Methodological Issues for Characterisation and Provenance Studies of Obsidian in Northeast Asia

Edited by

Akira Ono
Michael D. Glascock
Yaroslav V. Kuzmin
Yoshimitsu Suda

BAR International Series 2620
2014

Published in 2016 by
BAR Publishing, Oxford

BAR International Series 2620

Methodological Issues for Characterisation and Provenance Studies of Obsidian in Northeast Asia

ISBN 978 1 4073 1255 2

BAR Publishing is the trading name of British Archaeological Reports (Oxford) Ltd. British Archaeological Reports was first incorporated in 1974 to publish the BAR Series, International and British. In 1992 Hadrian Books Ltd became part of the BAR group. This volume was originally published by Archaeopress in conjunction with British Archaeological Reports (Oxford) Ltd / Hadrian Books Ltd, the Series principal publisher, in 2014. This present volume is published by BAR Publishing, 2016.

Printed in England

BAR
PUBLISHING

BAR titles are available from:

BAR Publishing
122 Banbury Rd, Oxford, OX2 7BP, UK
EMAIL info@barpublishing.com
PHONE +44 (0)1865 310431
FAX +44 (0)1865 316916
www.barpublishing.com

This volume is dedicated to the Golden Anniversary celebrating the beginning of scientific obsidian provenance studies in 1964, with Professor Colin Renfrew FBA FSA (a.k.a. The Lord Renfrew of Kaimsthorn) and Professor Johnson R. Cann FRS as pioneers

Table of Contents

List of Contributors

Fumito Akai	Kagoshima Board of Education, Shimofukumoto 3763–1, Kagoshima 91–0144, JAPAN (e-mail: fumito-akai@nifty.com)
Jeffrey R. Ferguson	Archaeometry Laboratory, Research Reactor Center, University of Missouri–Columbia, 1513 Research Park Dr., Columbia, MO 65211–3400, USA (e-mail: fergusonje@missouri.edu)
Michael D. Glascock	Archaeometry Laboratory, Research Reactor Center, University of Missouri–Columbia, 1513 Research Park Dr., Columbia, MO 65211–3400, USA (e-mail: glascockm@missouri.edu)
Andrei V. Grebennikov	Far Eastern Geological Institute, Far Eastern Branch of the Russian Academy of Sciences, 100-Letiya Vladivostoku Ave. 159, Vladivostok 690022, RUSSIA (e-mail: greandr@mail.ru)
Nobuyuki Ikeya	Numazu City Cultural Heritage Investigation Centre, Ohzuwa 46–1, Numazu City, Shizuoka Pref. 410–0873, JAPAN (e-mail: ikeya.nobuyuki@city.numazu.lg.jp)
Masami Izuho	Faculty of Social Sciences and Humanities, Tokyo Metropolitan University, Minami-Osawa 1–1, Hachioji-shi, Tokyo 192–0397, JAPAN (e-mail: izuhom@tmu.ac.jp)
Tarou Kannari	Centre for Obsidian and Lithic Studies, Meiji University, 1–1 Kanda–Surugadai, Chiyoda-ku, Tokyo 101–0064, JAPAN (e-mail: tarou.pbox@gmail.com)
Jong-Chan Kim	School of Physics, Seoul National University, Seoul 151–747, REPUBLIC OF KOREA (e-mails: marubosi@yahoo.co.kr; jckim@phya.snu.ac.kr)
Yaroslav V. Kuzmin	Institute of Geology and Mineralogy, Siberian Branch of the Russian Academy of Sciences, Koptyug Ave. 3, Novosibirsk 630090, RUSSIA (e-mail: kuzmin@fulbrightmail.org)
Masayuki Mukai	Asahikawa City Museum, Kagura 3–7, Asahikawa 070–8003, JAPAN (e-mail: mukaimasayuki2006@yahoo.co.jp)
Masashi Nagai	National Research Institute for Earth Science and Disaster Prevention, 3–1, Tennodai, Tsukuba City, Ibaraki Pref. 305–0006, JAPAN (e-mail: inagam@gmail.com)

Yuichi Nakazawa	Division of Human Evolution Studies, Department of Medicine, Hokkaido University, Kita 15, Nishi 7, Kita-ku, Sapporo 060–8638, JAPAN (e-mail: ynakazawa@med.hokudai.ac.jp)
Noriyoshi Oda	Archaeological Investigation Unit, Meiji University, Wakamatsu-cho 5–6–1, Fuchu City, Tokyo 183–0005, JAPAN (e-mail: odanori52@hotmail.com)
Akira Ono	Centre for Obsidian and Lithic Studies, Meiji University, 1–1 Kanda–Surugadai, Chiyoda-ku, Tokyo 101–0064, JAPAN (e-mail: onoak@meiji.ac.jp)
Vladimir K. Popov	Far Eastern Geological Institute, Far Eastern Branch of the Russian Academy of Sciences, 100-Letiya Vladivostoku Ave. 159, Vladivostok 690022, RUSSIA (e-mail: vladpov@fegi.ru)
Colin Renfrew	McDonald Institute for Archaeological Research, University of Cambridge, Downing St., Cambridge CB2 3ER, UK (e-mail: acr10@cam.ac.uk)
Kyohei Sano	Department of Earth Sciences, Hokkaido University of Education, Hokumon-cho 9, Asahikawa 070–8621, JAPAN (present address: Department of Earth and Planetary Sciences, Graduate School of Sciences, Kyushu University, Hakozaki 6–10–1, Higashi-ku, Fukuoka 812–8581, JAPAN) (e-mail: kyohei.sano.069@s.kyushu-u.ac.jp)
Hiroyuki Sato	Department of Archaeology, Faculty of Letters, University of Tokyo, 7–3–1 Hongo, Bunkyo-ku, Tokyo 113–0033, JAPAN (e-mail: hsato@l.u-tokyo.ac.jp)
Yoshimutsu Suda	Centre for Obsidian and Lithic Studies, Meiji University, 3670–8, Daimon, Nagawa-machi, Chiisagata-gun, Nagano Pref. 386–0601, JAPAN (e-mail: geosuda@gmail.com)
Shigeo Sugihara	Department of History and Geography, Meiji University (Professor Emeritus), 3–16–8, Sugano, Ichikawa City, Chiba Pref. 272–0824, JAPAN (e-mail: sugihara@kisc.meiji.ac.jp)
Keiji Wada	Department of Earth Sciences, Hokkaido University of Education, Hokumon-cho 9, Asahikawa 070–8621, JAPAN (e-mail: wada.keiji@a.hokkyodai.ac.jp)

List of Figures and Photos

List of Tables

Abbreviations

AD — *Anno Domini*, i.e., years after the birth of Jesus Christ
a.k.a. — also known as
a.s.l. — above sea level
BP — "before present", age in uncalibrated radiocarbon years as received from the laboratory
ca. — *circa*, i.e., approximately
cal BP — years before AD 1950, as applied to calibrated radiocarbon ages
CRDF — Civil Research and Development Foundation
e.g. — *exempli gratia*, i.e., for example
EDXRF, ED–XRF — Energy Dispersive X-ray Fluorescence
EPMA — Electron Probe Microanalysis
FEB RAS — Far Eastern Branch of the Russian Academy of Sciences
hXRF — handheld X-ray Fluorescence
i.e. — *id est*, that is to say
ICP–MS — Inductively Coupled Plasma – Mass Spectrometry
KAKENHI — Grant-in-Aid for Scientific Research; from Japan Society for Promotion of Science
LA–ICP–MS — Laser Ablation Inductively Coupled Plasma – Mass Spectrometry
Ma — million years ago
MEXT — Ministry of Education, Culture, Sport, Science, and Technology of Japan (also known in Japanese as Mombu Kagakusho)
NAA — Neutron Activation Analysis
OIS — Oxygen Isotope Stage
PCA — Principal Component Analysis
PI — Principal Investigator
PIXE — Proton-Induced X-ray Emission
ppm — parts-per-million, or per mille
Pref. — Prefecture
PXRF, pXRF — portable X-ray Fluorescence
REE — rare-earth elements
RFFI — Russian Foundation for Fundamental Investigations (also known as Russian Foundation for Basic Research [RFBR])
SB RAS — Siberian Branch of the Russian Academy of Sciences
SRM — standard reference material
US NSF — United States National Science Foundation
USGS — United States Geological Survey
vol.% — volume percent
vs. — *versus*, i.e., against
WDX, WDXRF — Wavelength Dispersive X-ray Fluorescence
wt.% — weight percent
WWII — Second World War
XRF — X-ray Fluorescence

Foreword

Colin RENFREW

It is an honour to be invited to introduce this volume, dedicated to the Golden Anniversary of our paper (Cann and Renfrew 1964) which the editors designate as 'the beginning of scientific obsidian provenance studies'. Joe Cann and I had the good fortune to discern at that time that trace element analysis of obsidian samples, both of artefacts and of geological specimens from obsidian outcrops, could answer a number of questions about early trade and exchange in obsidian, a material so widely used globally in prehistoric times. A way was now open to document early cases of mobility, and indeed very early seafaring. The application of trace element analysis to the obsidian of the Mediterranean region (and of central Europe) was soon followed by studies in the Near East (Renfrew *et al.* 1966, 1968).

In those early papers it was recognised that for successful characterisation studies it was desirable to identify specific properties which are 'homogeneous within sources and heterogeneous between them' (Cann and Renfrew 1964, 142–3). An important natural property of obsidian is its tendency to devitrify with time, so that eruptive activity before the Tertiary is not relevant to the issue, and in most cases the relevant obsidian sources are of Quaternary or sometimes late Tertiary date. In many parts of the world obsidian sources are relatively few in number, which facilitates the identification through trace element analysis of the original source of an artefact found in an archaeological context, since individual sources were soon shown to have consistent trace element signatures. The early use of Optical Emission Spectroscopy was soon supplemented and then superseded by Neutron Activation Analysis (Gordus *et al.* 1968; Aspinall *et al.* 1972), and the dating technique of fission-track analysis was also employed (Durrani *et al.* 1971). From about 1980 the technique of X-ray Fluorescence (XRF) Spectrometry became the most widely used, accessing all the possibilities offered by portable XRF (see Shackley 2010).

Systematic studies of obsidian exchange in Mesoamerica (Pires-Ferreira 1976) soon followed. Spatial analysis of obsidian finds from well-stratified archaeological contexts led to the investigation of different modes of exchange mechanism, with cases of "down-the-line" exchange recognised in widely differing area (Renfrew *et al.*, 1968, 328; Pires-Ferreira 1976; Renfrew 1975, 1977). Monograph studies of early obsidian exchange in different regions subsequently followed (Ericson 1981; Cauvin *et al.* 1998), along with systematic studies of specific sources (Torrence 1986; Tykot 1998).

In the 1990s there was an upsurge of obsidian studies (Shackley 1998; Glascock 2002). More recently the potential for productive work on obsidian procurement in Northeast Asia has been more clearly perceived (Kuzmin and Glascock 2010), and the present volume represents the culmination of that process, applying an international perspective to the research potentialities of the region. In the first place its obsidian sources are now much more thoroughly explored and documented, with an appropriate emphasis upon Japan.

Obsidian exchange in Japan has been systematically investigated for many years (e.g., Mochizuki 1997). A special feature in Japan, which is not seen in the Mediterranean area and little observed in Mesoamerica either, is the profusion of obsidian sources within quite localised areas, as the papers in the present volume now so thoroughly document. Thus Ferguson and colleagues (Chapter 2) refer to 13 primary sources on Hokkaido Island with eight further sources there in secondary deposits, while Wada and colleagues (Chapter 5) discern to 11 primary and ten secondary sources in Hokkaido. Kannari and colleagues (Chapter 4) assert that there are more than 100 geological sources of obsidian in Japan, and in his discussion focussing on the Kanto and Chibu regions of Central Japan; Ikeya (Chapter 8) lists 26 areas and sub-areas of obsidian sources. This information usefully supplements the documentation on source areas made available in the previous volume for Hokkaido Island (Izuho and Hirose 2010), Honshu Island (Tsutsumi 2010), and Kyushu Island (Obata *et al.* 2010) respectively.

The earlier work on the Kamchatka Peninsula (Russian Far East) (Grebennikov *et al.* 2010) is now carried forward (Grebennikov *et al.*, Chapter 7), and the important Paektusan source in Korea is now more fully investigated, with the correction of earlier errors (Kim, Chapter 11; see also Jia *et al.* 2010). Earlier work in the Russian Far East is summarised here by Kuzmin and Glascock (Chapter 6), and this work in the Russian Far East, in Sakhalin Island and the Kurile Islands is brought up to date in Kuzmin's comparative study (Chapter 10). So what previously seemed a rather confusingly complex picture on the Asiatic side of the North Pacific Rim has been admirably clarified in the papers in this volume and its predecessor, and the terrain has been mapped out with great clarity.

Looking across the perspective of fifty years of obsidian characterisation studies there are several major gains to record here. In the first place the quantitative standardisation of obsidian analyses was systematically addressed at the International Symposium which took place at the Centre for Obsidian and Lithic Studies at Meiji University in November 2011 as reported here by the editors (Ono *et al.*, Chapter 1). Although duplicate samples have previously been run to permit correlation and standardisation between analytical methods and laboratories (for instance, see Durrani *et al.* 1971, 242, for fission-track analysis compared with Optical Emission Spectroscopy), more systematic and wide-ranging standardisation has not, I think, been undertaken. So the international recognition and availability of a suite of standard samples is much to be welcomed. This promises to establish a new basis for inter-laboratory comparison and standardisation.

A further concern of the present volume is indeed the comparison of different analytical methods. Here the non-destructive analysis of obsidian artefacts is a particular focus, where both Wavelength Dispersive XRF (Suda, Chapter 3) and Energy Dispersive XRF (Kannari *et al.*, Chapter 4; Ikeya, Chapter 8) are discussed, along with studies employing Neutron Activation Analysis.

All of this lays the basis for some important archaeological studies, where the understanding of the distribution of obsidian sources and the appropriate analytical techniques give clear answers about the origin of the obsidian found, usually as artefacts, in secure archaeological contexts. Of particular interest is the emerging evidence for the transport and exchange of obsidian already during the Upper Palaeolithic period in this area. The present volume contains a particularly well-focussed study (Izuho *et al.*, Chapter 9) at the Upper Palaeolithic site of Ogachi-Kato 2 in Hokkaido Island.

All of this is important since it lays the foundations for a truly global comparative study of terrestrial and maritime trade and exchange, using obsidian as a convenient marker. As we noted already in 1964 (Cann and Renfrew 1964, 111) good quality obsidian from a wide variety of sources has broadly comparable physical properties all over the world (although that statement still requires detailed substantiation – a matter for future studies). Its exploitation to make effective tools requires considerable skill, but no further technical resources. So the comparative study of tool production and exchange across the continents is an interesting prospect.

Over the past fifty years there has been an understandable tendency for obsidian studies to be undertaken on a regional basis – in the Mediterranean, in the Near East, in North America, in Mesoamerica, in Japan etc. Indeed systematic studies do need first to be carried out on a regional basis. The present volume takes in a very wide geographical range – the entire North Pacific Rim. Its very success in doing so now prompts comparison with obsidian characterisation studies in other areas, for instance in New Guinea and the South Pacific (e.g., Spriggs *et al.* 2011; Torrence 2004; Torrence *et al.* 2009). Yet the very systematic presentation in this volume and in its predecessor (Kuzmin and Glascock 2010) now offers the possibility of transcending these regional concerns (which are indispensable, of course, as a preliminary to further discussion), and of moving from the regional to the global. Several major questions in world prehistory can now be answered through obsidian characterisation studies, as in perhaps no other way.

Among these key questions is the issue of seafaring (Anderson *et al.* 2010). When can the first seafaring be documented? The literature is extensive on this subject, much of it based on the evidence for the transport of obsidian, documented by characterisation studies. The marine transportation of obsidian during the late Upper Palaeolithic period was recognised for the Aegean source of Melos during the 1980s (Renfrew and Aspinall 1990), and is now much more fully exemplified for Japan, in particular for the Kozu-shima source, persuasively documented here by Ikeya (Chapter 8). The transportation of Hokkaido obsidian by sea to Sakhalin and the Kurile Islands in the early Holocene and possibly earlier (Kuzmin, Chapter 10) is of note.

The earlier exploitation of obsidian in East Asia by pre-sapiens hominins, perhaps *Homo erectus*, has yet to be documented. But such research is now underway in East Africa (Moutsiou 2012; Carter 2014). It remains to be seen whether the Paektusan source, near the border between China and North Korea, was exploited already by pre-sapient hominins. The characterisation studies presented here (Kim, Chapter 11) open the way to such inquiries.

The useful documentation of the obsidian sources of Northeast Asia offered here, with the accompanying information on early obsidian utilisation, is an important step forward for global obsidian studies. It also suggests interesting further avenues of inquiry. The work in Japan, not least on Honshu Island (Tsutsumi 2010) and on Hokkaido (Izuho and Hirose 2010; Izuho *et al.*, Chapter 9) underlines the intriguing question as to why obsidian was simultaneously sought from a number of different sources already as early as the Upper Palaeolithic period. It now seems clear that the nearest available source was not always the one utilised. For instance, at the site of Ogachi-Kato 2, three separate lithic industries are recognised, each with different patterns of source utilisation. In particular the microblade industry used obsidian from a variety of local and of non-local sources. The authors (Izuho *et al.*, Chapter 9), at the end of their paper, leave open to question the motivations of the obsidian workers and the circumstances of life on Hokkaido at that time which may have led to this variety in source utilisation. Was the obsidian from some sources preferred to that of others for its mechanical properties, as experienced in the process of artefact manufacture? Or were symbolic qualities assigned to the obsidian from specific sources (qualities which would be difficult now to measure effectively)? The technical precision in successful characterisation displayed in these papers now permits such questions to be posed. They focus, in a very interesting way, on issues of choice as they were experienced in Upper Palaeolithic times – issues which it is rarely easy to investigate today on the basis of the archaeological record.

By these means the techniques of archaeological science are being brought to bear in a particularly fruitful way to allow the effective study of human intentions and preferences in prehistoric times. Behavioural patterns, exchange relations, travel experience, and personal preferences are now open to investigation, already from Upper Palaeolithic times and onwards, in an explicit and verifiable manner.

References

Anderson, A., J. H. Barrett, and K. V. Boyle (eds). 2010. *The Global Origins and De-velopment of Seafaring*. Cambridge, McDonald Institute for Archaeological Research.

Aspinall, A., S. W. Feather, and C. Renfrew. 1972. Neutron Activation Analysis of Aegean Obsidians. *Nature* 237, 333–334.

Cann, J. R., and C. Renfrew. 1964. The Characterization of Obsidian and Its Application to the Mediterranean Region. *Proceedings of the Prehistoric Society* 30, 111–133.

Carter, T. 2014. The Contribution of Obsidian Characterization Studies to Early Prehistoric Archaeology. In *Lithic Raw Material Circulation and Exploitation in Prehistory. A Comparative Perspective in Diverse Environments* (E.R.A.U.L. 138), edited by M. Yamada and A. Ono, 23–33. Liège, Université de Liège.

Cauvin, M.-C., A. Gourgaud, B. Gratuze, N. Arnaud, G. Poupeau, J.-L. Poidevin, and C. Chataignier. 1998. *L'Obsidienne au Proche et Moyen Orient* (B.A.R. International Series 738). Oxford, BAR Publishing.

Durrani, S. A., H. A. Khan, N. Taj, and C. Renfrew. 1971. Obsidian Source Identification by Fission Track Analysis. *Nature* 233, 242–245.

Ericson, J. E. 1981. *Exchange and Production Systems in Californian Prehistory: The Results of Obsidian Hydration and Chemical Characterization of Obsidian Sources.* (British Archaeological Reports Series 110). Oxford, BAR Publishing.

Glascock, M. D. (ed.). 2002. *Geochemical Evidence for Long-Distance Exchange.* Westport, CT, Bergin & Garvey.

Gordus, A. A., G. A. Wright, and J. B. Griffin. 1968. Obsidian Sources Characterized by Neutron-Activation Analysis. *Science* 161, 382–384.

Grebennikov, A. V., V. K. Popov, M. D. Glascock, R. J. Speakman, Y. V. Kuzmin, and A. V. Ptashinsky. 2010. Obsidian Provenance Studies on Kamchatka Peninsula (Far Eastern Russia): 2003–9 Results. In *Crossing the Straits: Prehistoric Obsidian Source Exploitation in the North Pacific Rim* (B.A.R. International Series 2152), edited by Y. V. Kuzmin and M. D. Glascock, 89–120. Oxford, BAR Publishing.

IZUHO, M., and W. HIROSE. 2010. A Review of Archaeological Obsidian Studies on Hokkaido Island (Japan). In *Crossing the Straits: Prehistoric Obsidian Source Exploitation in the North Pacific Rim* (B.A.R. International Series 2152), edited by Y. V. Kuzmin and M. D. Glascock, 9–25. Oxford, BAR Publishing.

JIA, P. W., T. DOELMAN, C. CHEN, H. ZHAO, S. LIN, R. TORRENCE, and M. D. GLASCOCK. 2010. Moving Sources: A Preliminary Study of Volcanic Glass Artifact Distributions in Northeast China Using pXRF. *Journal of Archaeological Science* 37, 1670–1677.

KUZMIN, Y. V., and M. D. GLASCOCK (eds). 2010. *Crossing the Straits: Prehistoric Obsidian Source Exploitation in the North Pacific Rim* (B.A.R. International Series 2152). Oxford, BAR Publishing.

MOCHIZUKI, A. 1997. Identification of Sources of Obsidian Found in Chubu and Kanto Districts by X-ray Fluorescence Analysis. *Xsen Bunseki no Shinpo* 28, 157–168 (in Japanese with English Abstract).

MOUTSIOU, T. 2012. Changing Scales of Obsidian Movement and Social Networking. In *Unveiling the Palaeolithic. Ten Years of Research at the Centre for the Archaeology of Human Origins, University of Southampton* (B.A.R. International Series 2400), edited by K. Rubens, I. Romanowska and R. Bynoe, 85–96. Oxford, BAR Publishing.

OBATA, H., I. MORIMOTO, and S. KAKUBUCHI. 2010. Obsidian Trade between Sources on Northwestern Kyushu Island and the Ryukyu Archipelago (Japan) during the Jomon Period. In *Crossing the Straits: Prehistoric Obsidian Source Exploitation in the North Pacific Rim* (B.A.R. International Series 2152), edited by Y. V. Kuzmin and M. D. Glascock, 57–71. Oxford, BAR Publishing.

PIRES-FERREIRA, J. W. 1976. Obsidian Exchange in Formative Mesoamerica. In *The Early Mesoamerican Village*, edited by K. V. Flannery, 292–306. New York, Academic Press.

RENFREW, C. 1975. Trade as Action at a Distance: Questions of Integration and Communication. In *Ancient Civilization and Trade*, edited by J. A. Sabloff and C. C. Lamberg-Karlovsky, 3–59. Albuquerque, University of New Mexico Press.

RENFREW, C. 1977. Alternative Models for Exchange and Spatial Distribution. In *Exchange Systems in Prehistory*, edited by T. Earle and J. Ericson, 71–90. New York, Academic Press.

RENFREW, C., and A. ASPINALL. 1990. Aegean Obsidian and the Franchthi Cave. In Perlès, C., *Les Industries Lithiques Taillés de Franchthi (Argolide, Grèce)*, 257–270. Bloomington, Indiana University Press.

RENFREW, C., J. E. DIXON, and J. R. CANN. 1966. Obsidian and Early Cultural Contact in the Near East. *Proceedings of the Prehistoric Society* 32, 30–72.

RENFREW, C., J. E. DIXON, and J. R. CANN. 1968. Further Analysis of Near Eastern Obsidian. *Proceedings of the Prehistoric Society* 34, 319–331.

SHACKLEY, M. S. (ed.). 1998. *Archaeological Obsidian Studies: Method and Theory*. New York & London, Plenum Press.

SHACKLEY, M. S. 2010. Thoughts and Inference on Prehistoric Obsidian Source Exploitation in the Pacific Rim and Beyond. In *Crossing the Straits: Prehistoric Obsidian Source Exploitation in the North Pacific Rim* (B.A.R. International Series 2152), edited by Y. V. Kuzmin and M. D. Glascock, 219–223. Oxford, BAR Publishing.

SPRIGGS, M., C. REEPMEYER, ANGGRAENI, P. LAPE, L. NERI, W. P. RONQUILLO, T. SIMANJUNTAK, G. SUMMERHAYES, D. TANUDIRJO, and A. TIAUZONI. 2011. Obsidian Sources and Distribution Systems in Island Southeast Asia: A Review of Previous Research. *Journal of Archaeological Science* 38, 2873–2881.

TORRENCE, R. 1986. *Production and Exchange in Stone Tools: Prehistoric Obsidian in the Aegean.* Cambridge, Cambridge University Press.

TORRENCE, R. 2004. Now You See It, Now You Don't: Changing Obsidian Source Use in the Willaumez Peninsula, Papua New Guinea. In *Explaining Social Change: Studies in Honour of Colin Renfrew*, edited by J. Cherry, C. Scarre and S. Shennan, 115–125. Cambridge, McDonald Institute for Archaeological Research.

TORRENCE, R., P. SWADLING, W. AMBROSE, N. KONONENKO, P. RATH, and M. GLASCOCK. 2009. Obsidian Stemmed Tools and Mid-Holocene Interaction. *Asian Perspectives* 48, 118–147.

TSUTSUMI, T. 2010. Prehistoric Procurement of Obsidian from Sources on Honshu Island (Japan). In *Crossing the Straits: Prehistoric Obsidian Source Exploitation in the North Pacific Rim* (B.A.R. International Series 2152), edited by Y. V. Kuzmin and M. D. Glascock, 27–55. Oxford, BAR Publishing.

TYKOT, R. H. 1998. Mediterranean Islands and Multiple Flows: The Sources and Exploitation of Sardinian Obsidian. In *Archaeological Obsidian Studies: Method and Theory*, edited by M. S. Shackley, 67–82. New York & London, Plenum Press.

Preface and Acknowledgements

Akira ONO and Yaroslav V. KUZMIN

This volume appeared as the result of a collaboration between scholars of different institutions and countries who participated in an unprecedented Workshop in October–November 2011 in Japan (see Chapter 1). The Workshop consisted of two parts: 1) a fieldwork excursion in Hokkaido Island, with observations of several well-known sources of obsidian – the Shirataki and Oketo clusters, and the Tokachi-Mitsumata locale; and 2) the International Symposium "Methodological Issues of Obsidian Provenance Studies and the Standardisation of Geologic Obsidian", held at the Centre for Obsidian and Lithic Studies of Meiji University, at Nagawa Town (Nagano Prefecture). The second part of the Workshop was supported by the Strategic Research Foundation for Private Universities, by grant from the Ministry of Education, Culture, Sport, Science and Technology, Japan (MEXT), 2011–5 (No. S1101020). Participants revised their presentations delivered at the Symposium, and contributed papers to this volume.

Another reason for the creation of this collection of papers is that in 2014 we celebrate the Golden Anniversary of the beginning of obsidian provenance studies (in the modern understanding of the subject). The seminal article by Johnson R. Cann and Colin Renfrew "The Characterization of Obsidian and Its Application to the Mediterranean Region" was published in 1964, in Volume 30 of the *Proceedings of the Prehistoric Society*. This paper immediately triggered further research, in order to: a) establish securely the 'geological' sources of obsidian used by prehistoric people to acquire raw material for tool manufacture; and b) study in-depth the related cultural and social phenomena such as the origin of exchange and trade. It turned out that after 1964 *obsidian* became the most universal commodity for studying ancient contacts and migrations in many parts of the world! We dedicate this book to the pioneers in the obsidian provenance research, including Profs C. Renfrew and J. R. Cann. Prof. C. Renfrew has contributed a Foreword to this volume, and we are very grateful to him for this generous step.

In this Preface, we would like to point out mainly technical subjects. The spelling in the chapters follows British English, with reference to *The Concise Oxford Dictionary* (tenth edition, revised; J. Pearsall, editor; Oxford, Oxford University Press, 2001); and *The Shorter Oxford English Dictionary on Historical Principles* (sixth edition; Oxford, Oxford University Press, 2007). Synonyms are from *The Oxford American Writer's Thesaurus* (C. A. Lindberg, compiler; New York, Oxford University Press, 2004). For the creation of the Index, *The Chicago Manual of Style* (fifteenth edition; Chicago & London, University of Chicago Press, 2003) was used as a guideline. Geological terms in this volume follow the *Oxford Dictionary of Earth Sciences* (second edition; A. Allaby and M. Allaby, editors; Oxford, Oxford University Press, 2003); the *Dictionary of Geological Terms* (third edition; R. L. Bates and J. A. Jackson, editors; New York, Anchor Books, 1984); the *Oxford Companion to the Earth* (P. L. Hancock and B. J. Skinner, editors; New York, Oxford University Press, 2000); the *English–Russian Dictionary of Geology* (P. P. Timofeev and M. N. Alekseev, editors; Moscow, Russky Yazyk Publishers, 1988); and the *Russian–English Dictionary of Geology* (second edition; Y. G. Leonov, editor; Moscow, RUSSO Press, 2003). Geographic names and terms are mainly from *The Times Atlas of the World* (seventh comprehensive edition; London, Times Books, 1989); *Japan: A Bilingual Atlas* (Tokyo, Kodansha International, 1991); and the *Merriam Webster's Geographical Dictionary* (third edition; Springfield, MA, Merriam-Webster, Inc., 1997).

The style of references, written in non-Romance and non-Germanic languages, follows our previous volume "*Crossing the Straits: Prehistoric Obsidian Source Exploitation in the North Pacific Rim*"

(B.A.R. International Series 2152) published in 2010 by BAR Publishing (Oxford, UK). The original spelling of titles and sources (books, journals, edited volumes, etc.) written in Japanese, Russian, and Korean, is provided in the Roman alphabet along with the translation of the articles' and volume' titles only. Each chapter has its own list of references. The Index contains geographic and personal names, the main archaeological sites and cultural complexes mentioned in the text, and geological and archaeological terms important for the main topics of this book.

Numerous colleagues took part in the process of reviewing and editing the individual chapters of this volume. On behalf of all co-editors, we would like to thank them for the help they provided. Included are the following scholars (in alphabetical order, without titles or academic degrees): Robert E. Ackerman (Washington State University, Pullman, WA, USA); Anna Maria De Francesco (University of Calabria, Arcavacata CS, Italy); Scott M. Fitzpatrick (University of Oregon, Eugene, OR, USA); Ellery Frahm (University of Sheffield, Sheffield, UK); Andrei E. Izokh (Institute of Geology and Mineralogy, Siberian Branch of the Russian Academy of Sciences, Novosibirsk, Russia); Jun Kawai (Kyoto University, Kyoto, Japan); Susan G. Keates (Düsseldorf, Germany); Nikolai A. Klyuev (Institute of History, Archaeology, and Ethnography of the Far Eastern Nations, Far Eastern Branch of the Russian Academy of Sciences, Vladivostok, Russia); Vladimir L. Leonov (Institute of Volcanology and Seismology, Far Eastern Branch of the Russian Academy of Sciences, Petropavlovsk-Kamchatskyi, Russia); Candace C. Lindsey (Columbia, MO, USA); Eiichi Sato (Kobe University, Kobe, Japan); Christopher M. Stevenson (The College of William and Mary, Petersburg, VA, USA); Glenn Summerhayes (Otago University, Dunedin, New Zealand); Andrei V. Tabarev (Institute of Archaeology and Ethnography, Siberian Branch of the Russian Academy of Sciences, Novosibirsk, Russia); Toru Tateishi (Agency for Cultural Affairs, Tokyo, Japan); Karisa Terry (Central Washington University, Ellensburg, WA, USA); and Takashi Tsutsumi (Asama Jomon Museum, Miyota, Nagano Pref., Japan).

As always, Drs David Davison and Rajka Makjanić, and the staff of BAR as the Publisher of British Archaeological Reports (B.A.R.), warmly accepted our proposal to produce this volume as part of the B.A.R. International Series. We truly appreciate the assistance provided by all the staff of BAR with the final preparation of the book, especially Mr Darko Jerko (Zagreb, Croatia) for doing the layout and related activities.

Finally, we would like to express our sincere congratulations to all the 21 participants of this volume. We were working hard to get it done, and now we have it in hands!

Tokyo (Japan) and Novosibirsk (Russia),

31 October 2013

Chapter 1
INTRODUCTION: CHARACTERISATION AND PROVENANCE STUDIES OF OBSIDIAN IN NORTHEAST ASIA — THE VIEW FROM THE EARLY 2010s

Akira ONO, Yaroslav V. KUZMIN, Michael D. GLASCOCK, and Yoshimitsu SUDA

> '... two sharp peaks, with a saddle between them, and the whole steep side below was shining white — not with snow, but there were only a few patches of it to be seen in clefts, but with wet disintegrated pumice-stone, large lumps of which we had noticed on the banks of the Sungari [River] on our road through the forests.
>
> ... we go to the top and looked over the edge, and lo! at the bottom of a crater on whose brink we were standing, about three hundred and fifty feet below us, we saw a beautiful lake, its colours of the deepest, most pellucid blue, and, though the wind was howling above, its surface as still as Lake Leman, reflecting the crown of fantastic peaks with which the rugged top of the mountain was adorned.'
>
> *H. E. M. James (1888), The Long White Mountain* or *A Journey in Manchuria. London, Longmans, Green & Co.*

This volume originated from several long-term international collaborations in different fields of obsidian-related studies – geology, archaeology, and geochemistry, often combined under the term 'geoarchaeology'. Northeast Asia is a 'classical' region of the Pacific Rim where *obsidian* (i.e., high quality volcanic glass, practically waterless, and with extremely sharp edge resulting from conchoidal fracture) is widely distributed among volcanic rocks and is common at prehistoric sites, especially in Japan and some regions of the Russian Far East such as the Kamchatka Peninsula. Multidisciplinary studies of 'geological' (both primary and secondary) obsidian sources were initiated by geoarchaeologists who were anxious to answer questions coming from the archaeological community – which sources of obsidian were used by prehistoric humans to acquire the raw material? And, where were they located? As the result of decades of research, the long-distance distribution networks centred at the sources of obsidian in Northeast Asia have been established. The most important of these are shown in the Logo of this volume (see the cover page).

The border-crossing nature of obsidian provenance studies in Northeast Asia — including Japan, the Russian Federation (i.e., Russia), the Republic of Korea (i.e., South Korea), and the People's Republic of China (i.e., China) — required the coordination of international efforts to determine the obsidian sources which were exploited by prehistoric people because they are located in different modern countries. Because this was not done until recently, significant problems exist in the methodology, data acquisition, and even communication between scholars (if we consider that papers and reports are published in at least four-to-five unrelated languages). The standardisation of obsidian research methods in Northeast Asia is therefore one of the primary aims of this book.

Coincidentally, in 2014, the international scientific community celebrates the Golden Anniversary of scientific studies for obsidian provenance. In the early 1960s, an urgent need for the investigation of obsidian sources based on solid data from the geosciences became evident (e.g., Green 1962, 1964). The first comprehensive paper on obsidian sourcing (by modern standards), by Cann and Renfrew (1964), was published exactly 50 years ago, and more work soon followed (e.g., Renfrew *et al.* 1965, 1966, 1968), including in Japan (Suzuki 1969).

Before the early 1980s, obsidian sourcing for archaeological purposes was still a kind of 'exotic' area of research. Nowadays, it is almost impossible to find a volume where comprehensive information on obsidian source studies on a worldwide basis exists. A 1993 bibliography by the International Association of Obsidian Studies (IAOS) only partially covers publications released before the 1990s (Skinner and Tremaine 1993). The most recent progress and problems in obsidian provenance studies for archaeology are summarised by Freund (2013). Thanks to Prof. Colin Renfrew and other pioneers in the field, obsidian source investigations are now flourishing around the globe!

The obsidian provenance studies on geochemical and geological bases today allow archaeologists not only to understand very fine details of raw material acquisition but also to connect them to the social and environmental changes that affected the lives of early human societies, for example, in the Near East (e.g., Frahm and Feinberg 2013).

Studies of obsidian provenance are one of the most reliable ways to understand prehistoric migrations. They also represent an invaluable tool for detecting the earliest evidence of seafaring. The movement of obsidian across wide open seas during the Upper Palaeolithic (ca.

30,000–35,000 BP), as in the cases of the Kozu-shima source off coast of Honshu Island in the Pacific (see Ikeya, this volume) and in the Aegean region of the Mediterranean (e.g., Renfrew *et al.* 1965; see also: Laskaris *et al.* 2011; Simmons 2012), is an important (although still indirect!) evidence of early watercraft use.

Worldwide compendia of obsidian research are still quite rare. Some of the information has been presented in volumes edited by Shackley (1998), Glascock (2002), and Liritzis and Stevenson (2012), and in a monograph by Shackley (2005). We consider the collection of papers in this volume as a continuation of obsidian research in the Pacific Rim, following a recent volume edited by Kuzmin and Glascock (2010), which received a positive response (see Hirth 2011).

BACKGROUND OF THE 2011 WORKSHOP: THE 'MYSTERY' OF PAEKTUSAN OBSIDIAN

Following publication of the groundbreaking work by Cann and Renfrew (1964), obsidian provenance studies were initiated in different parts of the globe. Here we concentrate on our area of research — Northeast Asia — and present an overview of obsidian sourcing in historical perspective related to the issues of standardisation and international cooperation.

The Paektu [Baegdu] (a.k.a. Baitoushan and Changbaishan) Volcano, or Paektusan, on the modern border of the People's Democratic Republic of Korea (i.e., North Korea) and China, is a clear example of problems arising from the comparison and sharing of data on obsidian composition from this source, generated by researchers from Japan, South Korea, and Russia (e.g., Kuzmin, this volume; Kim, this volume). Initially explored by Europeans in the late nineteenth century (see Epigraph), Paektusan became almost unreachable since the 1950s. The lack of reference 'geological' obsidian samples from this source has hampered provenance studies of 'archaeological' volcanic glass in a large region covering North Korea, South Korea, far eastern Russia, and Northeast China (or Manchuria) (e.g., Kuzmin 2010). For decades, the crucial source region of the Paektusan Volcano remained practically inaccessible because of political tensions on and around the Korean Peninsula. Only very recently, an opportunity to obtain obsidian samples from the North Korean side of this large stratovolcano has materialised (Stone 2013; see also Kim, this volume).

The first scientific visits to the Paektusan Volcano took place at the end of the nineteenth century. In 1886–7, three British explorers — Sir Henry E. M. James, Sir Francis E. Younghusband, and Henry E. Fulford — travelled to Manchuria and the Paektusan Volcano region, but their journey resulted in a general geographic description only (James 1888). In 1897, the Russian geologist Eduard E. Anert studied the geology of the northern part of the Korean Peninsula and Manchuria, and visited the Paektusan Volcano where he described 'obsidian' near the lake shore within the volcano's caldera (on the modern North Korean side) (see Anert 1904, 275). As far as we are aware, this is the first reliable information regarding obsidian in mainland Northeast Asia obtained through geological reconnaissance.

The next stage of geological research at Paektusan belongs to the post-WWII period when Russian scholars studied the geology and petrology of the Korean Peninsula and Manchuria. In the late 1950s, the Russian geologist Evgeny P. Denisov examined the North Korean side of Paektusan, and confirmed the presence of obsidian near the southern rim of the caldera (Denisov 1965, 65). In 1974, the famous Russian scholar and Full Member of the USSR Academy of Sciences [Academician] Aleksei P. Okladnikov (e.g., Fagan 2003, 164) visited North Korea and acquired upon request (in a kind of semi-secret fashion because North Koreans were generally reluctant to give something to the foreigners, even from the friendly Soviet Union) four obsidian flakes, with a comment that they were from the Paektusan Volcano area. Fortunately, Okladnikov's assistant Aleksander K. Konopatsky preserved these samples, and later donated them for a geochemical study (see Kuzmin *et al.* 2002a, 513–4). In the 1980s, a joint Russian – North Korean expedition studied the Holocene volcanic ashes (tephra) on the southern slopes of Paektusan (Chichagov *et al.* 1991). However, the participants did not pay attention to the presence of obsidian and therefore did not collect any samples (V. P. Chichagov, personal communication 2000). In the 1990s, a joint German – North Korean team worked at the Paektusan Volcano, but their focus was on the so-called "Millennium Eruption" that took place sometime in the tenth century AD; those scholars did not gather any new information about the presence of obsidian on the southern part of Paektusan (see Horn and Schmincke 2000).

The first geochemical data on the obsidian from Paektusan were obtained in the 1980s (Warashina 1984; Sohn and Shin 1991). However, methodological inconsistencies hampered the secure determination of a chemical 'signature' for the Paektusan source. In the 1990s, Tetsuo Warashina and co-authors (see Warashina *et al.* 1998) analysed obsidian artefacts collected from the Kaineijodai site near the modern town of Onsong, North Hamgyong Province of North Korea, during the colonial period (1905–45) (Matsushita 1998; see also: Kuzmin 2010, 148). This collection is now stored at the Museum of Kyoto University, Japan (M. Tomii, personal communication 2003; Y. V. Kuzmin, personal observation 2003). However, these artefacts do not necessarily represent the source (i.e., 'geological') material from the Paektusan Volcano proper, because we now know that in this part of the Korean Peninsula and adjacent Manchuria obsidian was also brought by ancient people from the Basaltic Plateau source in Primorye Province of Russia (e.g., Jia *et al.* 2010, 2013).

Thus, without securely collected geological samples from the obsidian source itself it would not be absolutely correct to use the geochemical data on artefacts as a 'fingerprint' for the Paektusan source. Only in the 2000s, has the situation changed significantly (e.g., Kuzmin *et al.* 2002a; Popov *et al.* 2005; Kim *et al.* 2007). Until very recently we did not have solid data about the presence of volcanic glass on the Korean side of the Paektusan Volcano. Fortunately, a joint North Korean – British – US field trip to Paektusan was conducted in August 2013 (see Stone 2013), and volcanic glass was observed on the southern slope of the volcano (C. Oppenheimer, personal communication 2013). This gives us hope to obtain more reliable geochemical data on Paektusan obsidian in the near future.

THE CENTRE FOR OBSIDIAN AND LITHIC STUDIES, MEIJI UNIVERSITY: A SHORT OUTLINE

The Centre for Obsidian and Lithic Studies of Meiji University (Tokyo) was founded in April 2001, based on a long-term partnership between Meiji University and the town of Nagawa (Nagano Pref.). It was initially supported by a programme from the Ministry of Education, Culture, Sport, Science, and Technology of Japan (MEXT) in 2000–4. Upon completion of this first step, in 2006 the Centre became a branch of the Meiji University Museum. In 2010, the Centre was newly organised for the further enhancement of obsidian studies and international collaborative research networks, as one of the scholarly hubs comprising the Organisation for Strategic Coordination of Research and Intellectual Properties at Meiji University.

Photo 1.1. The building of the Centre for Obsidian and Lithic Studies, Meiji University (Nagawa Town, Nagano Pref.)

The Centre is situated near Nagawa Town in Nagano Pref. (Figure 1.1), and it includes an entrance hall, a research laboratory, archaeological research rooms, collection rooms, a library, and offices (Photo 1.1). The Centre's current Director is Prof. Dr Akira Ono; more information can be found on the Centre's website (in English): http://www.meiji.ac.jp/cols/english/

The Centre for Obsidian and Lithic Studies is a unique research facility in Japan devoted to various aspects of obsidian studies, both in the natural sciences and the humanities. Obsidian research in the field of human–environmental resources comprises the core axis of the Centre's activities, and includes research on the elucidation of human–environment interaction along with

Figure 1.1. The travel routes of the 2011 Workshop (by air and by land)

archaeological, geological, and palaeoenvironmental studies. The Centre's research topics include: 1) the advancement of archaeological research aimed toward enhancing knowledge of prehistoric obsidian exploitation in source areas, lithic tool production studies, and the reconstruction of obsidian circulation systems; 2) the reconstruction of obsidian formation mechanisms, volcanic eruption ages, and standardisation of obsidian analytical data through various geochemical analyses; 3) the palaeoenvironmental reconstruction for the age range of 50,000–10,000 BP (i.e., OIS 3–2) in Northeast Asia, with particular reference to Palaeolithic and Jomon subsistence practices; and 4) the establishment and development of international research networks.

The main project of the Centre, "Historical Variation in Interactions between Humans and Natural Resources: Towards the Construction of a Prehistoric Anthropography", received financial support from the MEXT for the period 2011–5. This project aims to integrate humans and their natural resource environment as a system, and to construct an anthropography (i.e., human geography) of historical variations. The range of issues with regard to people and their resource environment covers all aspects of human history. The main intention is to reconstruct the interaction between humans and their natural environment as a prehistoric anthropography, when people lived symbiotically with their surrounding environment. The 2011 Workshop was a vital part of the project's activity.

THE 2011 WORKSHOP: THE HOKKAIDO FIELDTRIP AND THE NAGANO SYMPOSIUM

In order to facilitate international cooperation in the field of obsidian research in Northeast Asia, the Centre for Obsidian and Lithic Studies came up with an initiative to conduct a Workshop on the standardisation of obsidian analyses and the results of obsidian provenance studies for prehistoric complexes in Northeast Asia. It was suggested to combine a scientific session with a field excursion to some of the most important obsidian sources in Japan located in Hokkaido Island which already were the subject of joint studies (Hall and Kimura 2002; Kuzmin and Glascock 2007; Kuzmin *et al.* 2002b, 2013; Sato *et al.* 2002; see also Ferguson *et al.*, this volume; and Izuho *et al.*, this volume). The Workshop took place on 28 October – 6 November 2011. Foreign participants from Russia, USA, and South Korea arrived in Japan, and all of the Workshop's members travelled to Hokkaido and Honshu islands (Figure 1.1).

The Hokkaido Fieldtrip

The first part of the Workshop was an extended field trip to Hokkaido Island on 29 October – 2 November 2013 (Figure 1.1). Japanese and foreign participants first flew to the Asahikawa City airport, and afterwards visited the most important sources of obsidian in Hokkaido Island. Before that, a lecture about the volcanic geology, petrology, and geochemistry of Hokkaido was given by Dr Keiji Wada on 29 October 2011 at the Department of Earth Sciences, Hokkaido University of Education (Asahikawa City).

On 30–31 October 2011, participants visited the Shirataki obsidian source area (Photos 1.2–1.4), and collected obsidian samples for further comparative study. On the morning of 1 November 2011, another cluster of obsidian sources — Oketo — was explored (Photo 1.5). Finally, in the afternoon of 1 November 2011 the obsidian source at Tokachi-Mitsumata was visited (Photo 1.6).

The Nagano International Symposium

The International Symposium "Methodological Issues of Obsidian Provenance Studies and the Standardisation of Geologic Obsidian" took place at the Centre for Obsidian and Lithic Studies on 5–6 November 2013. Before that, on 4 November 2011 participants visited archaeological excavations near the Centre. The large-scale obsidian mining site at Hoshikuso-toge [Hoshikuso Pass], dated to the Jomon period, has been under investigation since 1984, and scholars observed the ongoing excavations at one of the largest mining pits. Today, Hoshikuso-toge is a Registered National Archaeological Site. Afterwards, participants visited the exhibition of the Obsidian Museum of Nagawa Town located near the Centre.

The main topics of the Symposium were: 1) standardisation of obsidian analyses; and 2) current progress on obsidian provenance studies in Northeast Asia. As far as we know, these issues were not previously discussed in such detail, neither for Northeast Asia nor for other parts of the world.

The Scientific Programme of the Symposium was as follows:

5 November 2011

9:00 – 9:10 a.m. A. Ono: Opening Address.

9:10 – 9:40 a.m. V. K. Popov and A. V. Grebennikov: Results of Geochemical Study on the Basaltic Glasses from the Shkotovo Plateau and Obsidian Samples from the Paektusan Volcano by Various Methods, NAA, ICP–MS, and PIXE–PIGME, by Different Laboratories.

9:40 – 10:10 a.m. J.-C. Kim: The Paektusan Volcano Source and Geochemical Analysis of Archaeological Obsidian in Korea.

10:10 – 10:40 a.m. A. V. Grebennikov and V. K. Popov: Geochemistry of Volcanic Glasses and Search Strategy for Still Unknown Obsidian Sources in Kamchatka Peninsula (Russian Far East).

10.40 – 11.00 a.m. Coffee break.

11:00 – 11:30 a.m. J. R. Ferguson and M. D. Glascock: Determining Obsidian Source Groups in Complex Regions: the Use of Multiple Analytical Methods.

Photo 1.2. Participants of the 2011 Hokkaido Fieldtrip at the Akaishiyama (Summit) outcrop (Shirataki area), 30 October 2011. 1 – Keiji Wada; 2 – Jong-Chan Kim; 3 – Jeffrey R. Ferguson; 4 – Masami Izuho; 5 – Akira Ono; 6 – Yaroslav V. Kuzmin; 7 – Masayuki Mukai; 8 – Candace C. Lindsey; 9 – Michael D. Glascock; 10 – Vladimir K. Popov; 11 – Kyohei Sano; 12 – Satoru Yamada; 13 – Hiroyuki Sato; 14 – Kazutaka Shimada; 15 – Noriyoshi Oda; and 16 – Andrei V. Grebennikov

11:30 a.m. – 12:00 noon. N. Ikeya: Identifying the Sources of Archaeological Obsidian in Chubu and Kanto Regions, Japan, by EDXRF Analysis.

12:00 noon – 12:30 p.m. T. Kannari, M. Nagai, and S. Sugihara: Methods and Problems of Sourcing Obsidian Artefacts Using X-ray Fluorescence Analysis.

12:30 – 2:00 p.m. Lunch break.

2:00 – 2:30 p.m. Y. V. Kuzmin: Geoarchaeological Aspects of Obsidian Studies in the Russian Far East and Brief Comparison with Neighbouring Areas.

2:30 – 3:00 p.m. K. Wada, K. Sano, M. Mukai, M. Izuho, and H. Sato: Chemical Composition and Microstructure of Obsidians from Hokkaido Source Area, with Special Reference to Geological and Petrological Data for the Shirataki Obsidian Lava Complex.

3:00 – 3:30 p.m. M. Izuho: Reconstructing Hunters–Gatherers Obsidian Procurement Strategy: a Case Study from the Late Upper Palaeolithic Site of Kamihoronai-Moi, Hokkaido (Japan).

3:30 – 4:00 p.m. Coffee break.

Photo 1.3. Excursion to the Hachigozawa outcrop (Shirataki area, Hokkaido), 30 October 2011. K. Wada giving explanations

Photo 1.4. Excursion to the Ajisaitaki outcrop (Shirataki area, Hokkaido), 31 October 2011

4:00 – 4:30 p.m. Y. Suda: Characterisation and Standardisation of Geological Obsidian Using Physical and Chemical Appliances: Application to Obsidian Artefacts.

Photo 1.5. Excursion to the Tokoroyama outcrop (Oketo area, Hokkaido), 1 November 2011; participants collecting obsidian samples from the talus

Photo 1.6. Excursion to the Tokachi-Mitsumata source (Hokkaido), 1 November 2011; participants collecting obsidian samples from the Sannosawa River bed

4:30 – 5:00 p.m. M. D. Glascock and J. R. Ferguson: Obsidian Provenance by Trace Element Analysis: Analytical Techniques, Standardisation, and the Prospects for Data Sharing.

5:00 – 5:40 p.m. Discussion.

6 November 2011

9:00 – 10:00 a.m. Discussion of Future Tasks and Actions.

10.00 – 10.30 a.m. Splitting of Obsidian Sample for Inter-Comparison Analyses.

11:00 a.m. – 2:00 p.m. Return to Tokyo.

For the continuation of standardisation in obsidian research, specially selected samples from the Hachigozawa and Ajisaitaki sources (Shirataki area), and

Photo 1.7. Before the beginning of the ceremony of splitting the Shirataki obsidian sample for inter-comparison at the Centre for Obsidian and Lithic Studies on 6 November 2011. From left to right: M. D. Glascock, K. Wada, J.-C. Kim, V. K. Popov, A. Ono, and Akihiko Mochizuki; J. R. Ferguson sits in the background

the Oketo and Rubeshibe sources, were distributed among key participants. A special ceremony was organised for splitting a large obsidian specimen from the Shirataki source (Photo 1.7). The analytical work is underway (see Suda *et al.* 2013).

We hope that the 2011 Workshop will facilitate increased international cooperation in the field of obsidian studies in Northeast Asia and worldwide. This volume is a step in this direction.

THE CONTENT OF THE VOLUME

In this section, we briefly characterise the individual chapters which constitute this book. Technical details about the spelling of words and geographic names are given in the Preface and Acknowledgements. For some politically-sensitive regions of Northeast Asia like the Kurile Islands (see Stephan 1974, 1994) or the Korean Peninsula and adjacent areas, place names which are used in different countries and languages, are given: for example, the *Kunashir* or *Kunashiri* Island; *Sea of Japan* or *East Sea*; and *Soya* or *La Pérouse*, and *Korea* or *Tsushima* straits.

In ***Chapter*** **2**, J. R. Ferguson and co-authors report on the establishment of a comprehensive multi-elemental database for the obsidian sources in Hokkaido Island using both Energy Dispersive X-ray Fluorescence (EDXRF) and Neutron Activation Analysis (NAA). The multi-method approach establishes a foundation for highly accurate provenance studies of obsidian artefacts in Hokkaido requiring the least effort and expense, while minimising sample destruction. By identifying the most discriminating elements from the obsidian sources before analysing artefacts, the sources uniquely differentiated by non-destructive EDXRF are identified. The remaining sources which can only be separated by destructive NAA of a portion of an artefact are also identified. Out of the 21 sources in Hokkaido, only four sources were found to require NAA of the artefacts to accurately determine the primary locale. The multi-method approach described has the potential for broad application to provenance studies in other regions where multiple sources of obsidian are located.

In ***Chapter*** **3**, Y. Suda describes the development of a non-destructive Internal Standard Method for provenance identification of obsidian artefacts using Wavelength Dispersive XRF. The method was developed by employing the conventional Fusion Bead Method and internal standardisation to obtain quantitative data for the elements Ca, Mn, Rb, and Y for seven major Japanese sources. A variation diagram plotting the ratios of Y/Ca vs. Rb/Mn is identified as the most useful plot for source discrimination between the seven sources. More research will be necessary to demonstrate the utility of this method in complex regions such as Hokkaido.

In ***Chapter*** **4**, T. Kannari and colleagues examine the effectiveness of using intensity ratios for sourcing obsidian artefacts measured by EDXRF. The authors point out the problem applying EDXRF to artefacts of different shapes and sizes which produce inaccurate compositional data for small artefacts. By employing intensity ratios, the data for artefacts from the same source but with vastly different sizes are less likely to provide ambiguous sourcing results.

In ***Chapter*** **5**, K. Wada and colleagues perform microscopic observation and Electron Probe Microanalysis (EPMA) of the 'geological' obsidian in Hokkaido Island, and establish the standard composition for 24 volcanic glass groups. The variations of CaO/Al_2O_3 vs. TiO_2/K_2O ratios, and CaO vs. FeO* ratio allow them to discriminate different obsidian sources. Based on these data, the source identification for obsidian artefacts in Hokkaido and neighbouring Northeast Asia is now more secure than before.

In ***Chapter*** **6**, Y. V. Kuzmin and M. D. Glascock summarise the results of the systematic application of the NAA to volcanic glasses in the southern Russian Far East, including Primorye [Maritime] Province, the Amur River basin, Sakhalin Island, and the Kurile Islands. Several sources of obsidian were established, and each of them now has its own geochemical 'fingerprint'. This allows them to identify the provenance of obsidian artefacts not only from the Russian Far East but also from the adjacent regions of Northeast Asia, such as Manchuria and the Korean Peninsula (e.g., Jia *et al.* 2010, 2013).

In ***Chapter*** **7**, A. V. Grebennikov and co-authors describe the volcanic geology of the Kamchatka Peninsula (northern Russian Far East), and characterise each source of 'archaeological' obsidian in this region. Because of the abundance of obsidian sources in Kamchatka (no less than 30 are known to geologists) and very difficult logistics, at least six individual sources used by prehistoric people have not yet been pinpointed. The authors propose a strategy for the search and localisation of unknown obsidian sources, by analysing all the available information about their geology and geochemistry, including potassium–argon and argon–argon dating methods. After that, fieldwork verification is required, and now researchers have a greater potential to establish the exact position of these primary obsidian sources.

In ***Chapter*** **8**, N. Ikeya demonstrates the case of maritime transport of obsidian in the Upper Palaeolithic from the Kozushima [Kozu-shima] source to Honshu Island. Kozushima is a small volcanic island located ca. 50 km away from the main landmass in the open Pacific Ocean. This chapter is based on data generated from provenance identification of obsidian artefacts excavated in the Mount Ashitaka area (Chubu Region of Honshu Island). EDXRF was applied to artefacts, and the determination of sources is supported by independent NAA study. It seems that we now have solid evidence of sea voyages in Northeast Asia at ca. 33,000 BP (or ca. 36,000–38,500 cal BP).

In ***Chapter*** **9**, M. Izuho and co-authors discuss the obsidian raw material procurement system as reflected at the Upper Palaeolithic Ogachi-Kato 2 site in Hokkaido Island, as viewed from the integration of obsidian compositional studies and analysis of lithic artefact manufacture. The authors distinguish three different techniques, small flake industry, flake industry, and microblade industry; they evaluate these varieties as a reflection of different behavioural patterns of the Upper Palaeolithic hunter–gatherers, especially in terms of accessibility of obsidian sources in northeastern Hokkaido through time. The authors also propose the examination of research topics which are not yet solved.

In ***Chapter*** **10**, Y. V. Kuzmin gives an updated review of the geoarchaeological aspects of obsidian provenance studies in the greater Russian Far East, including its northeastern part (Kamchatka and Chukotka regions). A short historiographical overview of obsidian studies in this part of Northeast Asia is also included. Based on the current data, it is noted that while obsidian was one of the raw materials in the Palaeolithic and Neolithic in the mainland regions, it never dominated in the stone tool assemblages. Insular prehistoric populations (Sakhalin and the Kuriles) used obsidian as a raw material more widely than on the continent, with primary sources located outside of these regions – in the neighbouring Hokkaido Island and Kamchatka Peninsula. The existence of long-distance exchange/trade of obsidian in the later prehistory of the Russian Far East (Neolithic–Palaeometal) is now evident, with distances often exceeding 1000 km in a straight line. The problem of seafaring related to the transport of obsidian across the sea straits is also discussed.

In ***Chapter*** **11**, J.-C. Kim provides new information about the geochemical composition of obsidian from the Korean Peninsula, including the Paektusan Volcano. An interpretation based on carefully-prepared data on Korean obsidian (from both archaeological sites and primary outcrops) resulted in the conclusion that two out of three geochemical groups of obsidian from the Korean Peninsula and neighbouring parts of Northeast Asia (PNK2 and PNK3 groups) have their original sources at Paektusan. As for the remaining PNK1 Group, more data have now been generated, and it is possible to suggest that its primary source is also located in the Paektusan Volcano region. This, however, still requires verification by means of fieldwork and subsequent laboratory analyses. The issue of Palaeolithic seafaring is also touched upon, with the suggestion that people in the southernmost part of the Korean Peninsula were able to cross the Korea [Tsushima] Strait at ca. 25,000 BP by bringing obsidian from sources in Kyushu Island.

We hope that this collection of papers will be useful for all students of obsidian provenance and related activities in Northeast Asia and worldwide. Along with the previous volume (see Kuzmin and Glascock 2010), it constitutes a solid basis for in-depth research of obsidian and its role in ancient human societies.

References

ANERT, E. E. 1904. *Puteshestvie po Manchzurii* [À Travers la Mandjourie] (Memoirs of the Imperial Russian Geographic Society on General Geography. Vol. XXXV). St.-Petersburg, Imperatorskoe Russkoe Geograficheskoe Obshchestvo.

CANN, J. R., and C. RENFREW. 1964. The Characterization of Obsidian and Its Application to the Mediterranean Region. *Proceedings of the Prehistoric Society* 30, 111–133.

CHICHAGOV, V. P., R. K. MUK, A. E. CHERKINSKYI, and O. A. CHICHAGOVA. 1991. Radiocarbon Age of Tephra–Buried Trees on the Paektusan Volcano, North Korea. *Doklady of the USSR Academy of Sciences, Earth Science Sections* 306(1–6), 36–39.

DENISOV, E. P. 1965. *Noveishaya Tektonika i Podznekainozoisky Vulkanizm Yuzhnogo Primorya i Prilegayushchikh Oblastei* [The Modern Tectonics and Late Cenozoic Volcanism of Southern Primorye and Adjacent Regions]. Vladivostok, Dalnevostochnoe Knizhnoe Izdatelstvo.

FAGAN, B. 2003. *Archaeologists: Explorers of the Human Past*. New York, Oxford University Press.

FRAHM, E., and J. M. FEINBERG. 2013. Environment and Collapse: Eastern Anatolian Obsidians at Urkesh (Tell Mozan, Syria) and the Third-Millennium Mesopotamian Urban Crisis. *Journal of Archaeological Science* 40, 1866–1878.

FREUND, K. P. 2013. An Assessment of the Current Applications and Future Directions of Obsidian Sourcing Studies in Archaeological Research. *Archaeometry* 55, 779–793.

GLASCOCK, M. D. (ed.). 2002. *Geochemical Evidence for Long-Distance Exchange*. Westport, CT, Bergin & Garvey.

GREEN, R. C. 1962. Obsidian: Its Application to Archaeology. *Newsletter (New Zealand Archaeological Association)* 5, 8–16.

GREEN, R. C. 1964. Sources, Ages, and Exploitation of New Zealand Obsidian: An Interim Report. *Newsletter (New Zealand Archaeological Association)* 7, 133–143.

HALL, M. E., and H. KIMURA. 2002. Quantitative EDXRF Studies of Obsidian Sources in Northern Hokkaido. *Journal of Archaeological Science* 29, 259–266.

HIGASHIMURA, T. 1984. Osanri Yuzeok Chulto Hukyoseok ui Hyong-gwang Bunseok [X-ray Fluorescence Analysis of Obsidians Excavated at Osanni Site]. In *Osanri Yuzeok*, edited by H.-J. Im and H.-S. Kwon, 69–73. Seoul, Seoul National University Museum Press.

HIRTH, K., 2011. Review of Yaroslav V. Kuzmin & Michael D. Glascock (ed.). *Crossing the Straits: Prehistoric Obsidian Source Exploitation in the North Pacific Rim* (British Archaeological Reports International Series 2152). *Antiquity* 85, 1090–1091.

HORN, S., and H.-U. SCHMINCKE. 2000. Volatile Emission during the Eruption of Baitoushan Volcano (China/North Korea) ca. 969 AD. *Bulletin of Volcanology* 61, 537–555.

JAMES, H. E. M. 1888. *The Long White Mountain,* or *Journey in Manchuria: with Some Account of the History, People, Administration and Religion of that Country*. London, Longmans, Green & Co.

JIA, P. W., T. DOELMAN, C. CHEN, H. ZHAO, S. LIN, R. TORRENCE, and M. D. GLASCOCK. 2010. Moving Sources: A Preliminary Study of Volcanic Glass Artifact Distributions in Northeast China Using pXRF. *Journal of Archaeological Science* 37, 1670–1677.

JIA, P. W., T. DOELMAN, R. TORRENCE, and M. D. GLASCOCK. 2013. New Pieces: The Acquisition and Distribution of Volcanic Glass Sources in Northeast China during the Holocene. *Journal of Archaeological Science* 40, 971–982.

KIM, J.-C., D. K. KIM, M. YOON, C. C. YUN, G. PARK, H. J. WOO, M.-Y. HONG, and G. K. LEE. 2007. PIXE Provenancing of Obsidian Artefacts from Paleolithic Sites in Korea. *Bulletin of the Indo-Pacific Prehistory Association* 27, 122–128.

KUZMIN, Y. V. 2010. Crossing Mountains, Rivers, and Straits: A Review of the Current Evidence for Prehistoric Obsidian Exchange in Northeast Asia. In *Crossing the Straits: Prehistoric Obsidian Source Exploitation in the North Pacific Rim* (B.A.R. International Series 2152), edited by Y. V. Kuzmin and M. D. Glascock, 137–153. Oxford, BAR Publishing.

KUZMIN, Y. V., and M. D. GLASCOCK. 2007. Two Islands in the Ocean: Prehistoric Obsidian Exchange between Sakhalin and Hokkaido, Northeast Asia. *Journal of Island & Coastal Archaeology* 2, 99–120.

KUZMIN, Y. V., and M. D. GLASCOCK (eds). 2010. *Crossing the Straits: Prehistoric Obsidian Source Exploitation in the North Pacific Rim* (B.A.R. International Series 2152). Oxford, BAR Publishing.

KUZMIN, Y. V., M. D. GLASCOCK, and M. IZUHO. 2013. The Geochemistry of the Major Sources of Archaeological Obsidian on Hokkaido Island (Japan): Shirataki and Oketo. *Archaeometry* 55, 355–369.

KUZMIN, Y. V., M. D. GLASCOCK, and H. SATO. 2002b. Sources of Archaeological Obsidian on Sakhalin

Island (Russian Far East). *Journal of Archaeological Science* 29, 741–749.

KUZMIN, Y. V., V. K. POPOV, M. D. GLASCOCK, and M. S. SHACKLEY. 2002a. Sources of Archaeological Volcanic Glass in the Primorye (Maritime) Province, Russian Far East. *Archaeometry* 44, 505–515.

LASKARIS, N., A. SAMPSON, F. MAVRIDIS, and I. LIRITZIS. 2011. Late Pleistocene/Early Holocene Seafaring in the Aegean: New Obsidian Hydration Dates with the SIMS–SS Method. *Journal of Archaeological Science* 38, 2475–2479.

LIRITZIS, I., and C. M. STEVENSON (eds). 2012. *Obsidian and Ancient Manufactured Glasses*. Albuquerque: University of New Mexico Press.

MATSUSHITA, H. 1998. Chosen Ikyo Hokudo Kainei Jyo Fukin no Sekki Nitsuite [Stone Tools near Kainei Castle in North Korea]. In *Toki no Kizuna Michi wo Tadoru*, edited by Ishizuki Kimio Sensei wo Shinobu, 439–464. Sapporo, Hon Kanko Iinkai.

POPOV, V. K., V. G. SAKHNO, Y. V. KUZMIN, M. D. GLASCOCK, and B.-K. CHOI. 2005. Geochemistry of Volcanic Glasses of the Paektusan Volcano. *Doklady Earth Sciences* 403, 254–259.

RENFREW, C., J. R. CANN, and J. E. DIXON. 1965. Obsidian in the Aegean. *The Annual of the British School at Athens* 60, 225–247.

RENFREW, C., J. E. DIXON, and J. R. CANN. 1966. Obsidian and Early Cultural Contact in the Near East. *Proceedings of the Prehistoric Society* 32, 30–72.

RENFREW, C., J. E. DIXON, and J. R. CANN. 1968. Further Analysis of Near Eastern Obsidian. *Proceedings of the Prehistoric Society* 34, 319–331.

SATO, H., Y. V. KUZMIN, and M. D. GLASCOCK. 2002. Source Analysis of Obsidian in Prehistoric Sakhalin and the Assessment of Its Distribution in Northeast Asia. *Hokkaido Kokogaku* 38, 1–13 (in Japanese with English Title).

SHACKLEY, M. S. (ed.). 1998. *Archaeological Obsidian Studies: Method and Theory*. New York & London, Plenum Press.

SHACKLEY, M. S. 2005. *Obsidian: Geology and Archaeology in the North American Southwest*. Tucson, University of Arizona Press.

SIMMONS, A. 2012. Mediterranean Island Voyagers. *Science* 338, 987–988.

SKINNER, C. E., and K. J. TREMAINE. 1993. *Obsidian: An Interdisciplinary Bibliography* (International Association for Obsidian Studies Occasional Papers 1). San Jose, CA, San Jose State University.

SON, P., and S.-C. SHIN. 1991. A Shell-Midden on the Island Sangnodaedo: A Neolithic Site off the Southern Coast of Korea. *Bulletin of the Indo-Pacific Prehistory Association* 10, 109–117.

STEPHAN, J. J. 1974. *The Kuril Islands: Russo–Japanese Frontier in the Pacific*. Oxford, Clarendon Press.

STEPHAN, J. J. 1994. *The Russian Far East: A History*. Stanford, CA, Stanford University Press.

STONE, R. 2013. Sizing Up a Slumbering Giant. *Science* 341, 1060–1061.

SUDA, Y., J. FERGUSON, M. D. GLASCOCK, V. K. POPOV, S. V. RASSKAZOV, T. A. YASNYGINA, J.-C. KIM, N. SAITO, H. TAKEHARA, K. WADA, A. ONO, A. V. GREBENNIKOV, and Y. V. KUZMIN. 2013. Standardization of Obsidian Compositional Data for Provenance Studies: Petrology and Data Compilation of Intra-Laboratory Results for Obsidian from the Shirataki Source, Northern Japan. In *5-th Arheoinvest Symposium 'Stories Written in Stone'. International Symposium on Chert and Other Knappable Material. Iaşi, Romania, 20–24 August 2013. Programme and Abstracts*, edited by O. N. Crandell and V. Cotiugă, 94. Iaşi, Editura Universităţii "Alexandru Ioan Cuza" din Iaşi.

SUZUKI, M. 1969. Fission Track Dating and Uranium Contents of Obsidian (I). *Daiyonki Kenkyu* 8, 123–130 (in Japanese with English Abstract).

WARASHINA, T., T. HIGASHIMURA, H. SATO, and Z. S. LAPSHINA. 1998. Sekki Genzai no Sanchi Bunseki [Analysis of Lithic Artefact Sources]. In *Nihon Bunkazai Kagaku-kai dai 15 kai Taikai Kenkyu Happyo Yoshi*, edited by Nihon Bunkazai Kagaku-kai, 138–139. Nara, Nihon Bunkazai Kagaku-kai.

METHODOLOGICAL ISSUES ON CHARACTERISATION OF OBSIDIAN SOURCES IN NORTHEAST ASIA

Chapter 2
MULTI-METHOD CHARACTERISATION OF OBSIDIAN SOURCE COMPOSITIONAL GROUPS IN HOKKAIDO ISLAND (JAPAN)

Jeffrey R. FERGUSON, Michael D. GLASCOCK, Masami IZUHO, Masayuki MUKAI, Keiji WADA, and Hiroyuki SATO

Abstract: *A comprehensive study of all known obsidian sources on the island of Hokkaido was undertaken using two analytical methods: Neutron Activation Analysis and Energy Dispersive X-ray Fluorescence. Twenty-one unique geochemical source groups were identified. Using a collection of artefacts from the site of Ogachi-Kato 2, we demonstrate a systematic approach to artefact sourcing that minimises time, expense, and sample destruction while achieving a high rate of success.*

Keywords: *Obsidian, Hokkaido Island, Japan, Neutron Activation Analysis, Energy Dispersive X-ray Fluorescence, Sourcing*

INTRODUCTION

We describe a multi-method approach utilising both Energy Dispersive X-ray Fluorescence (hereafter – ED–XRF or XRF) and Neutron Activation Analysis (henceforth – NAA) to maximise our ability to trace obsidian artefacts to specific geologic sources while at the same time reducing analytical cost and damage to the artefacts. The advantages of NAA are: 1) its excellent sensitivity, precision, and accuracy for a large number of elements; 2) because NAA is a bulk technique it is possible to analyse samples ranging in size from a few milligrammes to several grammes; and 3) it is possible to analyse different suites of elements based on half-life such that samples do not necessarily remain radioactive for years. The advantages of ED–XRF are: 1) it can be performed non-destructively; 2) it can measure several of the key discriminating elements in obsidian; 3) the analysis can be performed in a few minutes; and 4) the cost is minimal. Until recently, few studies have given consideration to the application of multiple analytical methods with the goal of achieving higher efficiency and reduced cost (Glascock 2010; Glascock *et al.* 2010).

THE MULTI-METHOD APPROACH

Throughout the history of archaeological research on obsidian, a number of analytical techniques have been employed to trace obsidian artefacts to specific raw material sources. In the early years, the choice of analytical method was usually the most familiar or most available. As the experience from employing a variety of analytical techniques to obsidian accumulated, the advantages and disadvantages of each method with respect to sourcing obsidian were learned. The specific elements and sensitivities for each analytical method are almost always different. Most require a minimum amount of material or involve sample destruction which limits the possibility of analysing archaeological materials. The methods of analysis differ with respect to precision, accuracy, data compatibility between laboratories, time required to complete the analysis, and expense.

The two most often employed methods for obsidian sourcing studies have been NAA and ED–XRF. Until recently, these approaches were almost always used independently. The strengths of NAA are based on the large number of possible elements (approximately 30) and its superior precision and accuracy. With a large number of elements, it is possible to inspect several combinations and discover multiple groups of elements that discriminate between chemically similar obsidian sources. However, the requirement that samples be partially destroyed so that they could be exposed to neutrons in a reactor and that they are made radioactive for a long period of time is a restricting factor for some archaeological specimens. The limited availability and higher expense of NAA are other impediments. Improvements such as procedures to measure short-lived elements that differentiate between some difficult sources (Glascock *et al.* 1994) and reduce the amount of residual radioactivity make some of these issues less problematic.

The main strengths of ED–XRF are based on its ability to measure a few critical elements (especially Rb, Sr, Y, Zr, and Nb) without destroying the sample, and to do so at low cost. Furthermore, the development of relatively inexpensive handheld XRF instruments, capable of producing data comparable to laboratory-based XRF instruments (when used correctly; see Ferguson 2012), has increased the worldwide prevalence of ED–XRF applied in obsidian studies. But, the lower sensitivity and precision for ED–XRF and the reduced number of elements can be constraints on the ability to differentiate between some chemically similar sources. For some regions such as western New Mexico (USA) where the number of sources is modest, ED–XRF is usually adequate. However, there are other regions such as the Rift Valley of Kenya in Africa (Ferguson *et al.* 2010; Merrick and Brown 1984) and western Mexico (Glascock *et al.* 2010) where the density of sources is high and many sources are not discernible by ED–XRF alone.

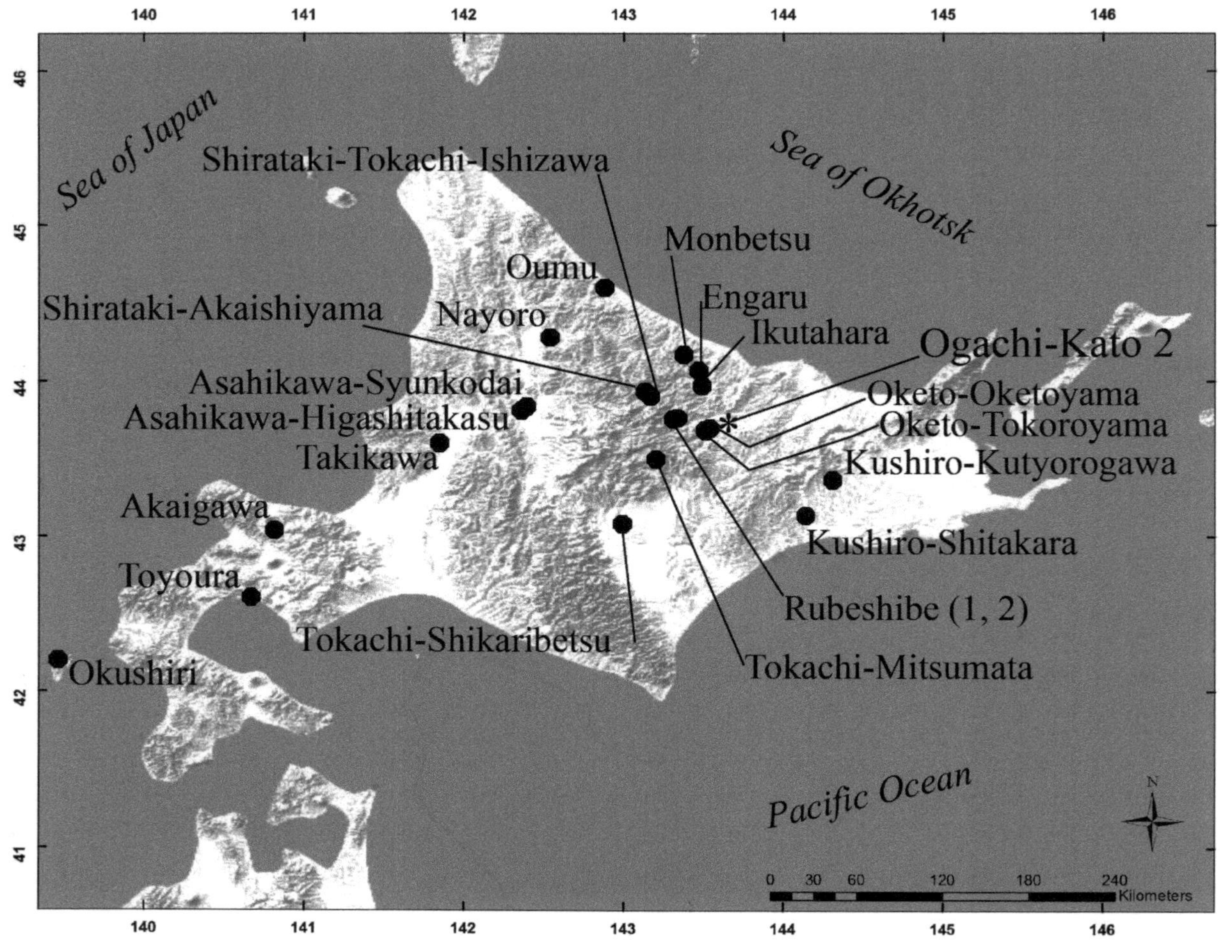

Figure 2.1. Map of Hokkaido obsidian sources (shown by circles) and the location of Ogachi-Kato 2 site (shown by asterisk)

GEOGRAPHY AND GEOLOGICAL SETTING

Located on the western margin of the Pacific Ocean, Hokkaido Island is the northernmost landmass of the Japanese Archipelago. The island is surrounded on the east by the Pacific Ocean, on the west by the Sea of Japan, and on the northeast by the Sea of Okhotsk. The nearest neighbouring lands are the Japanese island of Honshu (20 km to the south), the Russian island of Sakhalin (40 km to the north), and the island of Kunashir [Kunashiri] (15 km to the east).

Hokkaido is situated near the intersection of three tectonic plates, the Pacific, the Eurasian, and the North American. Due to the subduction of the Pacific Plate beneath the North American Plate, the region has been subject to a considerable amount of volcanism, including the formation of obsidian, since the Miocene. Volcanic activity on and around the island of Hokkaido took place in three chronological stages that can be distinguished based on the type of activity, composition of the magma, and types of eruption centres (Hirose and Nakagawa 1999). These stages include: 1) prior to the middle Miocene (> 12 Ma ago); 2) from the late Miocene until the early Pleistocene (ca. 12–1 Ma ago); and 3) from the early Pleistocene until the present (< 1 Ma ago).

During the Miocene period, large-scale alkaline volcanic activity occurred throughout Hokkaido producing obsidian with large felsic sediments (Hirose *et al.* 2000; Takanashi *et al.* 2011). From the end of the Miocene until the early Pleistocene, the intensity of volcanic activity decreased, with most of the activity occurring near the Kurile Arc in the form of rhyolitic lava domes and felsic pyroclastic flows on the eastern side of the island. The mountains associated with the Shirataki, Oketo, and Engaru obsidian sources were formed during this period. From the early Pleistocene until the present time, most of the obsidian was formed in the western part of Hokkaido and on Okushiri Island located just off the western coast of Hokkaido.

Obsidian sources are present throughout Hokkaido, but most are concentrated on the eastern side. The locations of all known sources in Hokkaido are listed in Table 2.1 and shown in Figure 2.1. Thirteen of these are primary sources, including: Akaigawa, Engaru, Ikutahara, Kushiro (Shitakara), Monbetsu (Kamimobetsu), Oketo (Oketoyama), Oketo (Tokoroyama), Okushiri (Katsumayama), Rubeshibe (Iwayama), Rubeshibe (Kayokozawa), Shirataki (Akaishiyama), Shirataki (Tokachi-Ishizawa) [in some publications and sources also Tokachiishizawa; see Wada *et al.*, this volume], and

Table 2.1. Obsidian sources in Hokkaido Island described in this study and their locations (see Figure 2.1)

Source Name	Source Type	Geographic Coordinates*	
		Latitude (N)	Longitude (E)
Akaigawa	Primary	43.0388	140.8155
Asahikawa (Higashitakasu)	Secondary	43.8342	142.3924
Asahikawa (Syunkodai)	Secondary	43.8111	142.3601
Engaru	Primary	44.065	143.4725
Ikutahara	Primary	43.9698	143.4923
Kushiro (Kutyorogawa)	Secondary	43.3644	144.3111
Kushiro (Shitakara)	Primary	43.1353	144.1402
Monbetsu (Kamimobetsu)	Primary (outcrop)	44.1689	143.3781
Oketo (Oketoyama)	Primary	43.6969	143.5403
Oketo (Tokoroyama)	Primary	43.685	143.5115
Okushiri (Katsumayama)	Primary (outcrop)	42.1989	139.4592
Oumu	Secondary	44.595	142.8818
Rubeshibe (Iwayama)	Primary	43.7607	143.3373
Rubeshibe (Kayokozawa)	Primary (outcrop)	43.756	143.3106
Shirataki (Akaishiyama)	Primary (outcrop)	43.9333	143.1333
Shirataki (Tokachi-Ishizawa)	Primary (outcrop)	43.9000	143.1666
Takikawa	Secondary	43.6006	141.8475
Tokachi-Mitsumata	Primary	43.4981	143.2026
Tokachi (Shikaribetsu)	Secondary	43.0801	142.9910
Toyoura	Secondary	42.6045	140.6698

* Coordinates are given in decimal degrees.

Tokachi-Mitsumata. The remaining eight sources are only found in secondary deposits created by transport of obsidian away from the primary source areas which are now covered or obscured by erosion, landslides, and other factors. The secondary sources include: Asahikawa (Higashitakasu), Asahikawa (Syunkodai), Kushiro (Kutyorogawa), Nayoro, Oumu, Takikawa, Tokachi (Shikaribetsu), and Toyoura. The large number of obsidian sources and the small geochemical differences between some of the geographically separated locales has made archaeological studies of Hokkaido obsidian challenging. Table 2.2 provides a comparison of previously used names for the Hokkaido sources.

PREVIOUS ARCHAEOLOGICAL AND GEOCHEMICAL RESEARCH IN HOKKAIDO

Procurement of obsidian by prehistoric groups in Hokkaido began by at least the Early Upper Palaeolithic (dated to ca. 30,000 BP) and continued throughout the prehistoric period. One of the earliest known sites showing evidence of obsidian use is Wakabano Mori located in eastern Hokkaido (Izuho and Akai 2005). Investigation of procurement and exchange of obsidian in Hokkaido began with the work of Kimura (1992) who studied the Shirataki obsidian sources. Follow-up studies by Kimura (1995, 1998, 2005) describe changes in obsidian procurement patterns during the Upper Palaeolithic.

A series of geochemical studies using Electron Microprobe Analysis (EMPA) were conducted on obsidian sources located throughout Hokkaido by M. Mukai and colleagues (Mukai 2005; Mukai and Wada 2001, 2003, 2004a, 2004b; Mukai *et al.* 2000, 2004). ED–XRF analyses were performed on various sources (Hall and Kimura 2002; Inoue 2003; Warashina 1999; Yoshitani *et al.* 2001). NAA was used on the Akaigawa, Oketo, Shirataki, and Tokachi-Mitsumata sources to establish the existence of a long-distance exchange network for obsidian from Hokkaido to Sakhalin Island (Kuzmin 2006; Kuzmin and Glascock 2007; Kuzmin *et al.* 2002, 2013; Sato 2004a, 2004b; Sato *et al.* 2002). Phillips and Speakman (2009) conducted an initial evaluation of obsidian on the Kurile Islands and used a portable ED–XRF spectrometer to determine that Hokkaido was the source of most obsidian to the southern Kuriles. However, as reported by Izuho and Sato (2007) no single comprehensive study of all sources in Hokkaido has been performed until very recently (see Wada *et al.*, this volume).

APPLICATION OF NAA AND ED–XRF TO OBSIDIAN FROM HOKKAIDO

In the following sections we describe the use of both NAA and ED–XRF to analyse source samples, in order to determine the analytical technique(s) necessary for accurate source assignments. We begin our work with a

Table 2.2. Current and previously published obsidian source names for Hokkaido Island

Source Name	Source Name	Source Name (Primary Locality*)
This Paper	**Mukai (2010)**	**Izuho and Sato (2007), Izuho and Hirose (2010)**
Asahikawa (Shunkodai)	Asahikawa-I	Chikabumidai (Unkown)
Asahikawa (Higashitakasu)	Asahikawa-II	Ubundai (Unkown)
Takikawa	Takikawa	Hokuryu (Unknown)
		Chippubetsu (Unkown)
Akaigawa	Akaigawa	Akaigawa (Dobokuzawa)
Nayoro	Nayoro	Nayoro (Unknown)
Engaru	Engaru	Engaru (Unkown)
Oumu	Oumu	Oumu (Unkown)
Okushiri	Okushiri	Okushiri (Katsumayama)
Tokachi (Shikaribetsu)	Tokachi-I	Shikaribetsu (Unknown)
Tokachi-Mitsumata	Tokachi-II	Tokachi-Mitsumata (Jyusan'nosawa)
Ikutahara	Ikutahara-I, Ikutahara-II	Ikutahara (Nitappugawa)
Rubeshibe 1	Rubeshibe-I	Keshomappu
Rubeshibe 2	Rubeshibe-II	
Toyoura	Toyoura	Toyoizumi (Unkown)
Monbetsu	Monbetsu	Monbetsu (Kamimobetsu)
Kushiro (Shitakara)	Kushiro-I	Kushiro (Unknown)
Kushiro (Kutyorogawa)	Kushiro-II	
Oketo (Tokoroyama)	Oketo-I	Oketo (Tokoroyama)
Oketo (Oketoyama)	Oketo-II	Oketo (Oketoyama)
Shirataki (Akaishiyama)	Shirataki-II	Shirataki (Akaishiyama)
Shirataki (Tokachi-Ishizawa)	Shirataki-I	Shirataki (Tokachi-Ishizawa)
—	—	Abashiri (Ponmoimisaki)

* Sources without indication as "Unknown" have primary locales.

comprehensive analysis of all available source samples by NAA, followed by analysis of the same samples by ED–XRF. The results from analysing the source samples are then used to identify the analytical method/ procedure(s) required for a successful analysis of the artefacts. Compositional data for obsidian from all known sources on the island of Hokkaido are presented to demonstrate the integration of data from multiple methods. For illustration, we describe the analysis of an assemblage of 129 obsidian artefacts from the site of Ogachi-Kato 2.

NAA Source Groups

The NAA at the Archaeometry Laboratory, University of Missouri Research Reactor Center (hereafter – MURR), consists of a combination of two separate irradiations and three counts. Full details on the irradiation and analytical procedures are available elsewhere (Glascock *et al.* 1998). The resulting dataset includes parts-per-million (ppm) concentrations for 28 elements (see Table 2.3). Only three scatterplots are needed to totally separate the source data into 21 groups. A plot of Rb vs. Sb (Figure 2.2) clearly isolates ten of the source groups. A second plot of Mn vs. Fe (Figure 2.3) separates another eight groups. The final three groups are shown in a plot of Cs vs. Co (Figure 2.4). While it is possible to use multivariate statistical methods (such as Principal Component Analysis and Mahalanobis distance) to separate and assign unknown samples, we typically use visual examination of numerous scatterplots, and Figures 2.2–2.4 demonstrate how the individual sources can be visually separated.

XRF Source Groups

While NAA is capable of readily distinguishing between all 21 known source groups in Hokkaido, restrictions on destructive analysis, as well as funding, often necessitate the use of an alternate technique for the analysis of archaeological samples. We have analysed the same source samples studied by NAA at MURR by ED–XRF and found that it is possible to separate a majority of the source groups. Here, we focus on six elements (Fe, Rb, Sr, Y, Zr, and Nb) reliably detected by ED–XRF in the obsidian source samples from Hokkaido. The

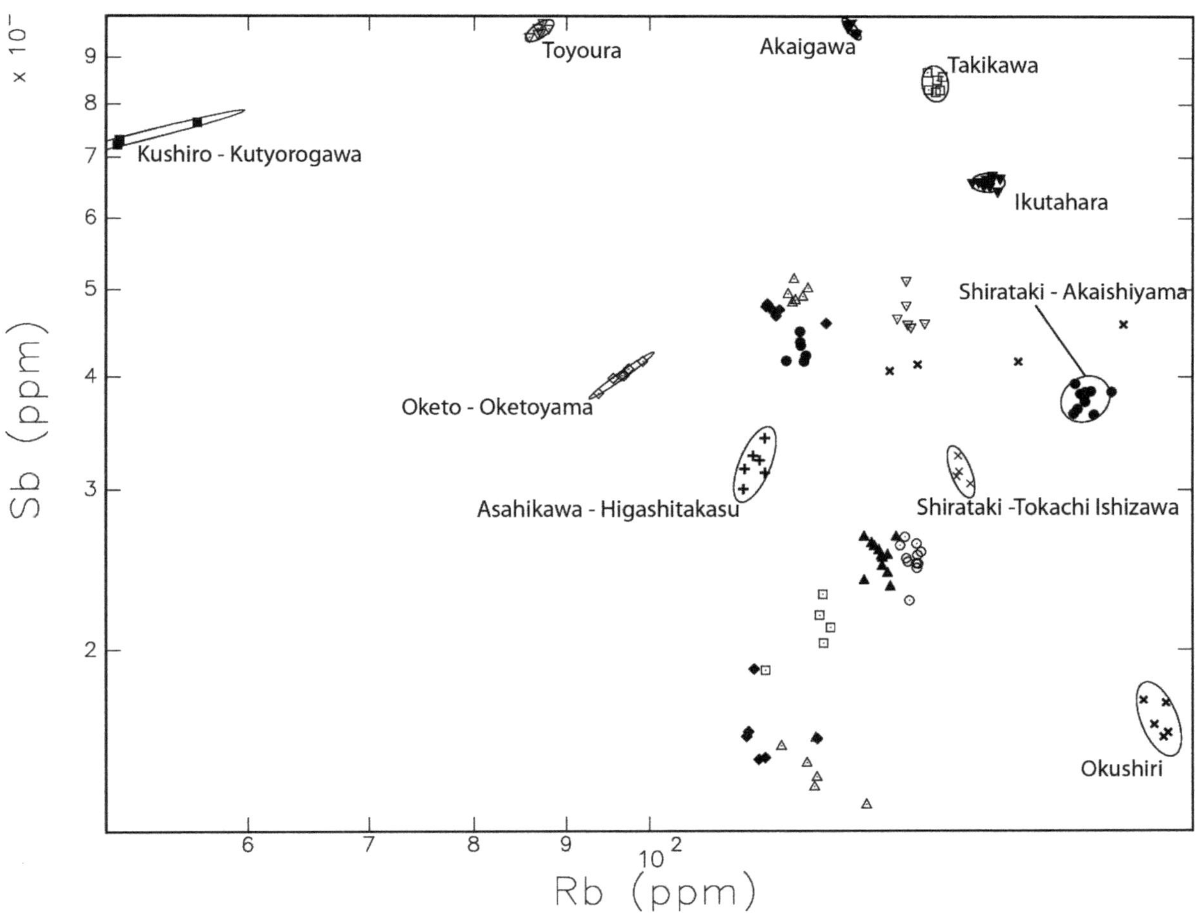

Figure 2.2. Scatterplot of Rb vs. Sb concentrations from NAA for all obsidian sources in Hokkaido. On Figures 2.2–2.11, only sources separated with 95% confidence ellipses are individually labelled

Table 2.3. Element concentration means and standard deviations by NAA and ED–XRF for obsidian sources in Hokkaido. Concentrations are listed in parts-per-million (ppm) of the element, unless % is noted

Element	**Akaigawa**	**Asahikawa (Higashitakasu)**	**Asahikaw (Syunkodai)**	**Engaru**	**Ikutahara**
NAA Results	(n = 5)	(n = 6)	(n = 6)	(n = 6)	(n = 12)
Na (%)	2.66 ± 0.02	2.78 ± 0.03	2.89 ± 0.05	3.10 ± 0.09	2.97 ± 0.05
Al (%)	6.39 ± 0.14	7.09 ± 0.15	7.60 ± 0.34	6.81 ± 0.36	6.99 ± 0.25
Cl	1202 ± 76	457 ± 31	394 ± 52	557 ± 30	592 ± 35
K (%)	3.68 ± 0.15	2.83 ± 0.15	2.74 ± 0.23	3.28 ± 0.27	3.69 ± 0.12
Sc	2.10 ± 0.02	2.79 ± 0.02	3.68 ± 0.07	4.71 ± 0.05	6.03 ± 0.05
Mn	483 ± 10	409 ± 4	554 ± 10	405 ± 3	241 ± 2
Fe (%)	0.71 ± 0.01	0.96 ± 0.01	1.35 ± 0.02	0.93 ± 0.01	1.11 ± 0.01
Co	0.46 ± 0.01	0.59 ± 0.03	1.54 ± 0.03	0.24 ± 0.02	0.50 ± 0.03
Zn	27.8 ± 0.7	40.3 ± 1.2	49.0 ± 1.8	56.1 ± 1.3	56.2 ± 2.4
Rb	129 ± 1	121 ± 1	114 ± 1	118 ± 3	154 ± 2
Sr	54 ± 5	128 ± 11	181 ± 29	50 ± 7	43 ± 9
Zr	120 ± 4	109 ± 4	122 ± 10	150 ± 7	203 ± 7
Sb	0.97 ± 0.01	0.43 ± 0.01	0.32 ± 0.01	0.47 ± 0.01	0.66 ± 0.01
Cs	10.5 ± 0.1	10.2 ± 0.1	8.7 ± 0.1	6.9 ± 0.1	9.1 ± 0.1
Ba	718 ± 13	724 ± 14	670 ± 20	761 ± 9	824 ± 23
La	27.2 ± 0.1	19.7 ± 0.2	18.0 ± 0.2	25.2 ± 0.2	29.4 ± 0.3

Element	Akaigawa	Asahikawa (Higashitakasu)	Asahikaw (Syunkodai)	Engaru	Ikutahara
Ce	53.6 ± 0.3	41.5 ± 0.4	38.7 ± 0.5	54.7 ± 0.6	64.7 ± 0.5
Nd	18.7 ± 1.2	15.3 ± 0.4	15.3 ± 0.7	23.7 ± 0.5	31.0 ± 7.6
Sm	3.68 ± 0.01	3.51 ± 0.02	3.51 ± 0.04	5.58 ± 0.03	6.29 ± 0.35
Eu	0.343 ± 0.003	0.490 ± 0.004	0.586 ± 0.010	0.667 ± 0.006	0.474 ± 0.005
Tb	0.51 ± 0.01	0.54 ± 0.01	0.57 ± 0.03	1.00 ± 0.03	1.01 ± 0.02
Dy	3.49 ± 0.34	3.12 ± 0.15	3.40 ± 0.23	6.51 ± 0.34	6.42 ± 0.28
Yb	3.12 ± 0.05	2.29 ± 0.03	2.34 ± 0.09	4.47 ± 0.02	4.01 ± 0.14
Lu	0.54 ± 0.01	0.37 ± 0.01	0.37 ± 0.01	0.65 ± 0.01	0.59 ± 0.02
Hf	3.42 ± 0.22	2.94 ± 0.03	3.26 ± 0.05	4.35 ± 0.09	6.66 ± 0.09
Ta	0.60 ± 0.01	0.58 ± 0.01	0.60 ± 0.01	0.64 ± 0.01	0.48 ± 0.01
Th	17.9 ± 0.1	10.2 ± 0.1	8.7 ± 0.1	10.2 ± 0.1	12.5 ± 0.1
U	5.2 ± 0.2	4.1 ± 0.2	3.8 ± 0.1	3.0 ± 0.2	3.1 ± 0.4
XRF Results	(n = 5)	(n = 6)	(n = 6)	(n = 5)	(n = 12)
Mn	487 ± 93	446 ± 57	553 ± 71	369 ± 12	256 ± 79
Fe (%)	0.80 ± 0.05	0.94 ± 0.01	1.32 ± 0.06	1.07 ± 0.05	1.19 ± 0.15
Rb	138 ± 6	122 ± 3	114 ± 5	127 ± 7	150 ± 16
Sr	55 ± 4	113 ± 6	152 ± 10	54 ± 4	50 ± 5
Y	26 ± 4	23 ± 1	25 ± 2	42 ± 2	41 ± 5
Zr	117 ± 3	113 ± 4	131 ± 4	168 ± 10	233 ± 24
Nb	7.4 ± 2.2	5.1 ± 1.4	6.4 ± 1.0	9.5 ± 1.3	6.7 ± 1.4

Element	Kushiro (Kutyorogawa)	Kushiro (Shitakara)	Monbetsu	Nayaro	Oketo (Oketoyama)
NAA Results	(n = 3)	(n = 5)	(n = 4)	(n = 6)	(n = 6)
Na (%)	3.33 ± 0.04	2.93 ± 0.03	2.47 ± 0.24	2.83 ± 0.01	3.13 ± 0.02
Al (%)	6.53 ± 0.16	6.50 ± 0.22	6.18 ± 0.11	6.93 ± 0.24	6.63 ± 0.18
Cl	760 ± 80	499 ± 27	111 ± 74	630 ± 31	473 ± 25
K (%)	1.76 ± 0.11	3.28 ± 0.07	2.74 ± 0.29	3.10 ± 0.09	2.89 ± 0.12
Sc	9.96 ± 0.24	3.95 ± 0.03	2.78 ± 0.01	2.64 ± 0.02	3.23 ± 0.06
Mn	690 ± 21	362 ± 5	239 ± 9	278 ± 1	374 ± 4
Fe (%)	1.53 ± 0.08	0.71 ± 0.01	0.72 ± 0.01	0.92 ± 0.01	0.87 ± 0.01
Co	0.84 ± 0.09	0.18 ± 0.01	0.32 ± 0.01	0.73 ± 0.01	0.45 ± 0.01
Zn	74.6 ± 3.5	36.0 ± 0.8	34.6 ± 1.9	34.6 ± 1.1	36.1 ± 0.6
Rb	53 ± 3	139 ± 1	155 ± 21	121 ± 1	97 ± 2
Sr	155 ± 21	50 ± 6	52 ± 9	115 ± 6	76 ± 4
Zr	183 ± 20	102 ± 7	92 ± 10	129 ± 3	130 ± 11
Sb	0.74 ± 0.02	0.26 ± 0.01	0.42 ± 0.02	0.50 ± 0.01	0.40 ± 0.01
Cs	3.7 ± 0.1	8.2 ± 0.1	12.2 ± 1.7	8.8 ± 0.1	5.2 ± 0.1
Ba	524 ± 20	872 ± 10	1043 ± 165	627 ± 14	698 ± 13
La	13.4 ± 0.2	21.8 ± 0.1	17.7 ± 0.2	21.8 ± 0.3	20.6 ± 0.4
Ce	32.9 ± 0.7	46.7 ± 0.5	38.5 ± 0.3	44.8 ± 0.5	41.0 ± 0.9
Nd	20.0 ± 2.0	18.3 ± 0.9	17.4 ± 0.7	16.7 ± 0.9	15.0 ± 0.5
Sm	5.72 ± 0.02	4.59 ± 0.03	4.69 ± 0.04	3.37 ± 0.03	3.22 ± 0.03
Eu	1.186 ± 0.039	0.468 ± 0.005	0.333 ± 0.006	0.523 ± 0.007	0.524 ± 0.010
Tb	1.16 ± 0.01	0.76 ± 0.01	0.87 ± 0.01	0.45 ± 0.01	0.51 ± 0.02

Element	Kushiro (Kutyorogawa)	Kushiro (Shitakara)	Monbetsu	Nayaro	Oketo (Oketoyama)
Dy	7.44 ± 0.89	4.63 ± 0.24	5.47 ± 0.21	2.89 ± 0.36	3.40 ± 0.08
Yb	5.49 ± 0.01	3.24 ± 0.03	4.16 ± 0.07	2.13 ± 0.03	2.46 ± 0.05
Lu	0.77 ± 0.02	0.49 ± 0.02	0.63 ± 0.01	0.36 ± 0.01	0.37 ± 0.01
Hf	5.84 ± 0.13	3.09 ± 0.05	3.41 ± 0.07	3.74 ± 0.04	3.51 ± 0.06
Ta	0.17 ± 0.01	0.58 ± 0.01	0.44 ± 0.01	0.42 ± 0.01	0.51 ± 0.01
Th	4.5 ± 0.1	12.1 ± 0.1	9.7 ± 0.1	12.0 ± 0.1	9.2 ± 0.2
U	1.6 ± 0.3	3.6 ± 0.1	3.3 ± 0.1	3.7 ± 0.3	2.5 ± 0.1
XRF Results	(n = 3)	(n = 5)	(n = 4)	(n = 6)	(n = 6)
Mn	716 ± 162	367 ± 92	268 ± 54	265 ± 62	395 ± 103
Fe (%)	1.85 ± 0.23	0.85 ± 0.11	0.80 ± 0.04	0.94 ± 0.03	0.94 ± 0.02
Rb	47 ± 8	142 ± 10	159 ± 24	122 ± 4	102 ± 3
Sr	130 ± 21	52 ± 3	61 ± 16	104 ± 6	84 ± 5
Y	35 ± 5	32 ± 4	36 ± 2	22 ± 3	25 ± 2
Zr	176 ± 43	108 ± 5	117 ± 2	139 ± 4	150 ± 3
Nb	5.8 ± 4.0	8.3 ± 1.5	4.7 ± 2.1	5.2 ± 1.3	7.5 ± 1.5

Element	Oketo (Tokoroyama and Kitatokoroyama)	Okushiri	Oumu	Rubeshibe 1 (Iwayama)	Rubeshibe 2 (Kayokozawa)
NAA Results	(n = 12)	(n = 5)	(n = 6)	(n = 6)	(n = 5)
Na (%)	2.71 ± 0.08	2.68 ± 0.03	2.80 ± 0.04	2.70 ± 0.12	2.81 ± 0.05
Al (%)	6.54 ± 0.29	7.38 ± 0.39	6.75 ± 0.31	7.07 ± 0.15	7.15 ± 0.22
Cl	467 ± 39	555 ± 60	519 ± 24	523 ± 30	511 ± 36
K (%)	3.47 ± 0.25	4.01 ± 0.15	3.67 ± 0.20	2.93 ± 0.18	3.02 ± 0.08
Sc	3.27 ± 0.04	1.96 ± 0.06	5.40 ± 0.05	2.92 ± 0.02	2.91 ± 0.02
Mn	318 ± 4	792 ± 21	177 ± 3	395 ± 5	433 ± 10
Fe (%)	0.72 ± 0.01	0.44 ± 0.04	0.85 ± 0.01	1.18 ± 0.01	1.29 ± 0.01
Co	0.53 ± 0.04	0.28 ± 0.04	0.52 ± 0.02	0.60 ± 0.01	0.71 ± 0.04
Zn	27.6 ± 1.6	20.0 ± 0.7	44.5 ± 0.8	45.9 ± 0.6	48.4 ± 0.3
Rb	134 ± 2	191 ± 2	139 ± 2	124 ± 4	114 ± 1
Sr	67 ± 8	131 ± 6	55 ± 9	123 ± 16	146 ± 13
Zr	114 ± 6	95 ± 5	127 ± 7	122 ± 7	132 ± 6
Sb	0.25 ± 0.01	0.17 ± 0.01	0.47 ± 0.02	0.15 ± 0.01	0.16 ± 0.02
Cs	6.7 ± 0.1	12.0 ± 0.2	9.0 ± 0.1	6.3 ± 0.1	6.0 ± 0.1
Ba	979 ± 18	870 ± 17	944 ± 15	779 ± 22	754 ± 19
La	21.8 ± 0.2	21.4 ± 1.9	28.4 ± 0.2	24.2 ± 0.2	23.2 ± 0.2
Ce	43.0 ± 0.5	40.8 ± 3.1	64.9 ± 0.5	52.4 ± 0.5	49.8 ± 0.3
Nd	15.0 ± 0.7	14.2 ± 0.6	29.8 ± 0.6	21.6 ± 1.9	21.0 ± 1.8
Sm	3.34 ± 0.02	2.95 ± 0.09	7.05 ± 0.08	4.59 ± 0.06	4.43 ± 0.03
Eu	0.358 ± 0.005	0.446 ± 0.007	0.403 ± 0.008	0.596 ± 0.007	0.627 ± 0.007
Tb	0.51 ± 0.01	0.37 ± 0.01	1.20 ± 0.02	0.67 ± 0.02	0.63 ± 0.01
Dy	3.38 ± 0.44	2.26 ± 0.36	8.22 ± 0.50	4.12 ± 0.30	3.89 ± 0.39
Yb	2.62 ± 0.02	2.09 ± 0.04	4.53 ± 0.06	2.81 ± 0.11	2.77 ± 0.11
Lu	0.42 ± 0.01	0.38 ± 0.01	0.65 ± 0.01	0.42 ± 0.01	0.42 ± 0.01
Hf	3.12 ± 0.03	2.05 ± 0.13	4.42 ± 0.05	3.74 ± 0.04	3.88 ± 0.01

Element	Oketo (Tokoroyama and Kitatokoroyama)	Okushiri	Oumu	Rubeshibe 1 (Iwayama)	Rubeshibe 2 (Kayokozawa)
Ta	0.57 ± 0.01	0.72 ± 0.01	0.54 ± 0.01	0.48 ± 0.01	0.46 ± 0.02
Th	11.8 ± 0.1	17.6 ± 0.7	11.7 ± 0.1	10.1 ± 0.1	9.5 ± 0.1
U	3.7 ± 0.3	5.4 ± 0.1	3.5 ± 0.1	2.9 ± 0.2	2.9 ± 0.1
XRF Results	(n = 12)	(n = 5)	(n = 6)	(n = 6)	(n = 6)
Mn	335 ± 106	735 ± 92	181 ± 58	445 ± 88	482 ± 118
Fe (%)	0.76 ± 0.05	0.50 ± 0.04	0.92 ± 0.08	1.21 ± 0.06	1.31 ± 0.10
Rb	134 ± 4	185 ± 11	137 ± 5	126 ± 8	118 ± 7
Sr	68 ± 3	121 ± 10	49 ± 3	121 ± 6	136 ± 11
Y	24 ± 3	21 ± 3	43 ± 2	28 ± 3	29 ± 2
Zr	123 ± 5	89 ± 5	143 ± 4	142 ± 6	150 ± 7
Nb	6.7 ± 1.4	8.6 ± 1.6	8.9 ± 2.3	6.5 ± 2.4	6.7 ± 1.8

Element	Shirataki (Akaishiyama)	Shirataki (Tokachi-Ishizawa)	Takikawa	Tokachi-Mitsumata	Tokachi (Shikaribetsu)	Toyoura
NAA Results	(n = 4)	(n = 10)	(n = 6)	(n = 6)	(n = 5)	(n = 6)
Na (%)	2.82 ± 0.04	2.92 ± 0.02	2.72 ± 0.03	3.00 ± 0.05	2.65 ± 0.04	2.73 ± 0.04
Al (%)	7.05 ± 0.15	7.24 ± 0.27	6.69 ± 0.28	7.13 ± 0.25	6.82 ± 0.29	6.40 ± 0.11
Cl	643 ± 16	523 ± 54	702 ± 65	453 ± 43	381 ± 32	1156 ± 80
K (%)	3.85 ± 0.22	3.77 ± 0.13	3.47 ± 0.10	3.67 ± 0.12	2.91 ± 0.36	2.96 ± 0.15
Sc	2.65 ± 0.02	2.92 ± 0.04	2.99 ± 0.03	4.07 ± 0.01	4.26 ± 0.24	2.07 ± 0.13
Mn	379 ± 7	448 ± 5	492 ± 9	374 ± 3	332 ± 11	464 ± 5
Fe (%)	0.79 ± 0.01	0.74 ± 0.01	0.58 ± 0.01	0.74 ± 0.01	0.96 ± 0.16	0.83 ± 0.04
Co	0.13 ± 0.01	0.06 ± 0.01	0.40 ± 0.01	0.25 ± 0.01	0.77 ± 0.34	0.86 ± 0.30
Zn	36.4 ± 0.9	36.6 ± 1.1	25.4 ± 0.8	39.7 ± 3.9	37.6 ± 1.6	29.7 ± 2.7
Rb	149 ± 1	174 ± 3	144 ± 1	140 ± 1	129 ± 9	87 ± 1
Sr	31 ± 6	20 ± 7	86 ± 55	53 ± 3	101 ± 24	118 ± 14
Zr	95 ± 8	84 ± 7	120 ± 6	102 ± 9	105 ± 11	124 ± 10
Sb	0.31 ± 0.01	0.38 ± 0.01	0.84 ± 0.02	0.25 ± 0.01	0.22 ± 0.02	0.96 ± 0.01
Cs	9.5 ± 0.1	11.9 ± 0.2	11.6 ± 0.1	8.4 ± 0.1	8.0 ± 0.3	7.6 ± 0.1
Ba	850 ± 6	183 ± 12	1025 ± 19	882 ± 35	777 ± 63	805 ± 15
La	20.3 ± 0.8	13.1 ± 0.2	25.9 ± 0.3	22.1 ± 0.4	21.1 ± 0.6	20.1 ± 0.2
Ce	43.1 ± 1.4	31.3 ± 0.5	54.5 ± 0.3	48.3 ± 0.3	45.0 ± 1.5	42.6 ± 0.3
Nd	16.3 ± 0.8	12.6 ± 0.5	19.8 ± 0.4	22.1 ± 4.6	17.2 ± 4.9	16.6 ± 1.5
Sm	3.97 ± 0.04	3.82 ± 0.06	4.22 ± 0.03	4.73 ± 0.21	4.06 ± 0.39	3.47 ± 0.01
Eu	0.279 ± 0.005	0.127 ± 0.004	0.512 ± 0.007	0.484 ± 0.007	0.501 ± 0.022	0.482 ± 0.005
Tb	0.65 ± 0.01	0.74 ± 0.02	0.59 ± 0.01	0.79 ± 0.02	0.63 ± 0.09	0.48 ± 0.01
Dy	4.15 ± 0.21	4.77 ± 0.31	3.83 ± 0.31	5.32 ± 0.52	3.91 ± 0.56	2.87 ± 0.16
Yb	3.03 ± 0.02	3.59 ± 0.08	3.04 ± 0.04	3.35 ± 0.02	2.60 ± 0.42	3.25 ± 0.11
Lu	0.45 ± 0.01	0.54 ± 0.01	0.51 ± 0.01	0.50 ± 0.04	0.41 ± 0.06	0.54 ± 0.01
Hf	2.75 ± 0.05	2.67 ± 0.06	3.38 ± 0.03	3.20 ± 0.02	3.06 ± 0.07	4.09 ± 0.12
Ta	0.53 ± 0.01	0.67 ± 0.01	0.57 ± 0.01	0.59 ± 0.01	0.54 ± 0.03	0.37 ± 0.01
Th	11.2 ± 0.2	9.7 ± 0.1	16.8 ± 0.1	12.3 ± 0.1	10.8 ± 1.0	9.9 ± 0.1
U	3.1 ± 0.3	3.8 ± 0.2	4.9 ± 0.2	3.8 ± 0.3	3.4 ± 0.5	3.7 ± 0.2

Element	Shirataki (Akaishiyama)	Shirataki (Tokachi-Ishizawa)	Takikawa	Tokachi-Mitsumata	Tokachi (Shikaribetsu)	Toyoura
XRF Results	(n = 3)	(n = 10)	(n = 6)	(n = 6)	(n = 4)	(n = 6)
Mn	390 ± 109	423 ± 111	509 ± 76	427 ± 89	393 ± 130	411 ± 84
Fe (%)	0.76 ± 0.03	0.75 ± 0.07	0.68 ± 0.02	0.77 ± 0.05	1.31 ± 0.27	0.80 ± 0.02
Rb	136 ± 2	173 ± 10	154 ± 4	143 ± 8	133 ± 9	84 ± 4
Sr	31 ± 3	12 ± 1	67 ± 4	54 ± 3	120 ± 16	104 ± 5
Y	29 ± 2	34 ± 2	29 ± 3	32 ± 1	26 ± 2	24 ± 3
Zr	95 ± 1	91 ± 3	120 ± 2	112 ± 4	124 ± 4	134 ± 8
Nb	5.7 ± 2.6	6.9 ± 1.2	7.2 ± 1.7	6.9 ± 1.9	5.9 ± 1.8	5.9 ± 2.0

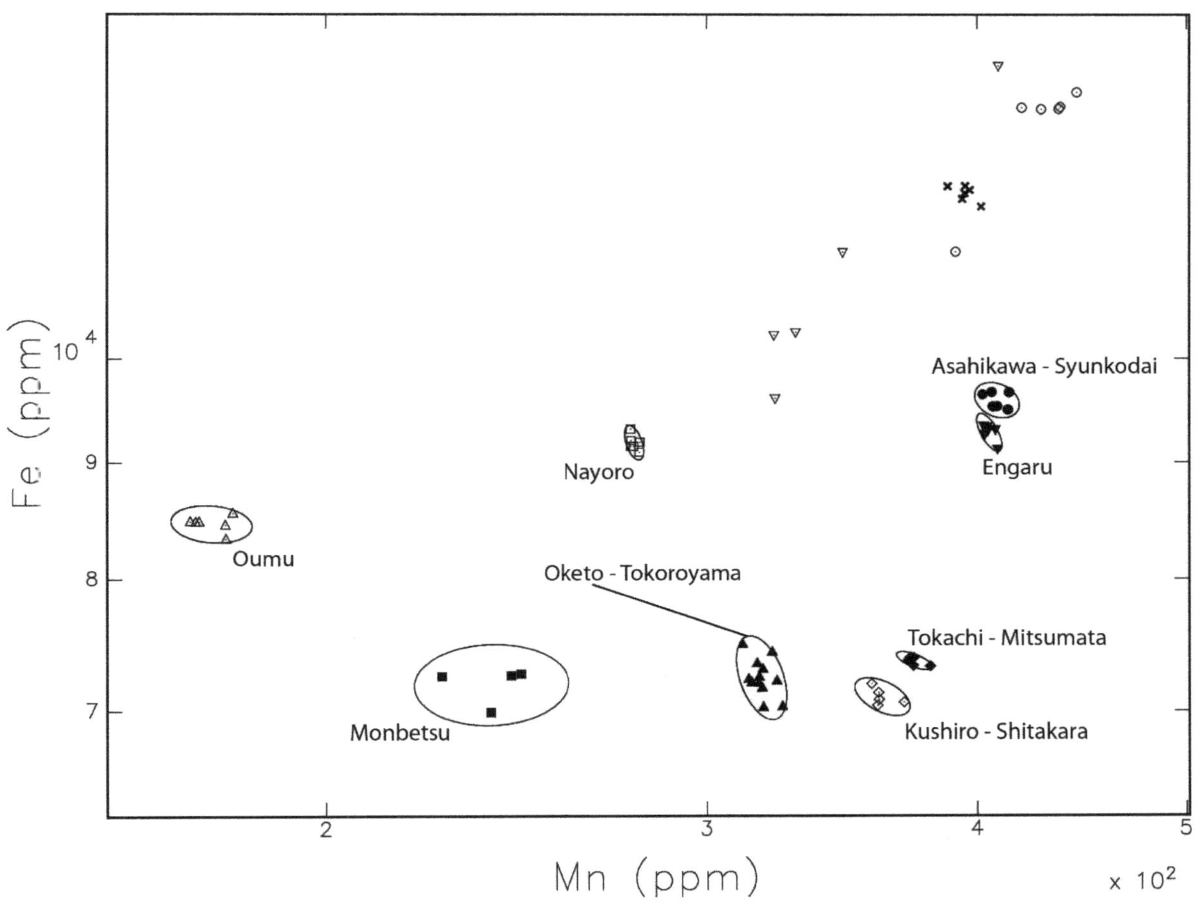

Figure 2.3. Scatterplot of Mn vs. Fe concentrations from NAA for obsidian sources in Hokkaido

concentration data are presented in Table 2.3. The analysis was conducted using a Bruker Tracer III-SD handheld ED–XRF apparatus, operated at 40 keV and 17 μA, with a filter consisting of 6 mil [one thousandth of an inch] Cu, 1 mil Ti, and 12 mil Al, and each sample was analysed for three minutes. The instrument was calibrated for obsidian using a set of 40 obsidian calibration samples developed at MURR (Glascock and Ferguson 2012).

The smaller number of elements for ED–XRF compared to NAA, coupled with reduced precision, necessitates using a greater number of bivariate plots to separate the source groups. Five source groups are distinguished in a plot of Rb vs. Sr (Figure 2.5). Four additional groups separate in a plot of Sr vs. Zr (Figure 2.6). The remaining samples split into two clusters (A and B) in a plot of Sr vs. Fe (Figure 2.7) and are plotted separately. A plot of Zr vs. Fe separates the first three source groups in Cluster A (Figure 2.8), and a plot of Fe vs. Rb (Figure 2.9) divides the last three groups. A plot of Rb vs. Sr (Figure 2.10) separates two of the source groups in Cluster B, but the remaining four groups — Monbetsu (Kamimobetsu), Kushiro (Shitakara), Tokachi-Mitsumata, and Akaigawa — are not distinguishable by ED–XRF (Figure 2.11). Laboratories using only ED–XRF (or another similarly

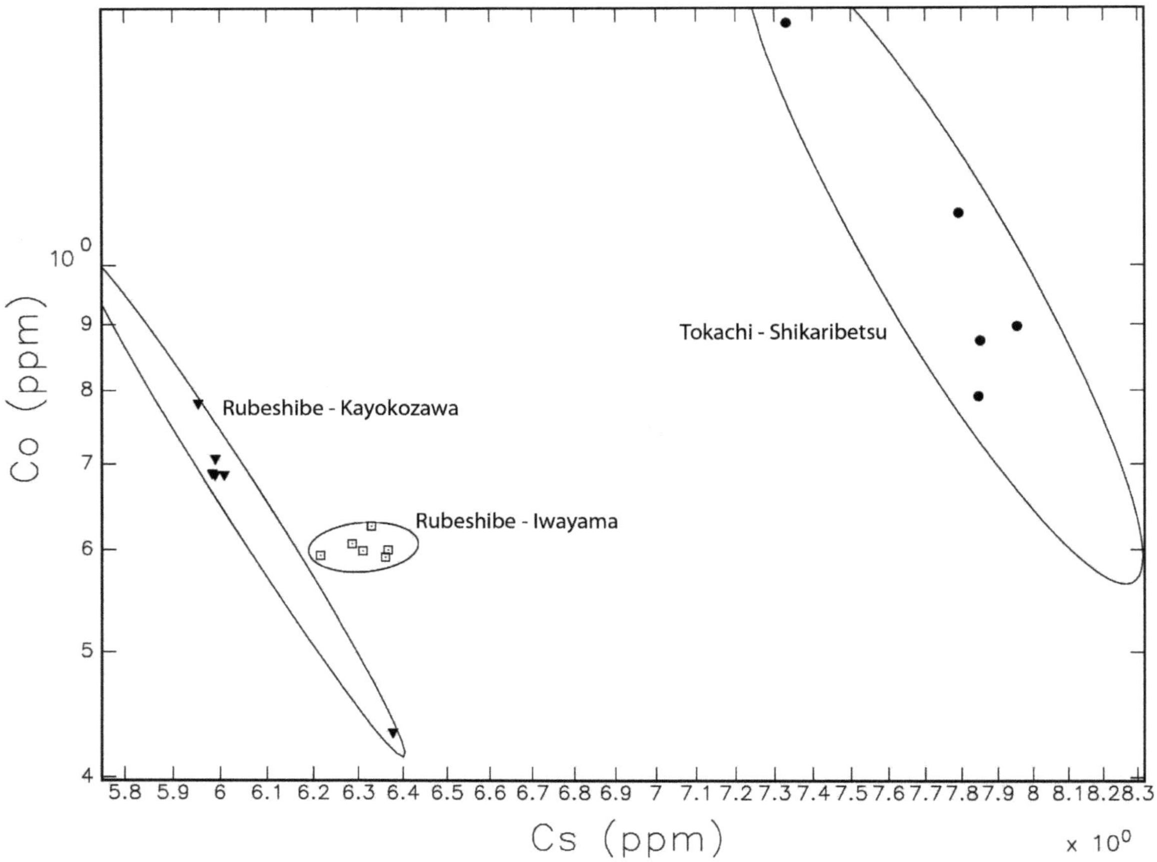

Figure 2.4. Scatterplot of Cs vs. Co concentrations from NAA for selected obsidian sources in Hokkaido

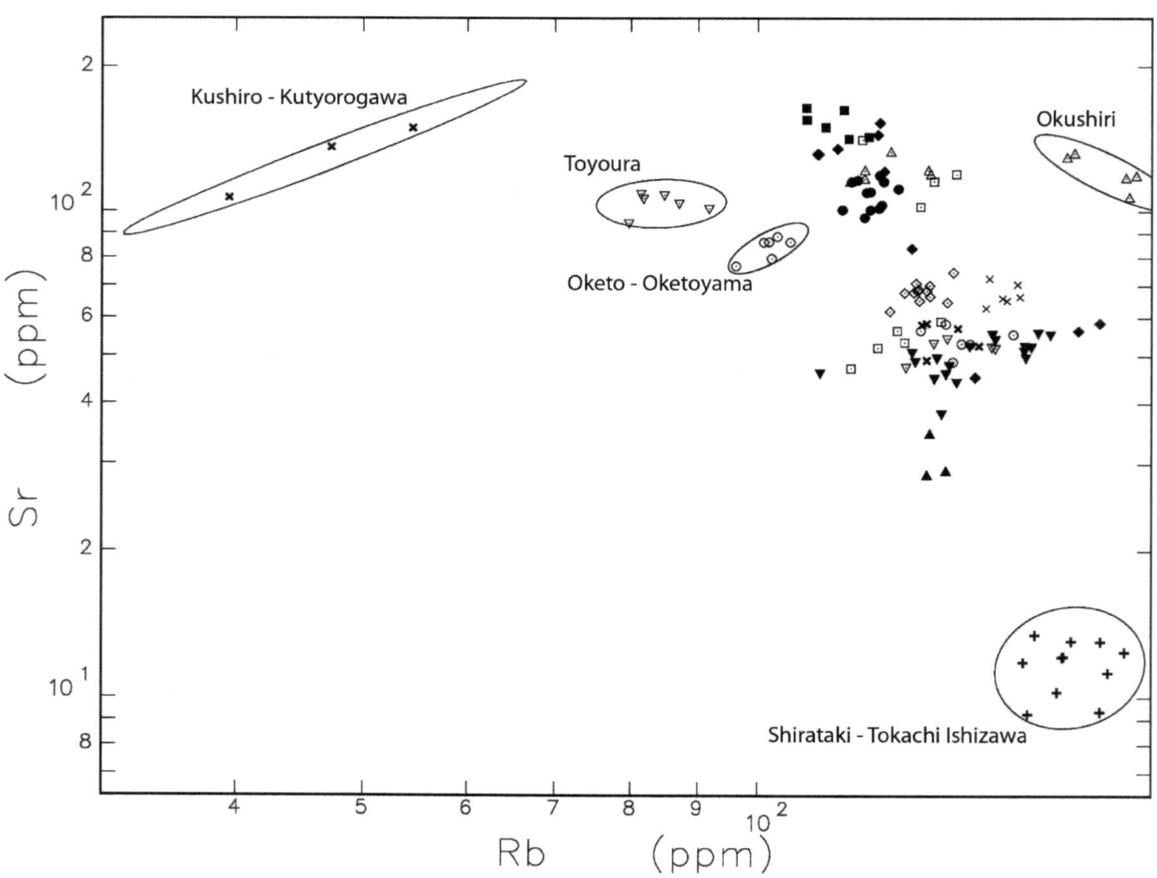

Figure 2.5. Scatterplot of Rb vs. Sr concentrations from ED–XRF for all sources in Hokkaido

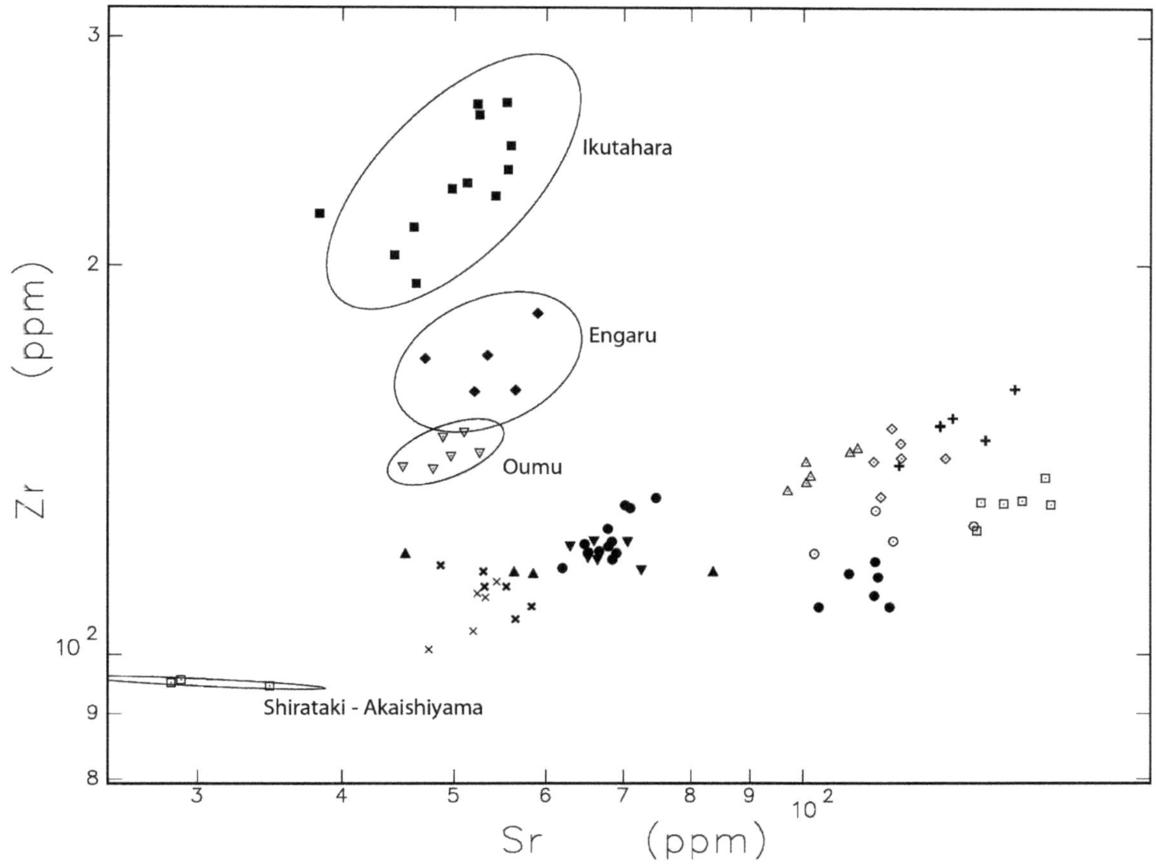

Figure 2.6. Scatterplot of Sr vs. Zr concentrations from ED–XRF for obsidian sources in Hokkaido

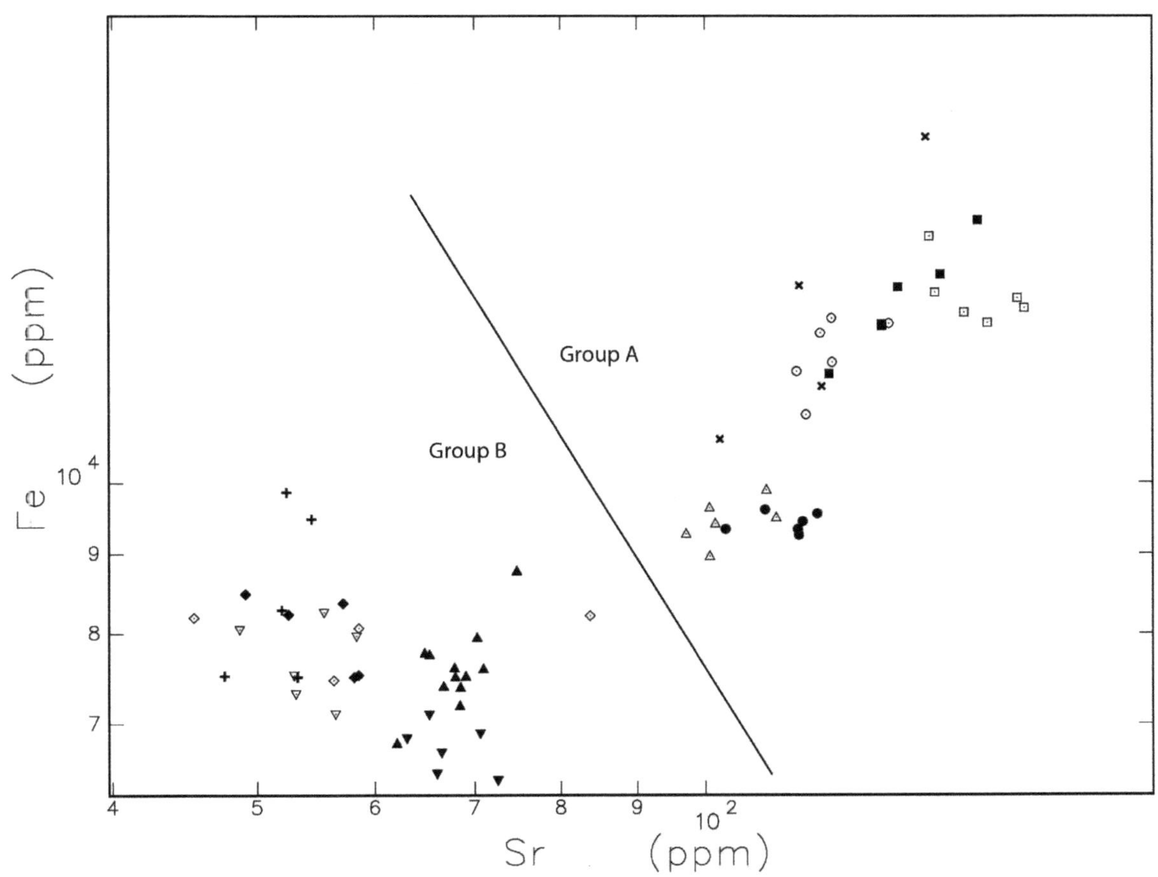

Figure 2.7. Scatterplot of Sr vs. Fe concentrations from ED–XRF used to separate groups A and B

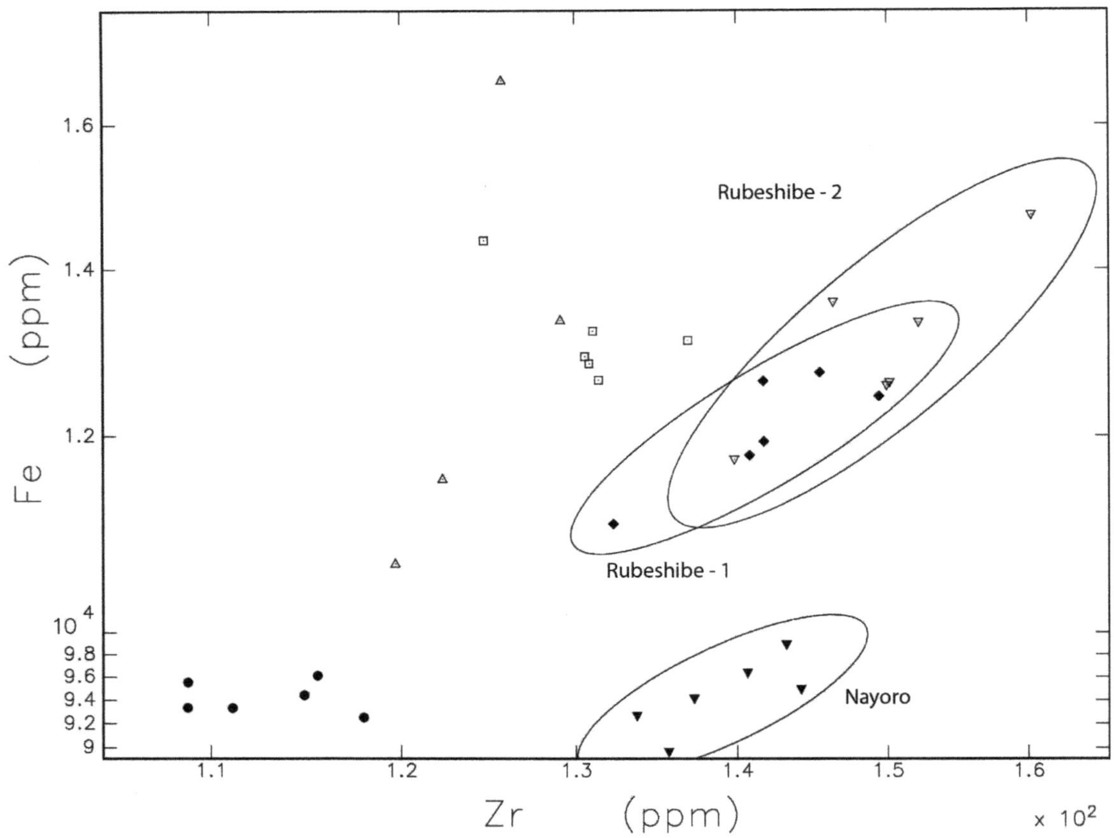

Figure 2.8. Scatterplot of Zr vs. Fe concentrations from ED–XRF used to separate the samples from Group A (see Figure 2.7)

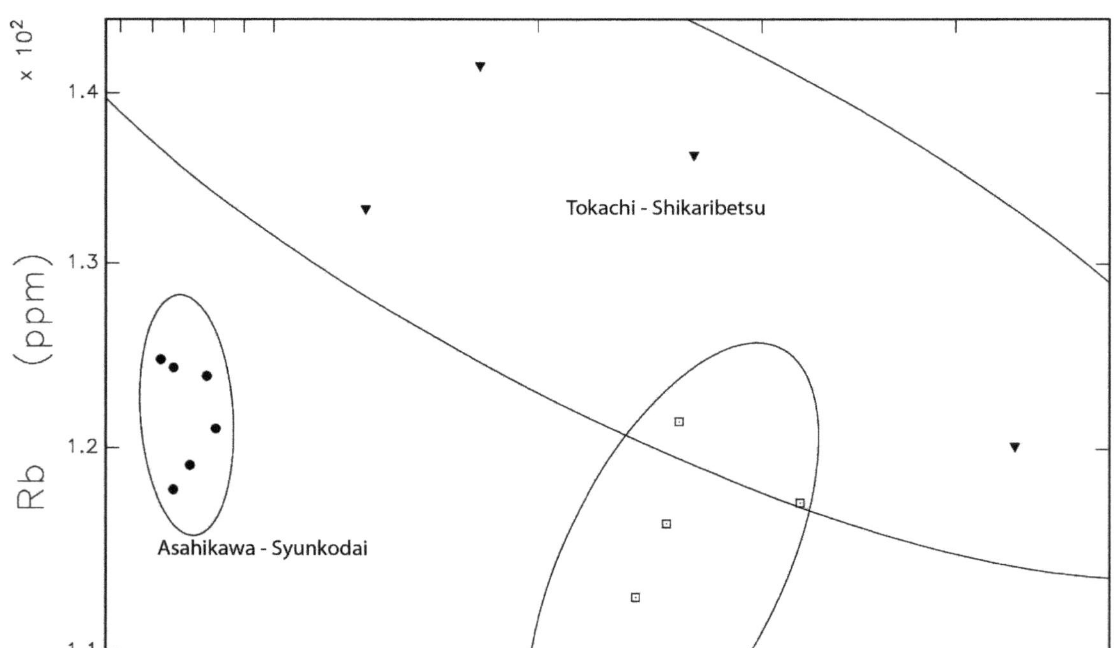

Figure 2.9. Scatterplot of Fe vs. Rb concentrations from ED–XRF used to separate the remaining samples (see Figure 2.8) from Group A in Figure 2.7

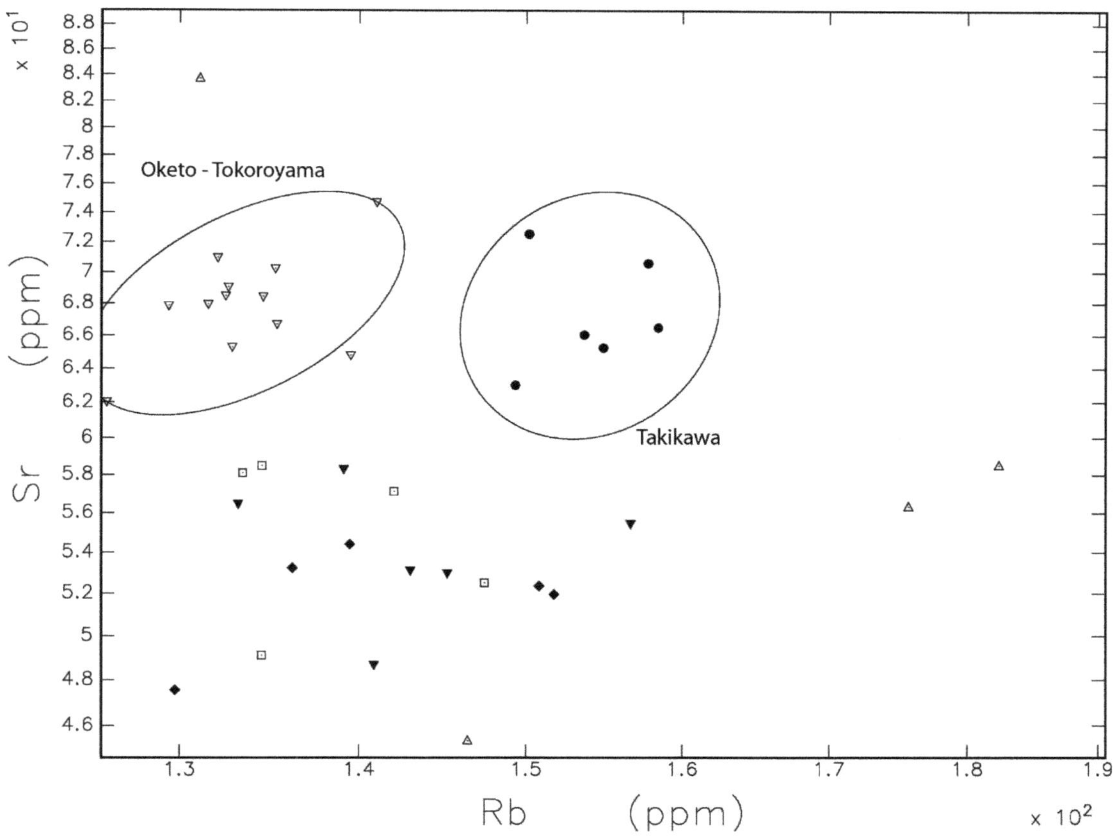

Figure 2.10. Scatterplot of Rb vs. Sr concentrations from ED–XRF used to separate the first two source groups from Group B (see Figure 2.7)

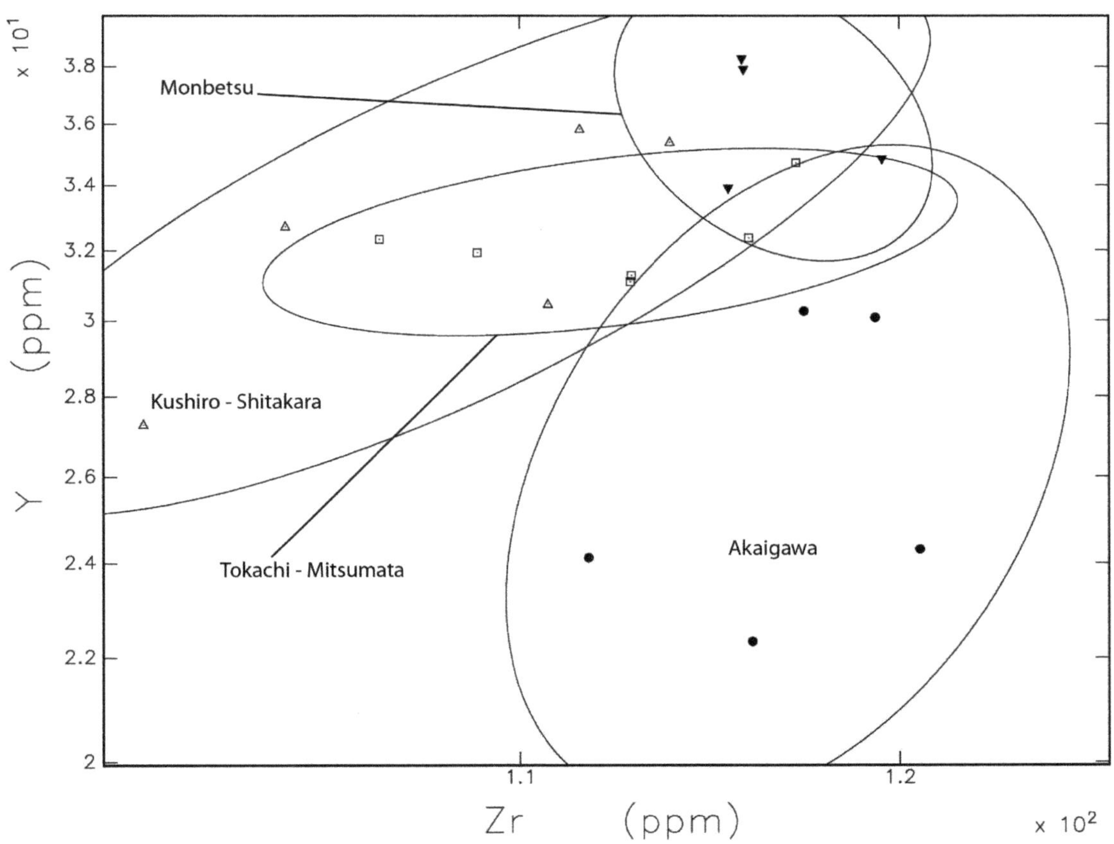

Figure 2.11. Scatterplot of Zr vs. Y concentrations from ED–XRF showing the overlap of the remaining source groups (see Figure 2.10) in Group B in Figure 2.7

limited technique) are not capable of accurately assigning sources to artefacts if any of the four overlapping sources are included in the sample.

Analysis of the Obsidian Artefacts from Ogachi-Kato 2 Site

NAA demonstrated that all 21 source groups are chemically unique; however, it is not possible to distinguish all of the groups by ED–XRF. We propose a two stage analysis that first maximises the speed, low cost, and non-destructive benefits of ED–XRF, and then applies a potentially non-destructive version of NAA to analyse any problematic samples. We have applied just such an approach to the analysis of 129 artefacts from the site of Ogachi-Kato 2 in Hokkaido (Figure 2.1; Table 2.4).

Ogachi-Kato 2 is located in the northeastern part of Hokkaido, on the second terrace of the Kunneppu River (see Izuho *et al.*, this volume). Upper Palaeolithic components are found in aeolian deposits which

Table 2.4. Element concentrations by ED–XRF (in ppm) of artefacts from the Ogachi-Kato 2 site (because many artefacts are extremely small, this causes some shift in concentrations, especially for Fe)

MURR ID	Artefact No.	Mn	Fe	Rb	Sr	Y	Zr	Nb	Source
MIH001	6	441	13,528	138	94	27	165	9	Oketo (Oketoyama)
MIH002	8	291	10,255	116	96	28	157	7	Oketo (Oketoyama)
MIH003	9	313	11,732	124	98	27	161	8	Oketo (Oketoyama)
MIH004	14	237	9069	105	86	22	152	10	Oketo (Oketoyama)
MIH005	29	368	11,394	119	88	23	148	9	Oketo (Oketoyama)
MIH006	31	299	6034	120	61	22	119	6	Oketo
MIH007	37	414	10,803	116	87	20	151	8	Oketo (Oketoyama)
MIH008	42	249	7002	140	75	27	130	6	Oketo
MIH009	45	274	7035	150	78	30	132	5	Oketo
MIH010	51	230	7920	106	90	23	154	8	Oketo (Oketoyama)
MIH011	54	387	8520	112	79	23	152	6	Oketo (Oketoyama)
MIH012	56	293	8396	97	73	26	153	7	Oketo (Oketoyama)
MIH013	68	729	13,659	129	134	27	152	5	Rubeshibe
MIH014	76	356	12,288	121	107	28	166	10	Oketo (Oketoyama)
MIH015	77	314	7021	84	75	21	150	6	Oketo (Oketoyama)
MIH016	78	349	6162	128	71	21	118	6	Oketo
MIH017	82	317	12,265	112	71	18	133	8	Oketo (Oketoyama)
MIH018	84	372	7206	94	71	22	138	6	Oketo (Oketoyama)
MIH019	85	626	9896	106	87	32	164	6	Oketo (Oketoyama)
MIH020	90	387	9554	111	92	25	154	9	Oketo (Oketoyama)
MIH021	91	460	10,836	111	91	23	167	6	Oketo (Oketoyama)
MIH022	100	371	8808	95	89	23	157	6	Oketo (Oketoyama)
MIH023	102	380	9737	119	95	26	164	8	Oketo (Oketoyama)
MIH024	109	522	11,413	117	94	29	146	9	Oketo (Oketoyama)
MIH025	116	438	11,057	112	96	26	149	10	Oketo (Oketoyama)
MIH026	117	511	11,488	110	93	25	149	9	Oketo (Oketoyama)
MIH027	118	353	5953	124	60	21	121	7	Oketo
MIH028	126	371	13,308	127	123	30	129	9	Tokachi (Shikaribetsu)
MIH029	128	292	6432	78	71	22	135	6	Oketo (Oketoyama)
MIH030	133	502	11,741	111	91	21	146	7	Oketo (Oketoyama)
MIH031	134	622	11,620	119	90	24	171	6	Oketo (Oketoyama)
MIH032	135	358	12,300	116	100	29	170	10	Oketo (Oketoyama)
MIH033	137	414	6212	132	63	23	116	5	Oketo
MIH034	141	416	8449	103	76	27	156	5	Oketo (Oketoyama)
MIH035	145	468	10,811	114	99	27	158	7	Oketo (Oketoyama)

MURR ID	Artefact No.	Mn	Fe	Rb	Sr	Y	Zr	Nb	Source
MIH036	150	497	12,024	128	97	24	169	9	Oketo (Oketoyama)
MIH037	156	327	10,922	109	96	25	146	9	Oketo (Oketoyama)
MIH038	158	261	8888	97	83	25	146	6	Oketo (Oketoyama)
MIH039	165	376	9252	115	92	23	151	10	Oketo (Oketoyama)
MIH040	166	468	11,802	96	90	25	138	7	Oketo (Oketoyama)
MIH041	170	421	11,177	127	100	27	172	10	Oketo (Oketoyama)
MIH042	173	426	10,338	119	99	27	160	3	Oketo (Oketoyama)
MIH043	181	430	8753	170	90	33	147	9	Oketo
MIH044	185	419	9595	110	95	23	147	9	Oketo (Oketoyama)
MIH045	188	265	9174	105	84	25	156	10	Oketo (Oketoyama)
MIH046	190	487	11,664	111	84	25	162	11	Oketo (Oketoyama)
MIH047	191	281	9243	110	97	23	165	11	Oketo (Oketoyama)
MIH048	192	550	10,646	108	83	27	153	8	Oketo (Oketoyama)
MIH049	194	234	7683	72	65	25	121	8	Oketo (Oketoyama)
MIH050	195	466	9813	97	82	25	136	4	Oketo (Oketoyama)
MIH051	199	602	8767	108	35	19	80	5	Tokachi-Mitsumata
MIH052	201	727	12,255	96	81	26	157	7	Oketo (Oketoyama)
MIH053	202	349	9594	97	78	20	140	6	Oketo (Oketoyama)
MIH054	205	368	12,181	131	98	28	168	12	Oketo (Oketoyama)
MIH055	206	341	11,277	115	80	30	160	8	Oketo (Oketoyama)
MIH056	210	360	9400	119	94	22	164	8	Oketo (Oketoyama)
MIH057	215	410	9441	121	91	27	155	11	Oketo (Oketoyama)
MIH058	217	284	9084	106	87	22	155	4	Oketo (Oketoyama)
MIH059	219	553	12,024	122	95	20	148	7	Oketo (Oketoyama)
MIH060	224	542	10,634	117	87	23	156	9	Oketo (Oketoyama)
MIH061	225	507	12,372	96	78	20	127	9	Oketo (Oketoyama)
MIH062	228	523	10,19	96	82	24	145	5	Oketo (Oketoyama)
MIH063	230	686	11,075	111	91	27	155	9	Oketo (Oketoyama)
MIH064	234	435	10,838	116	98	21	146	7	Oketo (Oketoyama)
MIH065	236	338	17,677	160	49	32	215	11	Ikutahara
MIH066	253	367	10,069	118	93	26	161	7	Oketo (Oketoyama)
MIH067	254	120	12,555	187	59	37	234	9	Ikutahara
MIH068	259	223	9804	106	91	26	155	5	Oketo (Oketoyama)
MIH069	260	441	12,473	100	81	23	122	12	Oketo (Oketoyama)
MIH070	261	341	10,229	113	90	29	152	7	Oketo (Oketoyama)
MIH071	264	427	12,993	93	68	16	114	7	Oketo (Oketoyama)
MIH072	266	411	12,442	120	92	28	137	8	Oketo (Oketoyama)
MIH073	274	576	11,199	95	80	23	134	5	Oketo (Oketoyama)
MIH074	276	520	13,168	124	99	26	146	8	Oketo (Oketoyama)
MIH075	280	362	9244	114	83	26	142	7	Oketo (Oketoyama)
MIH076	282	301	11,812	136	43	37	203	6	Ikutahara
MIH077	285	602	12,429	105	96	23	155	15	Oketo (Oketoyama)
MIH078	286	223	11,988	163	44	43	229	7	Ikutahara
MIH079	287	252	7065	136	66	22	121	8	Oketo
MIH080	294	281	12,649	161	60	41	209	10	Ikutahara
MIH081	295	529	10,712	113	85	23	138	4	Oketo (Oketoyama)
MIH082	298	612	12,004	114	89	22	167	11	Oketo (Oketoyama)

MURR ID	Artefact No.	Mn	Fe	Rb	Sr	Y	Zr	Nb	Source
MIH083	301	383	7500	89	72	26	136	5	Oketo (Oketoyama)
MIH084	305	473	12,107	108	94	26	148	10	Oketo (Oketoyama)
MIH085	309	289	5944	120	71	22	113	6	Oketo
MIH086	313	410	7880	160	27	28	108	12	Tokachi-Mitsumata?
MIH087	317	498	13,269	108	138	29	155	7	Rubeshibe
MIH088	319	410	9796	103	88	30	155	7	Oketo (Oketoyama)
MIH089	325	230	7338	88	74	25	134	4	Oketo (Oketoyama)
MIH090	333	383	8267	107	83	25	158	7	Oketo (Oketoyama)
MIH091	334	95	10,431	168	50	37	228	5	Ikutahara
MIH092	345	371	12,144	135	134	30	150	6	Rubeshibe
MIH093	352	338	6912	131	72	28	127	6	Oketo
MIH094	353	526	10,265	117	129	28	135	5	Rubeshibe
MIH095	354	218	6857	132	69	23	126	6	Oketo
MIH096	359	292	7180	83	74	27	136	5	Oketo (Oketoyama)
MIH097	361	308	7515	96	71	19	139	8	Oketo (Oketoyama)
MIH098	362	280	11,641	128	102	30	170	11	Oketo (Oketoyama)
MIH099	365	383	12,874	100	83	23	135	5	Oketo (Oketoyama)
MIH100	367	208	8381	101	85	23	158	5	Oketo (Oketoyama)
MIH101	369	471	9501	91	81	23	149	5	Oketo (Oketoyama)
MIH102	375	296	10,085	115	97	25	165	10	Oketo (Oketoyama)
MIH103	376	513	12,446	115	98	21	154	10	Oketo (Oketoyama)
MIH104	379	420	9176	105	98	29	166	9	Oketo (Oketoyama)
MIH105	381	372	10,722	120	99	29	161	8	Oketo (Oketoyama)
MIH106	382	664	12,312	130	101	25	164	9	Oketo (Oketoyama)
MIH107	389	413	13,274	104	85	20	147	9	Oketo (Oketoyama)
MIH108	391	395	10,881	100	58	25	123	8	Oketo (Oketoyama)
MIH109	395	340	11,669	123	108	31	160	10	Oketo (Oketoyama)
MIH110	399	324	12,958	115	84	26	163	10	Oketo (Oketoyama)
MIH111	405	623	10,611	60	49	8	97	5	Unknown*
MIH112	412	530	10,361	112	91	26	164	4	Oketo (Oketoyama)
MIH113	422	260	8884	107	88	25	158	5	Oketo (Oketoyama)
MIH114	424	338	9368	106	82	23	151	8	Oketo (Oketoyama)
MIH115	426	216	12,213	171	56	39	228	5	Ikutahara
MIH116	427	325	9668	105	71	26	145	9	Oketo (Oketoyama)
MIH117	428	496	10,776	106	89	30	153	3	Oketo (Oketoyama)
MIH118	429	535	11,114	114	94	31	157	7	Oketo (Oketoyama)
MIH119	442	303	7599	89	74	23	145	4	Oketo (Oketoyama)
MIH120	443	274	5583	118	58	18	118	5	Oketo
MIH121	447	288	8446	94	74	24	146	6	Oketo (Oketoyama)
MIH122	450	314	9682	157	71	18	126	9	Oketo
MIH123	453	449	12,308	129	95	29	167	6	Oketo (Oketoyama)
MIH124	454	259	11,285	111	87	21	188	5	Oketo (Oketoyama)
MIH125	455	158	12,626	170	55	36	233	8	Ikutahara
MIH126	456	571	13,489	119	107	27	160	11	Oketo (Oketoyama)
MIH127	458	424	12,139	120	99	22	154	12	Oketo (Oketoyama)
MIH128	459	972	13,727	113	87	20	144	7	Oketo (Oketoyama)
MIH129	474	322	7901	87	69	22	136	4	Oketo (Oketoyama)

* It is probably from the Oketo (Oketoyama) source (see text).

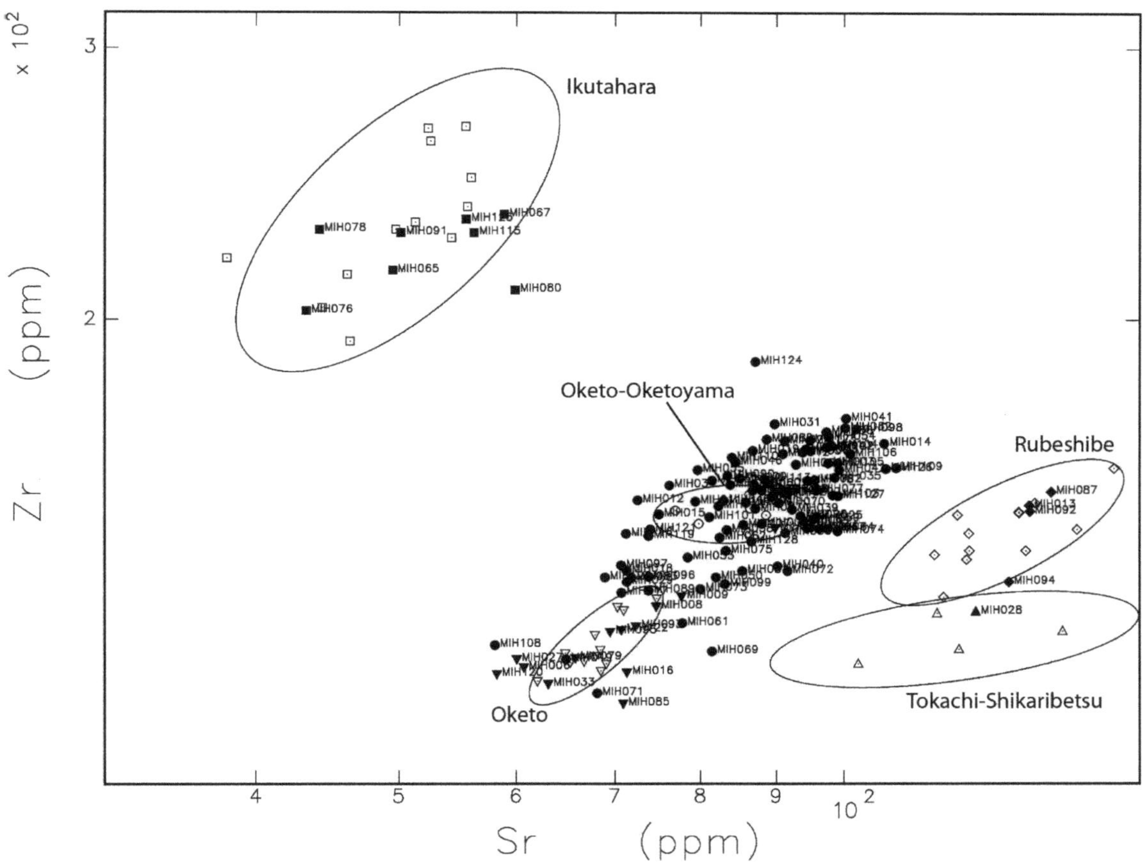

Figure 2.12. Bivariate plot of Sr vs. Zr concentrations showing all of the likely source reference groups [except for Tokachi (Mitsumata)] along with all of the artefacts from the Ogachi-Kato 2 site. On Figures 2.12–2.13, ellipses represent 90% confidence intervals for membership in the source group

underwent post-depositional disturbance by periglacial processes. Although no reliable geochronological data are available for this site, the Upper Palaeolithic components include the small flake, flake, and microblade assemblages which suggest an early-to-middle Upper Palaeolithic date (Izuho 2012).

Figure 2.12 shows the separation of most of the relevant source groups and assigned artefacts. The separation of the two Oketo groups requires a second plot (Figure 2.13). The artefacts do not always fit well into the confidence ellipses for the sources primarily due to variability in shape and thickness. Sample thickness has variable effects on the data and tends to spread the artefacts along a correlation line when compared to the analysis of large source specimens (Ferguson 2012).

The three artefacts with questionable source assignment by ED–XRF were subjected to NAA by short irradiation (Glascock *et al.* 1994) for further clarification. Short irradiation can accommodate some whole samples if they are small enough to fit in the polyvinyl tube. The radioactivity generated is very low, and the samples are no longer detectibly radioactive after approximately three months. The short irradiation only provides precise data for some elements (including Ba, Dy, Mn, and Na), but these are enough to separate the four sources difficult to distinguish by ED–XRF.

Sample MIH051 (see Table 2.4) is confirmed to be from Tokachi-Mitsumata source, although it is difficult to differentiate this source from Kushiro (Shitakara). Sample MIH086 seems closest to Tokachi-Mitsumata, but this is not clear. For now, a provisional assignment to Tokachi-Mitsumata is considered. Sample MIH111 is similar to Oketo (Oketoyama), but there are sufficient differences to consider it unassigned. A long irradiation NAA might help resolve this artefact's source assignment, although that would require sample destruction. A full discussion of the archaeologically relevant details for the Ogachi-Kato 2 site and artefact analysis are presented in Izuho *et al.* (this volume).

CONCLUSIONS

Obsidian sourcing is a powerful approach to understanding human behaviour that should be undertaken with a systematic methodology and proper selection of analytical techniques. It is important to carefully sample all potential source locations and characterise them with sufficient precision and breadth to determine which techniques allow for source identification while also minimising time, expense, and sample destruction. We propose a thorough analysis of all source samples by NAA and ED–XRF and then analysing artefacts by the easiest and least costly method possible.

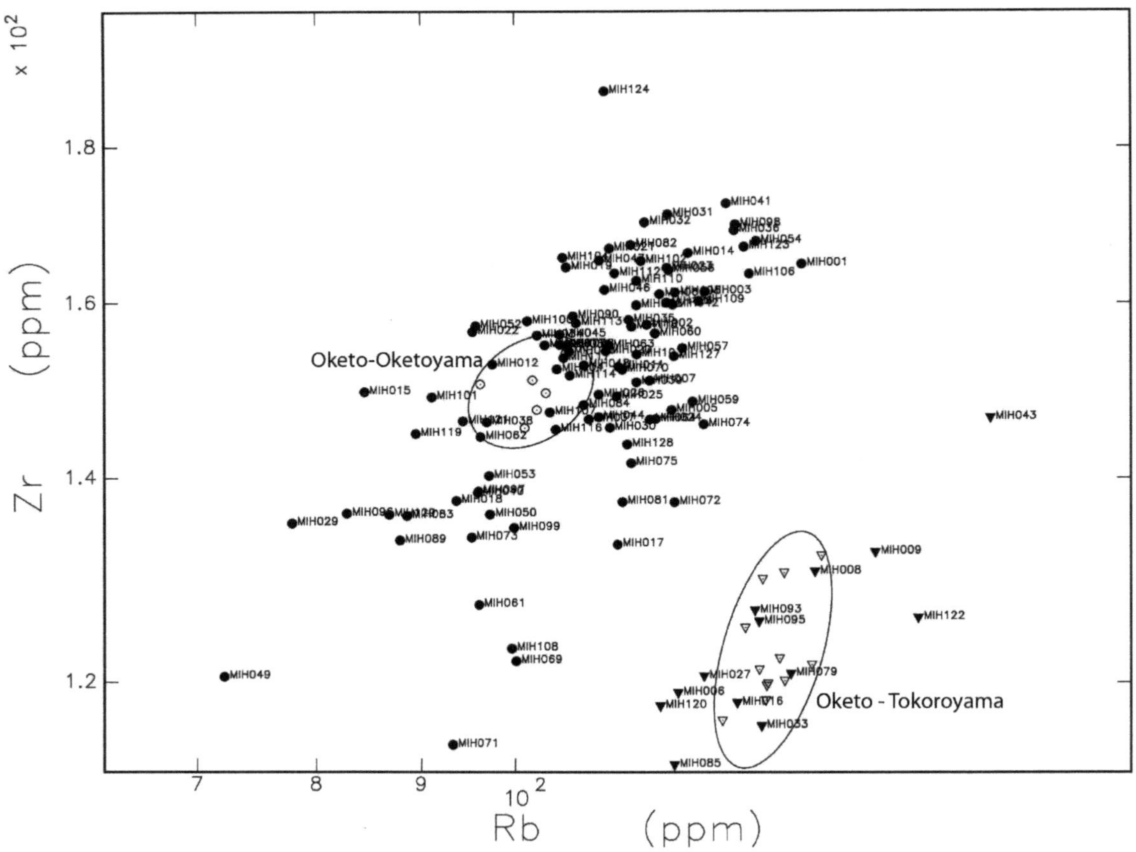

Figure 2.13. Bivariate plot of Rb vs. Zr concentrations showing the separation of the two Oketo source groups at the Ogachi-Kato 2 site

Although ED–XRF is widely available, not all researchers have access to NAA or another equally sensitive technique, and thus it might be necessary to submit the samples to a commercial laboratory.

The use of multiple methods has shown great advantages in most areas of the world, and Hokkaido is no exception. We have used ED–XRF and NAA to determine the unique chemical profiles of 21 obsidian sources on the island (see also Kuzmin *et al.* 2013). In our archaeological case study from Ogachi-Kato 2 it is possible to assign over 98% of the obsidian artefact assemblage to a specific source using only ED–XRF. The remaining pieces can be subjected to non-destructive NAA to distinguish between compositionally similar sources.

Acknowledgements

We acknowledge the assistance of Alex Brechbuhler (Archaeometry Laboratory, MURR) who prepared the samples for NAA, and the helpful comments and suggestions made by the participants of the 2011 Nagano Workshop. The Archaeometry Laboratory at MURR is supported by US NSF (Grant BCS-1110793). This research was also funded by the MEXT (Grant-in-Aid for Scientific Research 21242026); and the Japan Society for Promotion of Science (Grant-in-Aid for Scientific Research 24320157).

References

Ferguson, J. R. 2012. X-ray Fluorescence of Obsidian: Approaches to Calibration and the Analysis of Small Samples. In *Handheld XRF for Art and Archaeology*, edited by A. N. Shugar and J. L. Mass, 410–422. Leuven, Leuven University Press.

Ferguson, J. R., S. H. Ambrose, M. D. Glascock, A. S. Brooks, and J. E. Yellen. 2010. A Multi-Tier and Multi-Method Approach to the Analysis of Obsidian Source Data from the Central Rift Valley of Kenya. Poster presented at the 38th International Symposium on Archaeometry, Tampa, FL.

Glascock, M. D. 2010. Comparison and Contrast between XRF and NAA: Uses for Characterization of Obsidian Sources in Central Mexico. In *X-Ray Fluorescence Spectrometry (XRF) in Geoarchaeology*, edited by M. S. Shackley, 161–192. New York, Springer.

Glascock, M. D., G. E. Braswell, and R. H. Cobean. 1998. A Systematic Approach to Obsidian Source Characterization. In *Archaeological Obsidian Studies: Method and Theory*, edited by M. S. Shackley, 15–65. New York & London, Plenum Press.

Glascock, M. D., and J. R. Ferguson. 2012. Report on the Analysis of Obsidian Source Samples by Multiple Analytical Methods. Report on File. Missouri University Research Reactor, Archaeometry Laboratory, Columbia, MO, USA.

GLASCOCK, M. D., H. NEFF, K. S. STRYKER, and T. N. JOHNSON. 1994. Sourcing Archaeological Obsidian by an Abbreviated NAA Procedure. *Journal of Radioanalytical and Nuclear Chemistry* 180, 29–35.

GLASCOCK, M. D., P. C. WEIGAND, R. ESPARZA LÓPEZ, M. A. OHNERSORGEN, M. GARDUÑO AMBRIZ, J. B. MOUNTJOY, and J. A. DARLING. 2010. Geochemical Characterisation of Obsidian in Western Mexico: The Sources in Jalisco, Nayarit, and Zacatecas. In *Crossing the Straits: Prehistoric Obsidian Source Exploitation in the North Pacific Rim* (B.A.R. International Series 2152), edited by Y. V. Kuzmin and M. D. Glascock, 201–217. Oxford, BAR Publishing.

HALL, M., and H. KIMURA. 2002. Quantitative EDXRF Studies of Obsidian Sources in Northern Hokkaido. *Journal of Archaeological Science* 29, 259–266.

HIROSE, W., M. IWASAKI, and M. NAKAGAWA. 2000. Transition of Neogene Arc Volcanism in Central-Western Hokkaido Viewed from K-Ar Ages, Study of Volcanic Activity, and Bulk Rock Chemistry. *Journal of the Geological Society of Japan* 106, 120–135 (in Japanese with English Abstract).

HIROSE, W., and M. NAKAGAWA. 1999. Neogene Volcanism in Central-Eastern Hokkaido: Beginning and Evolution of Arc Volcanism Inferred from Volcanological Parameters and Geochemistry. *Journal of the Geological Society of Japan* 105, 247–265 (in Japanese with English Abstract).

INOUE, I. 2003. Shirataki Dai 30 Chiten Shutsudo Kokuyoseki no Kagaku Bunseki [Geochemical Analysis of Obsidian from Shirataki No. 30 Site]. In *Shirataki Dai 30 Chiten Iseki*, 245–258. Shirataki, Japan, Shirataki-mura Kyoiku Iinkai.

IZUHO, M. 2012. Matome [Conclusion]. In *Kokuyoseki no Ryutsu to Shouhi Karamita Kan-Nihonkai Hokubu Chiiki Niokeru Koushinsei Shakai no Keisei to Henyou (I)*, edited by H. Sato, 159–163. Tokyo, University of Tokyo Press.

IZUHO, M., and F. AKAI. 2005. Hokkaido no Kyusekki Hennen: Iseki Keisei Kateiron no Tekiyo [Geochronology of Palaeolithic Sites on Hokkaido]. *Kyusekki Kenkyu* 1, 39–55.

IZUHO, M., and W. HIROSE. 2010. A Review of Archaeological Obsidian Studies on Hokkaido Island (Japan). In *Crossing the Straits: Prehistoric Obsidian Source Exploitation in the North Pacific Rim* (B.A.R. International Series 2152), edited by Y. V. Kuzmin and M. D. Glascock, 9–25. Oxford, BAR Publishing.

IZUHO, M., and H. SATO. 2007. Archaeological Obsidian Studies in Hokkaido, Japan: Retrospect and Prospects. *Bulletin of the Indo-Pacific Prehistory Association* 27, 114–121.

KIMURA, H. (ed.). 1992. *Reexamination of the Yubetsu Technique and Study of the Horokazawa Toma Lithic Culture: Archaeological Survey on the Toma Lithic Culture*. Sapporo, Sapporo University Press.

KIMURA, H. 1995. Kokuyouseki, Hito, Gijutsu [Obsidian, Humans, and Technology]. *Hokkaido Kokogaku* 31, 3–63.

KIMURA, H. 1998. Obsidian, Humans, Technology. In *Paleoekologiya Pleistotseha i Kultury Kamennogo Veka Severnoi Azii i Sopredelnyikh Territorii. Tom 2*, edited by A. P. Derevianko, 302–314. Novosibirsk, Izdatelstvo Instituta Arkheologii i Etnografii.

KIMURA, H. 2005. *Kita no Kokuyoseki no Michi: Shirataki Isekigun* [The Obsidian Road of the North: Shirataki Site Cluster]. Tokyo, Shinsensha Publishers.

KUZMIN, Y. V. 2006. Recent Studies of Obsidian Exchange Networks in Prehistoric Northeast Asia. In *Archaeology in Northeast Asia: On the Pathway to Bering Strait*, edited by D. E. Dumond and R. L. Bland, 61–71. Eugene, University of Oregon Museum of Natural and Cultural History.

KUZMIN, Y. V., and M. D. GLASCOCK. 2007. Two Islands in the Ocean: Prehistoric Obsidian Exchange between Sakhalin and Hokkaido, Northeast Asia. *Journal of Island & Coastal Archaeology* 2, 99–120.

KUZMIN, Y. V., M. D. GLASCOCK, and M. IZUHO. 2013. The Geochemistry of the Major Sources of Archaeological Obsidian on Hokkaido Island (Japan): Shirataki and Oketo. *Archaeometry* 55, 355–369.

KUZMIN, Y. V., M. D. GLASCOCK, and H. SATO. 2002. Sources of Archaeological Obsidian on Sakhalin Island (Russian Far East). *Journal of Archaeological Science* 29, 741–749.

MERRICK, H. V., and F. H. BROWN. 1984. Obsidian Sources and Patterns of Source Utilization in Kenya and Northern Tanzania: Some Initial Findings. *African Archaeological Review* 2, 129–152.

MUKAI, M. 2005. Monbetsu Chiiki, Rubeshibe Chiiki, Toyoura Chiiki kara Sanshutsu suru Kokuyoseki Glass no Kagaku Sosei [Chemical Compositions of the Obsidian Glasses from the Monbetsu, Rubeshibe, and the Toyoura Areas]. *Asahikawa-shi Hakubutsukan Kenkyu Hokoku* 11, 9–20.

MUKAI, M. 2010. Hokkaido kara Sanshutsu suru Kokuyoseki Glass no Kagaku Sosei [Chemical Composition of Obsidian Glasses from Hokkaido Island]. *Asahikawa-shi Hakubutsu Kagakukan Kenkyu Hokoku* 2:1–33.

MUKAI, M., H. HASEGAWA, and K. WADA. 2000. Asahikawa Shuhen Chiiki niokeru Kokuyoseki Glass no Kagaku Sosei [Chemical Compositions of the Obsidian Glasses around Asahikawa District]. *Asahikawa-shi Hakubutsukan Kenkyu Hokoku* 6, 51–64.

MUKAI, M., R. SHIBUYA, and K. WADA. 2004. Ikutahara Chiiki kara Sanshutsu suru Kokuyoseki Glass no Kagaku Sosei [Chemical Compositions of the Obsidian Glasses from Ikutahara District]. *Asahikawa-shi Hakubutsukan Kenkyu Hokoku* 10, 35–40.

MUKAI, M., and K. WADA. 2001. Asahikawa Seiho, Chippubetsu–Hokuryu Chiiki kara Sanshutsu suru Kokuyoseki Glass no Kagaku Sosei [Chemical Compositions of the Obsidian Glasses from Chippubetsu and Hokuryu District, Western Area of

Asahikawa]. *Asahikawa-shi Hakubutsukan Kenkyu Hokoku* 7, 23–30.

MUKAI, M., and K. WADA. 2003. Engaru Chiiki Oumu Chiiki kara Sanshutsu suru Kokuyoseki Glass no Kagaku Sosei [Chemical Compositions of the Obsidian Glasses from Engaru and Oumu Districts]. *Asahikawa-shi Hakubutsukan Kenkyu Hokoku* 9, 19–26.

MUKAI, M., and K. WADA. 2004a. Okushiri To kara Sanshutsu suru Kokuyoseki Glass no Kagaku Sosei [Chemical Compositions of the Obsidian Glasses from Okushiri Island]. *Asahikawa-shi Hakubutsukan Kenkyu Hokoku* 10, 41–46.

MUKAI, M., and K. WADA. 2004b. Tokachi Chiho kara Sanshutsu suru Kokuyoseki Glass no Kagaku Sosei [Chemical Compositions of the Obsidian Glasses from Tokachi District]. *Asahikawa-shi Hakubutsukan Kenkyu Hokoku* 10, 47–56.

PHILLIPS, S. C., and R. J. SPEAKMAN. 2009. Initial Source Evaluation of Archaeological Obsidian from the Kuril Islands of the Russian Far East Using Portable XRF. *Journal of Archaeological Science* 36, 1256–1263.

SATO, H. 2004a. Russia Kyokuto Niokeru Senshi Jidai no Kokuyoseki no Riyo [Prehistoric Obsidian Exploitation in the Russian Far East]. *Kokuyouseki Bunka Kenkyu* 3, 45–55.

SATO, H. 2004b. Lithic Procurement and Reduction Strategy of Hirosato Industry in the Japan Sea Rim Area. *Seonsa Wa Kodae* 20, 205–221.

SATO, H., Y. V. KUZMIN, and M. D. GLASCOCK. 2002. Source Analysis of Prehistoric Obsidians and Assessment of Its Distribution in Northeast Asia. *Hokkaido Kokogaku* 38, 1–13 (in Japanese with English Title).

TAKANASHI, K., K. SHUTO, and M. SATO. 2011. Origin of Late Paleogene to Neogene Basalts and Associated Coeval Felsic Volcanic Rocks in Southwest Hokkaido, Northern NE Japan Arc: Constraints from Sr and Nd Isotopes and Major- and Trace-Element Chemistry. *Lithos* 125, 369–392.

WARASHINA, T. 1999. Sekki oyobi Tama Rui Genzairyo no Sanchi Bunseki [Source Identification of Raw Material for Stone Tools and Beads]. In *Kokogaku to Nendai Sokuteigaku, Chikyu Kagaku*, edited by S. Matsu'ura, Y. Uesugi and T. Warashina, 259–293. Tokyo, Doseisha Publishers.

YOSHITANI, A., M. KAWABE, O. SUDA, I. MIZUFUNE, and K. NAGAHATA. 2001. Oribe 17 Iseki kara Shutsudo shita Kokuyogan no Gensanchi Nituite [Approach on Resource of Obsidian from Oribe 17 Site]. *Kamishihoro-cho Higashi Taisetsu Hakubutukan Kenkyu Hokoku* 23, 1–19.

Chapter 3
APPLICATION OF AN INTERNAL STANDARD METHOD FOR NON-DESTRUCTIVE ANALYSIS OF OBSIDIAN ARTEFACTS BY WAVELENGTH DISPERSIVE X-RAY FLUORESCENCE SPECTROMETRY

Yoshimitsu SUDA

Abstract: *A method of non-destructive analysis suitable for provenance identification of obsidian artefacts was established using a Wavelength Dispersive X-ray Fluorescence (WDXRF) spectrometer. Obsidian nodules from seven major sources in Japan and one source in Peru were used as the standard materials for this analysis. Their qualified values were estimated with the analysis of the conventional Fusion Bead Method by the WDXRF. Internal Standard Method using the scattering X-rays was applied to the non-destructive analysis. The measurements of the quantitative values of Ca, Mn, Rb, and Y were possible using this method. These elements are adequate to discriminate between obsidian from different sources. The variation of the Y/Ca ratio with respect to the Rb/Mn ratio was quite useful in this discrimination. The provenance of archaeological obsidian is determined using a combination of the results of non-destructive analysis by the Internal Standard Method, and the Y/Ca vs. Rb/Mn ratios discrimination diagram.*

Keywords: *Obsidian; X-Ray Fluorescence Spectrometry, Internal Standard Method, Scattering X-rays, Non-Destructive Quantitative Analysis, Artefacts, Provenance Identification*

INTRODUCTION

Non-destructive analysis is required for many types of archaeological investigations, especially when dealing with nationally-designated cultural properties. X-ray Fluorescence (hereafter – XRF) analysis is commonly applied to the study of archaeological and cultural items (e.g., Shackley 2011) because the XRF spectrometer is able to perform elemental analysis without causing any damage to the artefact. The XRF spectrometers are roughly divided into the Wavelength Dispersive XRF (henceforth – WDXRF) and the Energy Dispersive XRF (hereafter – EDXRF). The EDXRF is able to reduce the damage to samples during analysis, because it offers higher detection efficiency than the WDXRF. The EDXRF has larger sample compartments. For this reason, the EDXRF is more widely used to perform non-destructive analyses on obsidian artefacts. The WDXRF, however, is suitable for performing the quantitative measurements with higher precision and accuracy compared to the EDXRF, even though its relatively smaller sample compartment is disadvantageous.

The WDXRF is equipped with the analysing crystals for apportioning the X-rays with a specific wavelength (λ) at θ degrees on the basis of Bragg's law:

$$2d\sin\theta = n\lambda$$

(where d is the lattice spacing of the analysing crystal, and n is an integer). On the other hand, the EDXRF is equipped with a semiconductor drift detector that measures the intensity of X-rays according to their energy. The measured X-rays' intensity is expressed by the counts of electronic signals per second. The relationship between the energy (E) and wavelength (λ) of the X-rays is given by the following formula (Bertin 1978, 7):

$$E\ (\text{keV: kiloelectronvolts}) = 12{,}396/\lambda\ (\text{Å: ångstroms})$$

The geochemistry of the obsidian artefacts is an appropriate indicator to validate the existence of long-distance exchange networks across mountain ranges, rivers, and seas, particularly during the Palaeolithic period (e.g., Burger and Glascock 2000; Kuzmin 2011; Ikeya 2009; see also Ikeya, this volume). The main research objective of the present research is to establish a methodology for obsidian artefact provenance studies on the basis of the geochemical analysis using a non-destructive method. The WDXRF spectrometer employed in this study was a Rigaku ZSX Primus III+ , equipped with the 3.0 kW type Rh anode X-ray tube and installed at the Centre for Obsidian and Lithic Studies of Meiji University (Nagawa, Nagano Pref.). The techniques discussed in this chapter are illustrated in Figure 3.1: 1) obsidian specimens from seven major sources in Japan and one source in Peru were prepared; 2) a portion of the specimen was pulverised, and another portion was sawed and fractured into a slab and a flake, respectively; 3) the powder was used to perform the quantitative analysis using the conventional 'Fusion Bead Method' (i.e., a destructive type of analysis); 4) the slab was used as the 'standard reference material' for non-destructive analysis; 5) the quantitative values after the Fusion Bead Method were used as 'certified values'; 6) the Internal Standard Method using scattering X-rays was applied to the non-destructive analysis of artefacts, in which the flake of obsidian was considered to be the equivalent of an artefact; 7) geochemical characterisation of obsidian in major sources was performed on the basis of the results after the Fusion Bead Method; and 8) the method preferred for artefact

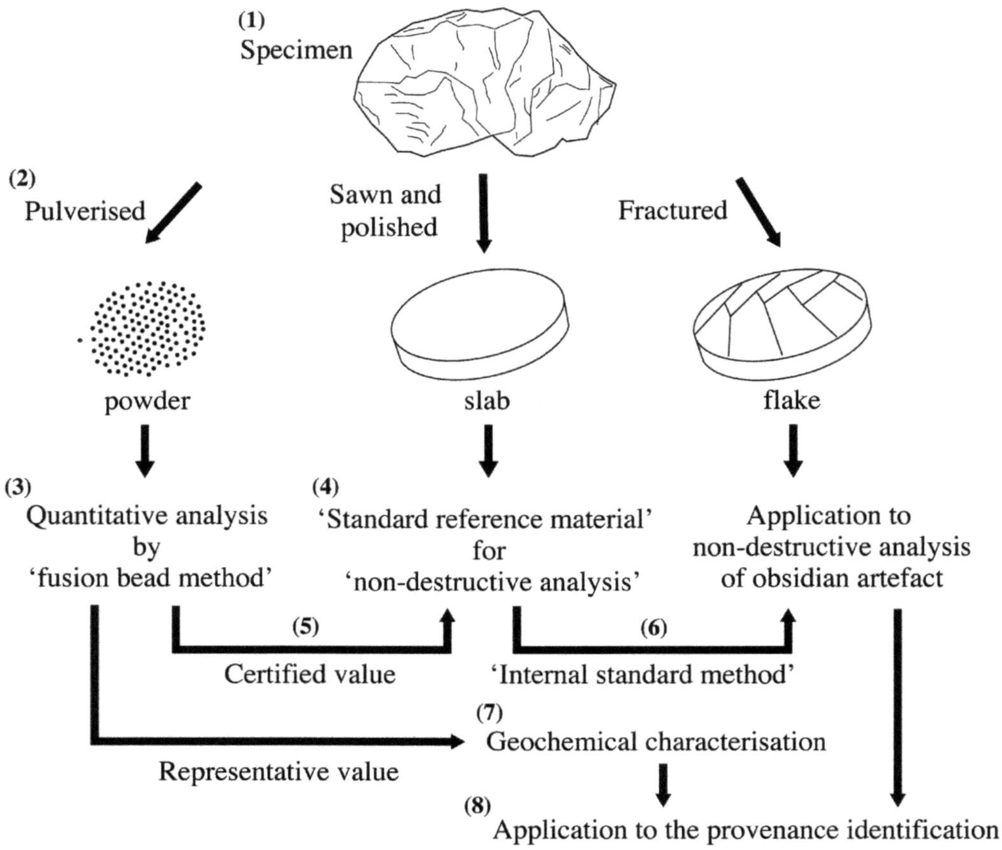

Figure 3.1. Techniques and materials used in the present study

provenance identification was examined on the basis of the geochemical characteristics of obsidian from the major sources.

DESCRIPTION OF SAMPLES

The Kirigamine–Yatsugatake region represents a major cluster of obsidian sources in central Japan (Figure 3.2). Many archaeological sites belonging to the Palaeolithic, Jomon, and Yayoi cultural complexes are located in this region. The Takayama, Hoshikuso-toge, Omegura, Higashimochiya, and Wada-toge obsidian sources are all located in the Kirigamine area, while the Tsumetayama is known as one of the major obsidian sources in the Yatsugatake area. Geological investigations indicate that obsidian in the Kirigamine area belongs to the Enrei volcanic rock class, while the obsidian in the Yatsugatake area is part of the Yatsugatake Volcanic Chain (Nakano *et al.* 1998). Both volcanic formations were formed during the Pleistocene, and they correspond to the calc-alkaline and high-potassic volcanic rocks series (Takahashi and Nishiki 2006).

Obsidian samples from the Wada-toge (WT-2) and Higashimochiya (HM-1) sources were extracted for the present study from outcrops, and thus represent the primary deposits. Specimens from the Takayama (TY-1), Omegura (OM-1), and Tsumetayama (TS-1) sources came from the beds of the Takayamagawa, Takamatsuzawa, and Sakasagawa rivers, respectively. These samples, therefore, originated from the secondary deposits. Obsidian from the Hoshikuso-toge (HT-1) source comes from a pyroclastic flow deposit, which is not an *in situ* product but geologically is considered as a primary locale. Shirataki in northern Japan and Quispisisa (Ayacucho) in Peru (South America) are major obsidian sources (Figure 3.2). Specimens from the Shirataki (obstd-1) and Quispisisa (PAY-1) sources were purchased from a mineral dealer. Thus, the details regarding their collection in the field are unknown.

The most representative types of obsidian are shown in Figure 3.3. Microphotographs of the obstd-1 sample are presented in Figure 3.4. The TY-1, HT-1, OM-1, HM-1, and TS-1 specimens are characterised by a brownish colour and high transparency. The WT-2, obstd-1, and PAY-1 specimens are blackish with low transparency. The major components of obsidian are a glassy matrix, spherulites, and microphenocrysts. The glassy matrix is observed in all obsidians. The spherulites with millimetre-scale are found in the TY-1, OM-1, TS-1, HM-1, and WT-2 specimens, and those of several centimetres in size are found in the obstd-1 specimen. Plagioclase microphenocrysts are detected in the HM-1 and PAY-1 obsidians, while biotite microphenocrysts are observed in the WT-2 specimen. Zircon micropheno-crysts and Fe–Ti oxides are found in the obstd-1 sample. The alignment and aggregation of crystallites and/or microphenocrysts form the foliation texture.

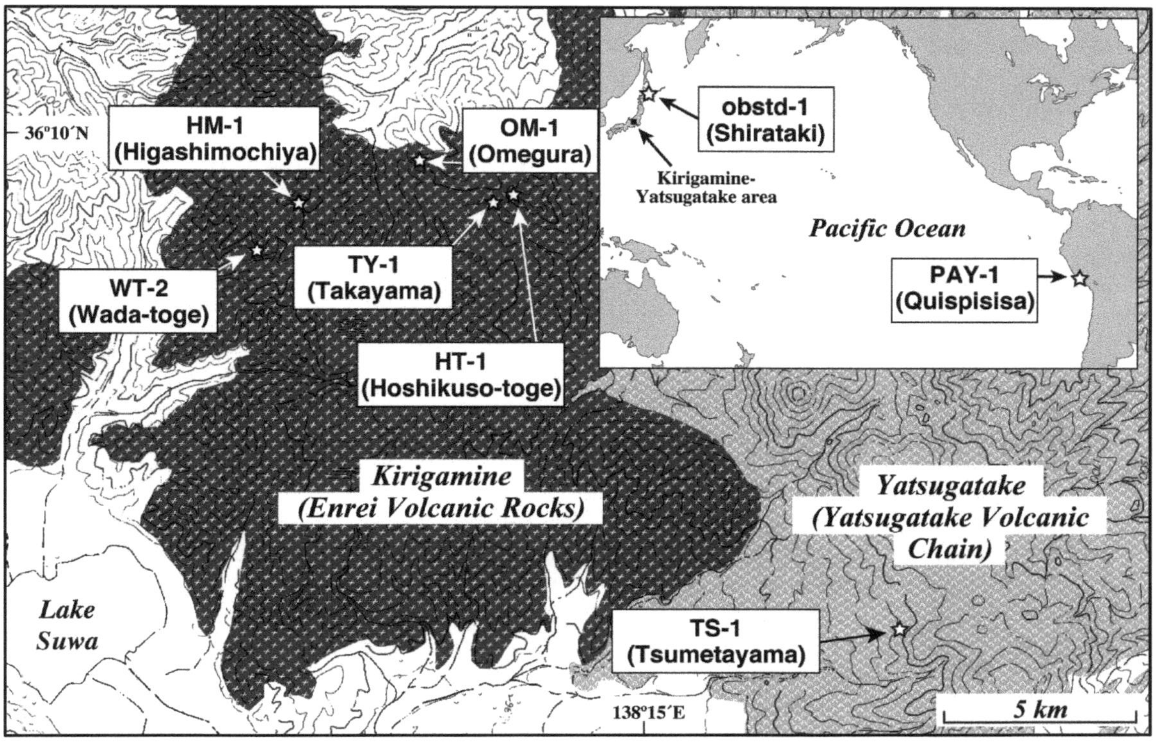

Figure 3.2. Locations of the major obsidian sources analysed in this study; shaded fields on the map show the exposure areas of Pleistocene volcanic rocks (after Nakano et al. *1998)*

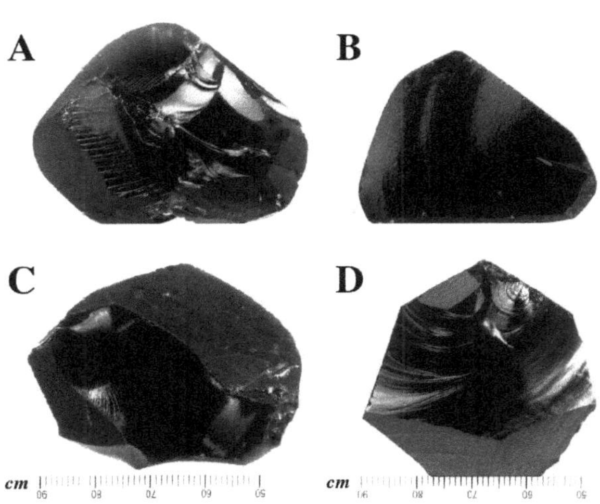

Figure 3.3. Flaked obsidian from the Hoshikuso-toge (A), Wada-toge (B), Shirataki (C), and Quispisisa (D) sources

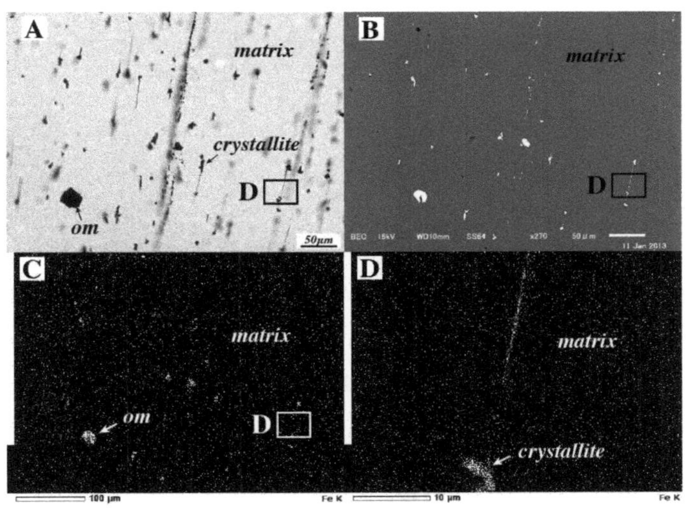

Figure 3.4. Microphotograph (A), backscattered electron image (B), and element mappings (C) of Fe for obsidian from the Shirataki source; on A–C parts, the area labelled as "D" is enlarged in D part

QUANTITATIVE ANALYSIS BY FUSION BEAD METHOD

Quantitative analysis of major elements as oxides (SiO_2, TiO_2, Al_2O_3, T-Fe_2O_3, MnO, MgO, CaO, Na_2O, K_2O, and P_2O_5) and trace elements (Rb, Sr, Y, and Zr) was performed to establish certified values of the reference material for the non-destructive study, and representative values of obsidian from the major sources. Geochemical standard materials from the Geological Survey of Japan (JA-1, JA-2, JA-3, JB-1a, JB-2, JB-3, JR-1, JR-2, JR-3, JG-1a, JG-2, JG-3, JGb-1, JGb-2, JSy-1, JH-1, JF-1, JF-2, and JP-1) were used for this analysis. Their qualified values were excerpted from Imai *et al.* (1995).

The Fusion Bead Method is widely applied to the quantitative analysis by WDXRF (Bertin 1978, 411). The bead is prepared by fusing a mixture of the powdered sample and a lithium-based flux. The fusion bead is able to homogenise the elemental composition of a sample to be measured, and to minimise matrix effects. In the present study, the bead with dilution ratio (flux weight/sample weight) of 5.000 (± 0.0001) was used to measure the major elements, while a bead with a dilution ratio of 2.000 (± 0.0002) was employed to measure the trace elements. Merck Spectromelt A12 (di-lithium tetraborate 66% + lithium metaborate 34%) was used as the flux. The total weight of the glass beads became 5.4 g. Additional details regarding the sample preparation methods and processing are available in Suda (2012, 2013), Suda and Motoyoshi (2011), and Suda *et al.* (2010, 2011).

The characteristic X-rays from the Kα line were chosen as the measurement lines for all elements. The X-ray tube was operated at 50 kV and 50 mA for all measurements. The diameter of the diaphragm (i.e., the measurement region on each specimen) was 3 cm. A PET analysing crystal ($2d$ = 8.76 Å) was chosen for the measurement of Si and Al. RX25, which is a multilayer film equivalent of the TAP ($2d$ = 25.7626 Å), was selected for the measurement of Mg and Na. Ge(111) ($2d$ = 6.53272 Å) was chosen for the measurement of P. LiF(200) ($2d$ = 4.0273 Å) was used to measure Ti, Fe, Mn, Ca, K, Sr, and Y. LiF(220) ($2d$ = 2.848 Å) was chosen for the measurements of Rb and Zr. A gas flow proportional counter was used for the measurements of Na, Mg, Al, Si, P, K, and Ca, while a scintillation counter was employed for the measurement of Ti, Mn, Fe, Rb, Sr, Y, and Zr. A mixed gas (PR gas: Ar 90% + CH_4 10%) was used as the counter gas. Details of the instrumental conditions can be found in Suda (2012, 2013). The values of the lattice distance ($2d$) of analysing crystal were excerpted from Rigaku (1982, 24).

The Semi-Fundamental Parameter (S–FP) method was applied for quantitative analysis. The resulting quantitative values (W_i) for an element (i) are given by the following Equation (1):

$$W_i = (a \times I_i + b) \times (1 + \sum D_j \times W_j) - A \times W_k \quad (1)$$

The S–FP method is fundamentally based on the calibration-standard method. The "a" and "b" indicate the constant numbers for the calibration line, and the "I_i" indicates the measured intensity. The absorption-enhancement effect (i.e., matrix effect) was calibrated by the correction coefficients (D_j) of a matrix element (j). The "D_j" is calculated after the Fundamental Parameter formula taken from the *Rigaku* software (ZSX version 6.48). The "W_j" is the quantitative value of matrix element (j). Spectral interferences from the Rb-Kβ and Sr-Kβ lines were taken into account for measurement of the Y-Kα and Zr-Kα lines, respectively. The factor "A" indicates the correction coefficient for the spectral interference taken from the Rigaku software. The factor "W_k" is the quantitative value for element (k) containing the interference lines (i.e., Rb and Sr). Results of the quantitative analysis are compiled in Table 3.1. The analysis was performed using two glass beads, repeated five times each. The result of the analysis indicates the mean value. Fe was assumed to be total trivalent iron (T-Fe_2O_3).

Table 3.1. Results of quantitative analysis of obsidian using the Fusion Bead Method

Source	***Takayama***		***Hoshikuso-toge***		***Omegura***		***Higashimochiya***	
	TY-1 (n = 10)		***HT-1 (n = 10)***		***OM-1 (n = 10)***		***HM-1 (n = 10)***	
	Value	***± 2σ***	***Value***	***± 2σ***	***Value***	***± 2σ***	***Value***	***± 2σ***
wt.%								
SiO_2	76.21	0.26	76.31	0.22	76.21	0.28	75.90	0.22
TiO_2	0.07	0.00	0.07	0.00	0.08	0.00	0.07	0.00
Al_2O_3	12.54	0.06	12.55	0.06	12.51	0.10	12.49	0.06
T-Fe_2O_3	0.66	0.01	0.65	0.01	0.76	0.01	0.66	0.01
MnO	0.10	0.00	0.10	0.00	0.09	0.00	0.10	0.00
MgO	0.06	0.02	005	0.02	0.07	0.02	0.05	0.01
CaO	0.48	0.01	0.48	0.01	0.53	0.01	0.48	0.01
Na_2O	3.89	0.02	4.02	0.03	3.90	0.02	3.83	0.04
K_2O	4.71	0.03	4.55	0.01	4.69	0.01	4.71	0.01
P_2O_5	0.01	0.00	0.01	0.00	0.01	0.00	0.01	0.00
Total	98.73		98.79		98.85		98.30	
ppm								
Rb	279	3	283	2	266	2	273	1
Sr	8.0	0.7	8.1	0.7	13.7	0.6	8.1	0.4
Y	46.3	1.0	47.2	1.0	40.1	0.7	45.9	1.0
Zr	91.9	1.5	93.6	1.7	98.6	0.9	91.9	2.2

Source	*Wada-toge*		*Tsumetayama*		*Shirataki*		*Quispisisa*	
	WT-2 (n = 10)		*TS-1 (n = 10)*		*obstd-1 (n = 10)*		*PAY-1 (n = 10)*	
	Value	$\pm 2\sigma$	*Value*	$\pm 2\sigma$	*Value*	$\pm 2\sigma$	*Value*	$\pm 2\sigma$
wt.%								
SiO_2	75.68	0.16	76.27	0.09	75.78	0.20	75.29	0.16
TiO_2	0.05	0.01	0.15	0.00	0.04	0.00	0.13	0.01
Al_2O_3	12.53	0.04	12.32	0.05	12.70	0.06	13.00	0.07
T-Fe_2O_3	0.72	0.01	0.89	0.01	1.15	0.01	0.82	0.01
MnO	0.11	0.00	0.05	0.00	0.05	0.00	0.05	0.00
MgO	0.05	0.01	0.14	0.01	0.05	0.02	0.16	0.02
CaO	0.48	0.00	0.72	001	0.53	0.01	0.84	0.01
Na_2O	4.04	0.05	3.90	0.04	3.85	0.03	3.90	0.03
K_2O	4.48	0.02	413	0.03	4.53	0.01	4.39	0.02
P_2O_5	0.01	0.00	0.02	0.00	0.02	0.00	0.02	0.00
Total	98.15		98.59		98.70		98.60	
ppm								
Rb	325	4	105	2	158	2	183	3
Sr	7.0	0.6	116	2	30.0	0.6	132	1
Y	52.0	1.1	17.3	0.7	30.1	0.9	13.0	0.8
Zr	89.0	1.5	102	2	74.8	1.3	93.1	2.1

METHODOLOGY FOR NON-DESTRUCTIVE ANALYSIS OF OBSIDIAN ARTEFACTS

Sample Preparation and Instrumental Conditions

Sawed and polished obsidian slabs from the eight major sources (Figure 3.1) were prepared to make the 'standard reference materials' for non-destructive analysis. The quantitative value after the Fusion Bead Method in Table 3.1 was used as the 'certified value'. The slabs are 0.5–0.8 cm thick, and were polished with a diamond paste (3 μm type). The elements for analysis were chosen from the major elements (Si, Ti, Al, Fe, Mn, Mg, Ca, Na, K, and P) and trace elements (Rb, Sr, Y, and Zr). Measurements were performed using the Kα line for all elements. The examinations were repeated six times using the same and/or different portions of the surface. The dwell times were duplicated, and the diameter of the diaphragm (i.e., measurement region on each specimen) was reduced to 1.0 cm compared to the instrumental conditions for the Fusion Bead Method. A sample spin was cut during the analysis to improve the stability of the measurement intensity because most obsidian artefacts have an irregular form and surface.

The spectral interferences of Rb-Kβ and Sr-Kβ lines were taken into account for the measurements of the Y-Kα and Zr-Kα lines, respectively. The intensity of the Y-Kα and Zr-Kα lines (I_i) were calibrated using the following Equation (2):

$$I_i = X_i - A \times X_j \quad (2)$$

The factor "A" is the correction coefficient of the interference lines (i.e., Rb-Kβ and Sr-Kβ lines) taken from the Rigaku software. The value "X_i" is the net intensity of the Y-Kα and Zr-Kα lines. The value "X_j" is the net intensity of the Rb-Kα and Sr-Kα lines. The net intensity means the gross intensity minus the background intensity of a given measurement line.

Selection of Elements

A variation diagrams of measurement intensity (net intensity) vs. theoretical intensity for a selected element are shown in Figure 3.5. The theoretical intensity was calculated using the Fundamental Parameter formula (e.g., Bertin 1976, 377) from the Rigaku software. Results of the correlation coefficient (R) for the least squares approximation lines indicate that the lowest correlations ($R < 0.99$) are found in the SiO_2 ($R = 0.66471$), Al_2O_3 ($R = 0.93118$), MgO (not determined), Na_2O ($R = 0.72449$), K_2O ($R = 0.95719$), P_2O_5 ($R = 0.51273$), and Zr ($R = 0.74711$) diagrams. On the other hand, the highest correlations ($R > 0.99$) are found in the TiO_2 ($R = 0.99586$), T-Fe_2O_3 ($R = 0.99402$), MnO ($R = 0.99351$), CaO ($R = 0.99954$), Rb ($R = 0.99884$), Sr ($R = 0.99943$), and Y ($R = 0.99783$) diagrams.

Microscopic investigation revealed that many opaque minerals and crystallites are found in some types of obsidian. Analysis with a Scanning Electron Microscope – Energy Dispersive X-ray Spectrometer (model JEOL JSM–6610LA) indicates that Fe is especially concentrated in these portions with respect to the glassy matrix part (Figure 3.4). This suggests that the

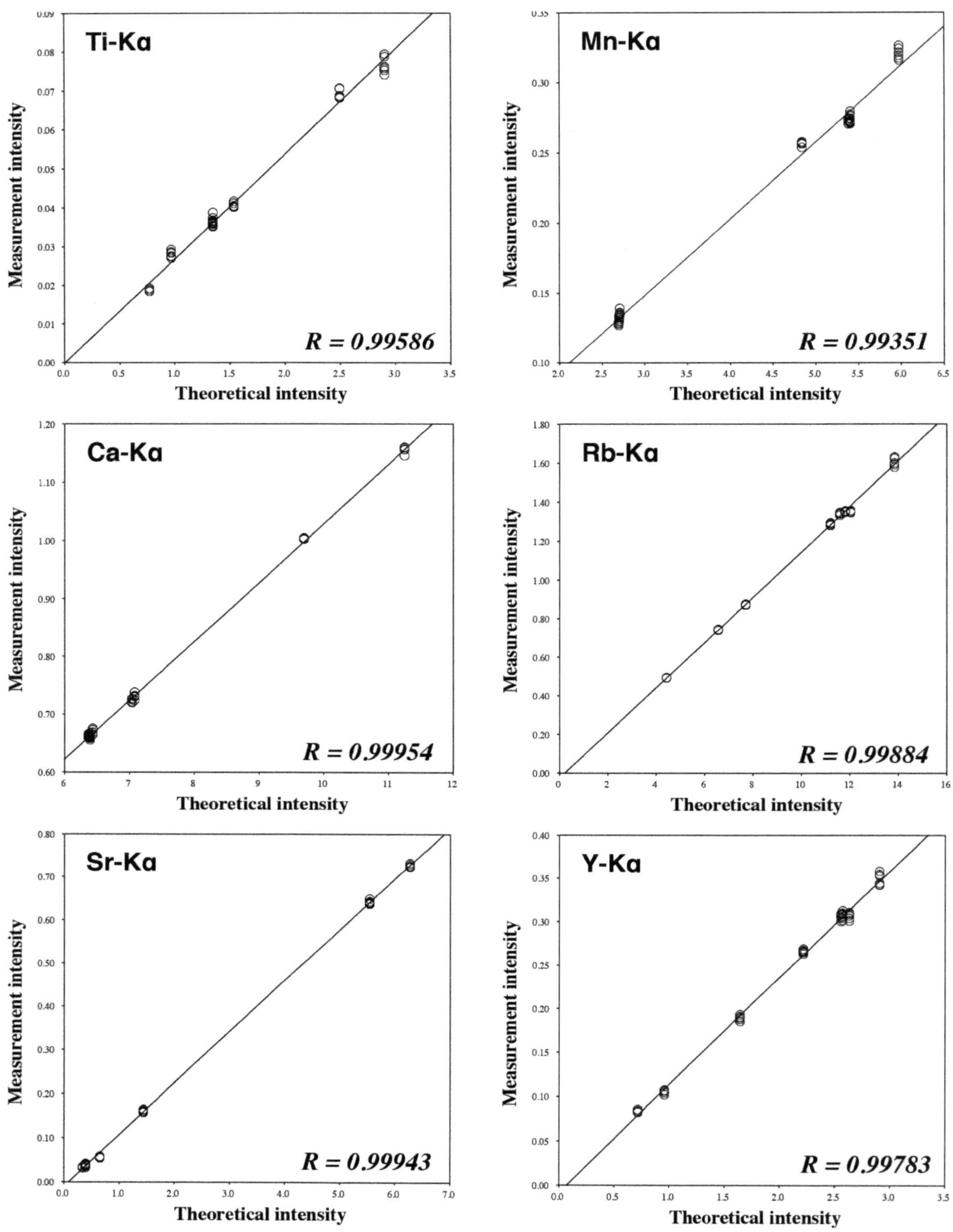

Figure 3.5. Variation diagrams of measurement intensity vs. theoretical intensity for selective elements

quantitative values for Fe are easily affected by the distribution and the concentration of the opaque minerals and/or the crystallites, which would become especially significant in non-destructive analysis. Furthermore, XRF analysis has shown that the value for Fe is determined either by the total amount of FeO (Fe^{2+}) or by the total amount of Fe_2O_3 (Fe^{3+}). This analytical technique is based on the assumption that the FeO/Fe_2O_3 ratios are generally conformable between the standard reference material and the measured specimens. Namely, the results of the quantitative values for Fe have some uncertainty unless the FeO and Fe_2O_3 are individually measured using a different method. Hence, the present study concludes that measurements of Si, Al, Fe, Mg, Na, K, P, and Zr are not suitable for non-destructive analysis, as opposed to measurements of Ti, Mn, Ca, Rb, Sr, and Y, which are deemed suitable.

Application of the Internal Standard Method

The Internal Standard Method using scattering X-rays was applied to the non-destructive analysis. In this method, the intensities of the Thomson scattered line (Rh-T), the Compton scattered line (Rh-C), the background of a measurement line (PB), and the background of the continuous spectrum at a given point (Rh-BG), are used as internal standard factors (F). The intensity of the measurement lines (X_E) is calibrated using the following Equation (3):

$$I_E = X_E / F \quad (3)$$

The matrix effect for a measurement line is calibrated with this method. Calibration of the spectral interferences is performed using the Equation (2) by substituting the I_i with X_E. Furthermore, theoretically, the Internal Standard Method using scattering X-rays is useful for the non-destructive analysis of obsidian artefacts due to the fact that some of the primary X-rays initiated by the X-ray tube are absorbed in the sample and contribute to the emission of fluorescent X-rays, and others are scattered and detected by the proportional counter. This means that the intensity of the scattering X-rays is affected by the condition of the sample. In other words, the intensity of these lines is correlated with the surface texture, density, and grain size and thickness of the sample. The surface of most obsidian artefacts is generally not flat but quite irregular because it suffered flaking and fracturing. Thus, this method would be applicable for the calibration between the slab (i.e., the form of reference material) and the flake (i.e., the form for most of archaeological obsidians).

The analysing crystal of LiF(200) was used for the measurements of the Rh-C, Rh-T, and Rh-BG lines, in which the θ angle were confirmed to be 18.36°, 17.55°, and 30.00°, respectively. The dwell times for the measurement of the lines were 40 seconds. Quantitative values were calculated on the basis of the Equation (1), in which the constant numbers (i.e., "a" and "b") of calibration lines for the Ti, Mn, Ca, Rb, Sr, and Y were calculated using the calibrated intensities (i.e., I_E) and the certified values (Table 3.1) of obsidian from the major sources (i.e., the TY-1, OM-1, HT-1, HM-1, WT-2, obstd-1, TS-1, and PAY-1 specimens). The measurements were performed using the polished slab specimens, and the analyses were repeated six times using different portions of the surface. The instrumental conditions for the elements are exactly the same as those in the analysis for the drawing of the diagrams in Figure 3.5.

A normalised multi-element diagrams compiling the calculated quantitative values for the reference materials are presented in Figure 3.6. The calculated quantitative value means the quantitative value of the calibration lines for each element. The results are presented in Table 3.2. The certified values expressed in wt.% (Table 3.1) were recalculated into parts-per-million (ppm) using the following equations: Ca = 10,000 x CaO / 1.39921; Ti = 10,000 x TiO_2 / 1.66856; and Mn = 10,000 x MnO / 1.29123. The multi-element diagram indicates that the normalisation values (i.e., value / qualified value ratio) are generally within the range of 0.9 to 1.1. Some results for Ti, after the internal standard method using the PB lines, and the result for Sr in OM-1 specimen, are beyond this range. Hence, the present study concludes that the results for Ti after the Internal Standard Method using the PB line are not reliable, and the results for Sr also present some uncertainty when using this method.

Application to Obsidian Artefacts

Quantitative analyses of Ca, Mn, Rb, and Y were performed using the replicas of obsidian artefacts (i.e., flaked obsidian). The specimens were made up of obsidian from the Hoshikuso-toge, Wada-toge, Shirataki, and Quispisisa sources, and they are shown in Figure 3.3. The measurements were repeated three times using the same surface. The results of the quantitative analysis by the Internal Standard Method are presented in Table 3.3, and compiled into the qualified value normalised multi-element diagrams (Figure 3.7). The qualified values are taken from Table 3.1. The results indicate that the normalisation values (i.e., value / qualified value ratio) are predominantly within the range of 0.9 to 1.1. The quantitative values obtained by the Internal Standard Method using the Rh-B, -C, and -T lines show almost the same pattern, which makes them inconclusive as to which method is best suited for non-destructive analysis. The results of the Internal Standard Method using the PB line have completely different patterns than the others, which also do not show any superiority using the application of the non-destructive analysis. Thus, in the present study, the mean values of all the results are considered to be the representative quantitative values after conducting non-destructive analysis by the Internal Standard Method.

GEOCHEMICAL CHARACTERISATION OF OBSIDIAN FROM MAJOR SOURCES

Representative values for obsidian from the major sources taken from Table 3.1 are presented in the JR-1 normalised multi-element diagram (Figure 3.8). The normalisation values of the JR-1 are taken from Imai *et al.* (1995): Ca, 4788 ppm; Ti, 659 ppm; Mn, 774 ppm; Rb, 258 ppm; Sr, 29.2 ppm; and Y, 45.2 ppm. Using this diagram, one can evaluate whether or not the quantitative values for the elements by the non-destructive analysis (i.e., Ca, Mn, Rb, and Y) are adequate for the geochemical characterisation of obsidian from the sources, and useful for the provenance identification of obsidian artefacts.

The profiles of obsidian from the Kirigamine area, namely from the Wada-toge, Takayama, Hoshikuso-toge, and Higashimochiya sources, are quite similar. Suda (2012) suggests that the obsidian in the Kirigamine area is derived from the same magmatic source, and that the fractional crystallisation of plagioclase must be

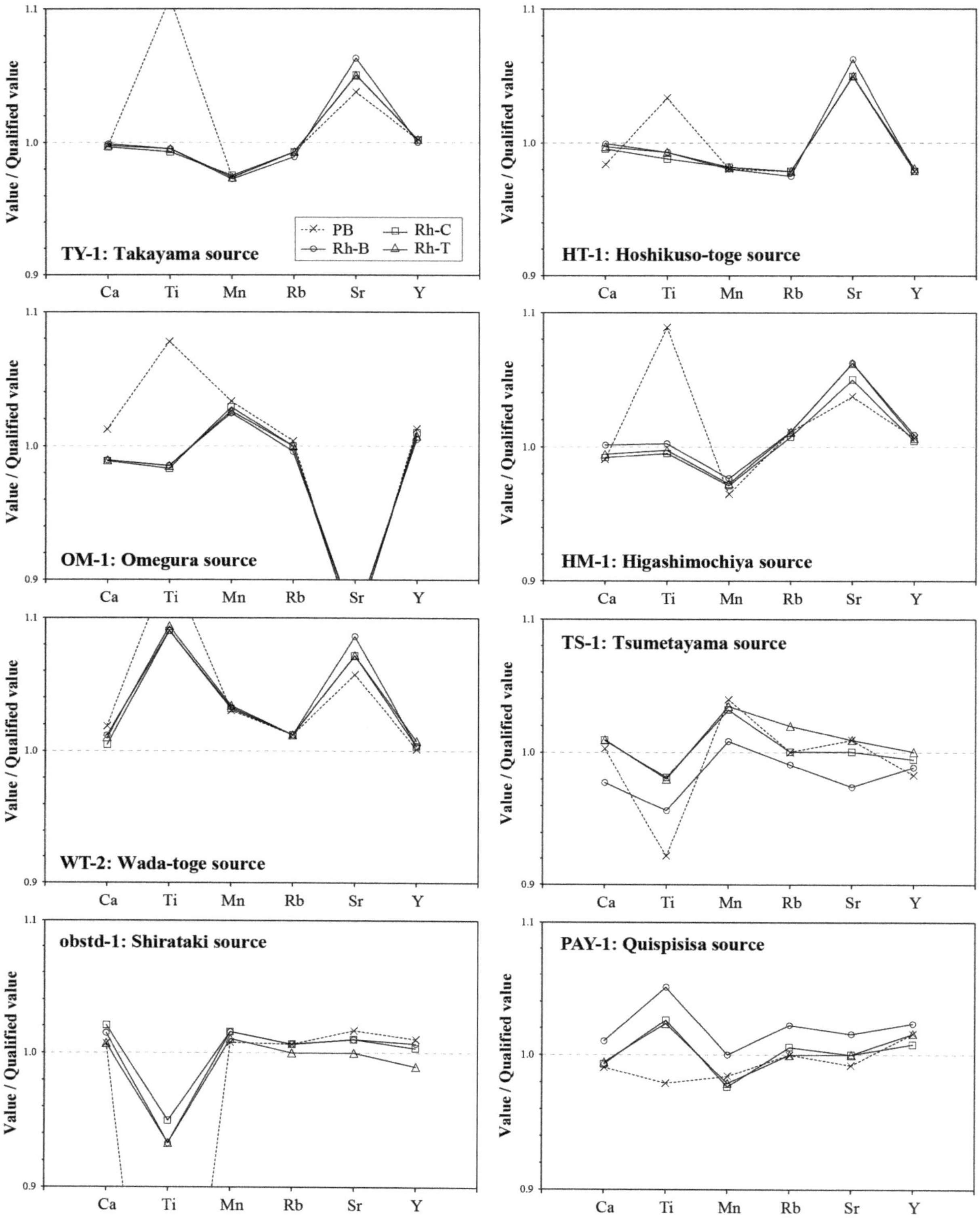

Figure 3.6. Qualified values normalised multi-element diagrams compiling the calculated values of the reference materials of obsidian for the non-destructive analysis. The PB, Rh-BG, Rh-C, and Rh-T indicate the lines of scattering X-rays used in the Internal Standard Method (see text for the abbreviations)

predominantly associated with their magmatic evolution.The degree of fractionation is the highest in the area of the Wada-toge source. The resulting patterns clearly indicate that the concentrations of the incompatible elements (i.e., Rb and Y) are most abundant in the WT-2 specimen. In the context of magmatic theory, the concentrations of the incompatible elements grow with increasing the degree of the fractionation.

The profiles of obsidian from the Tsumetayama, Shirataki, and Quispisisa sources indicate that these specimens are relatively depleted in Mn, Rb, and Y, but

Table 3.2. Results of quantitative analysis using the Internal Standard Method on polished slab obsidian specimens; see text for the abbreviations of scattering X-rays (PB, Rh-B, -C, and -T) used as the internal standard

	PB (n = 6)		***Rh-B (n = 6)***		***Rh-C (n = 6)***		***Rh-T (n = 6)***	
	Value	***± 2σ***	***Value***	***± 2σ***	***Value***	***± 2σ***	***Value***	***± 2σ***
TY-1 (Takayama)								
Ca	3423	140	3427	57	3419	21	3423	32
Ti	465	42	418	17	417	15	418	17
Mn	754	29	753	11	755	7	754	9
Rb	277	3	276	2	277	2	277	1
Sr	8.3	1.1	8.5	10	8.4	1.1	8.4	1.1
Y	46.4	1.4	46.3	1.6	46.4	1.1	46.4	1.3
HT-1 (Hoshikuso-toge)								
Ca	3376	218	3429	56	3415	37	3421	28
Ti	434	68	417	13	415	12	417	13
Mn	759	39	759	18	760	16	760	16
Rb	277	5	276	5	277	2	277	2
Sr	8.5	0.6	8.6	0.5	8.5	0.5	8.5	0.5
Y	46.2	1.4	46.2	1.8	46.2	1.2	46.3	1.2
OM-1 (Omegura)								
Ca	3834	163	3748	65	3746	37	3748	24
Ti	516	38	472	16	471	12	472	14
Mn	720	29	714	15	717	9	715	7
Rb	267	3	265	3	266	3	266	2
Sr	11.8	0.6	11.9	0.7	11.8	0.5	11.9	0.5
Y	40.6	0.6	40.3	0.6	40.5	0.6	40.4	0.6
HM-1 (Higashimochiya)								
Ca	3400	184	3436	56	3404	30	3413	30
Ti	457	74	421	30	418	31	419	31
Mn	747	33	756	15	752	9	753	7
Rb	276	2	276	5	275	1	276	2
Sr	8.4	0.6	8.6	0.6	8.5	0.6	8.6	0.5
Y	46.2	1.5	46.3	1.7	46.1	1.4	46.2	1.3
WT-2 (Wada-toge)								
Ca	3496	130	3472	59	3448	39	3466	45
Ti	342	36	327	16	327	18	328	19
Mn	878	45	880	31	879	22	881	19
Rb	329	7	329	13	329	9	329	8
Sr	7.4	0.5	7.6	0.5	7.5	0.5	7.5	0.4
Y	52.1	2.2	52.2	2.6	52.2	2.1	52.4	2.1
TS-1 (Tsumetayama)								
Ca	5159	320	5029	148	5189	19	5192	16
Ti	829	79	860	60	882	43	881	48
Mn	402	12	390	12	399	10	400	9
Rb	105	1	104	3	105	1	107	1
Sr	117	3	113	3	116	2	117	2
Y	17.0	0.5	17.1	0.6	17.2	0.6	17.3	0.5
obstd-1 (Shirataki)								
Ca	3814	143	3843	112	3865	49	3815	40
Ti	152	22	224	11	228	7	224	7

	PB (n = 6)		***Rh-B (n = 6)***		***Rh-C (n = 6)***		***Rh-T (n = 6)***	
	Value	***± 2σ***	***Value***	***± 2σ***	***Value***	***± 2σ***	***Value***	***± 2σ***
obstd-1 (Shirataki) *(continued)*								
Mn	390	27	393	3	393	8	391	8
Rb	159	2	159	3	159	1	158	1
Sr	31.1	1.1	30.9	1.2	30.9	0.9	30.6	1.0
Y	30.4	0.9	30.3	1.3	30.2	0.9	29.8	0.9
PAY-1 (Quispisisa)								
Ca	5949	331	6064	110	5964	69	5971	52
Ti	763	105	818	30	799	26	797	25
Mn	381	14	387	9	378	4	379	4
Rb	183	2	187	4	184	1	183	1
Sr	131	2	134	3	132	2	132	1
Y	13.2	0.4	13.3	0.6	13.1	0.3	13.2	0.4

Table 3.3. Results of quantitative analysis using the Internal Standard Method on flaked obsidian specimens; see text for the abbreviations of scattering X-rays (PB, Rh-B, -C, and -T) used as the internal standard

	PB (n = 3)		***Rh-B (n = 3)***		***Rh-C (n = 3)***		***Rh-T (n = 3)***	
	Value	***± 2σ***	***Value***	***± 2σ***	***Value***	***± 2σ***	***Value***	***± 2σ***
HT-1 (Hoshikuso-toge)								
Ca	3376	51	3376	51	3326	24	3328	30
Mn	753	12	753	12	746	6	745	7
Rb	278	4	278	4	276	1	275	2
Y	46.6	1.1	46.6	1.1	46.1	0.9	46.1	0.8
WT-2 (Wada-toge)								
Ca	3379	57	3379	57	3462	17	3440	32
Mn	843	14	843	14	869	6	860	2
Rb	315	4	315	4	325	1	322	1
Y	50.1	1.8	50.1	1.8	51.7	1.7	51.2	1.9
obstd-1 (Shirataki)								
Ca	3894	92	3894	92	3945	31	3859	23
Mn	403	5	403	5	405	11	399	12
Rb	161	3	161	3	162	1	160	1
Y	30.6	1.1	30.6	1.1	30.8	0.8	30.1	0.9
PAY-1 (Quispisisa)								
Ca	5930	14	5930	14	5864	41	5850	41
Mn	383	11	383	11	376	9	377	8
Rb	184	1	184	1	182	3	181	3
Y	13.9	0.6	13.9	0.6	13.8	0.6	13.8	0.7

enriched in Ca compared to obsidian from the Kirigamine area. This may reflect the petrogenetic diversity among different types of obsidian in each area.

These results indicate that such elements as Ca, Mn, Rb, and Y, are generally sufficient for the geochemical characterisation of obsidian from the major sources. The provenance identification of obsidian artefacts would be possible by comparing the profiles of geologic obsidian and archaeological samples using this multi-element diagram. However, the comparison method has some uncertainty. Thus, the combination of the multi-element diagram and a discrimination diagram drawn by a scatterplot is essential for the provenance identification of an obsidian artefact. This will be discussed in the next section.

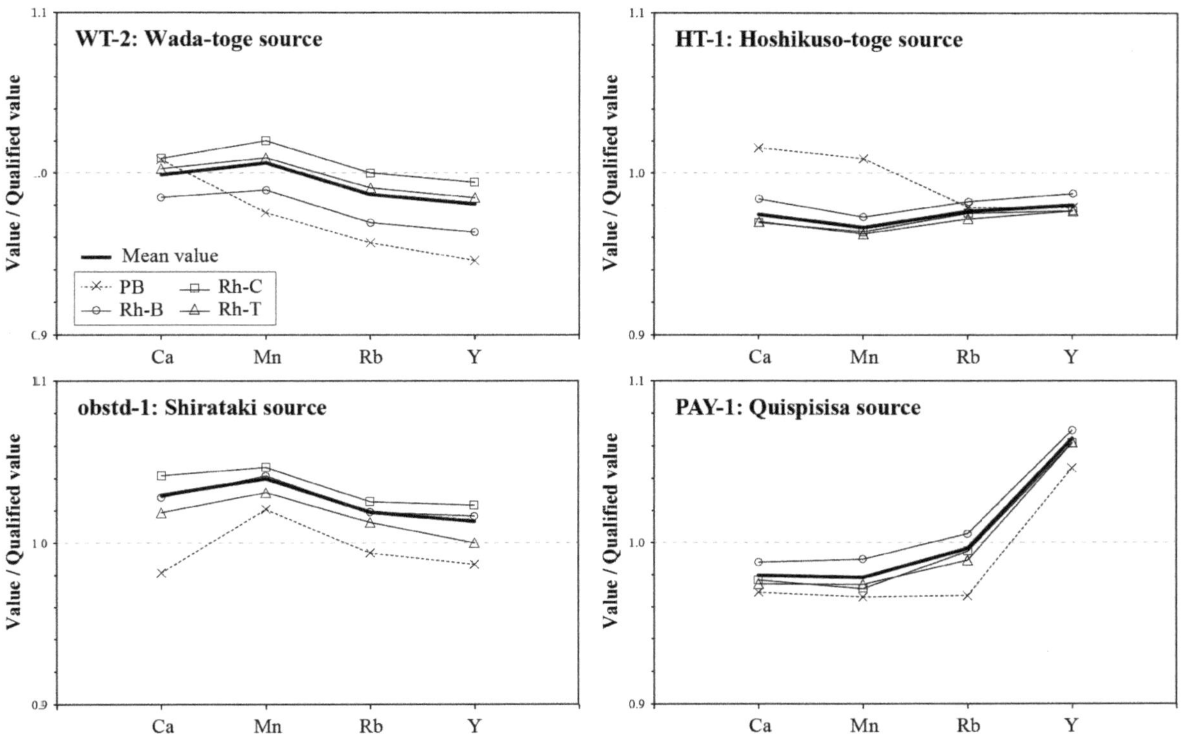

Figure 3.7. Qualified values normalised multi-element diagrams compiling the results of quantitative analysis of flaked obsidian using the non-destructive method. The PB, Rh-BG, Rh-C, and Rh-T indicate the lines used on the analysis in the Internal Standard Method (see text for the abbreviations)

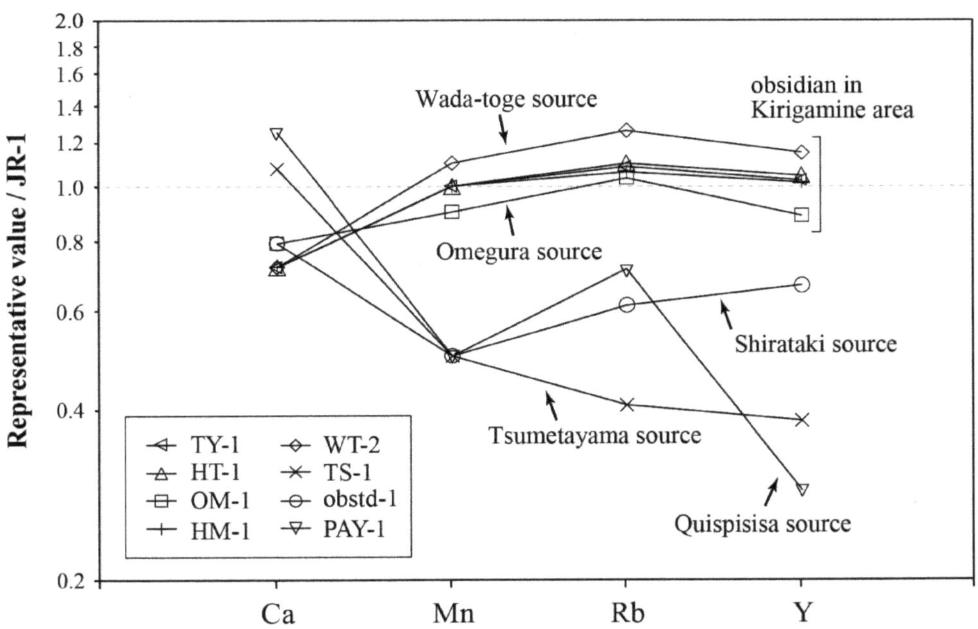

Figure 3.8. The JR-1 normalised multi-element diagram compiling the representative values of the obsidian in major sources analysed by the Fusion Bead Method (see text for the data sources and sample abbreviations)

PROVENANCE OF ARCHAEOLOGICAL OBSIDIAN

Profiles of the JR-1 normalised multi-patterns of obsidian from the Tsumetayama, Shirataki, and Quispisisa sources are completely different from those in the Kirigamine area (Figure 3.8). The profiles are generally characterised by the slopes between Ca and Y, and Mn and Rb. This suggests that the Y/Ca and Rb/Mn ratios are efficient indicators for characterising the geochemistry of obsidian from each source.

The JR-1 normalised Y/Ca vs. Rb/Mn ratios diagram is shown in Figure 3.9 (the Y_N/Ca_N vs. Rb_N/Mn_N diagram:

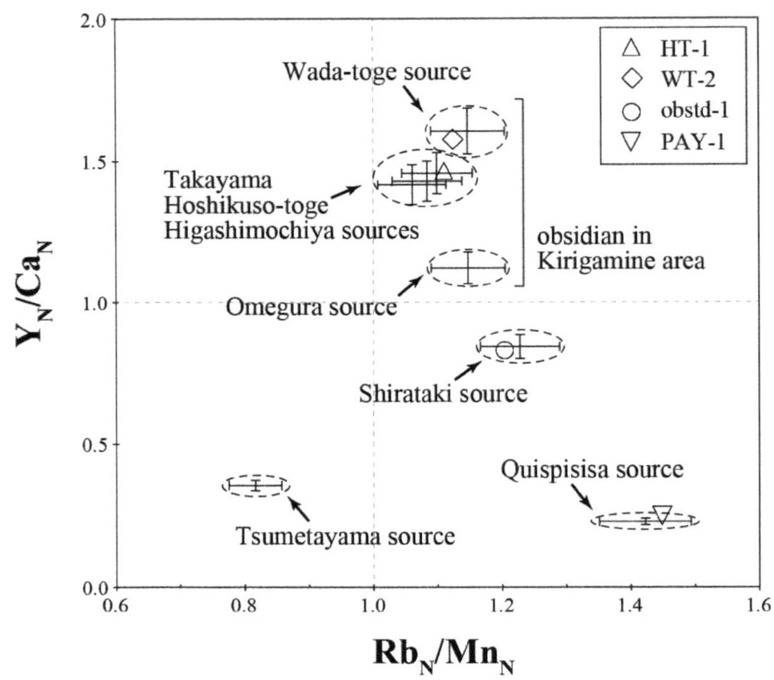

Figure 3.9. The JR-1 normalised Y/Ca vs. Rb/Mn ratios variation diagram for obsidian from major sources analysed in the present study. The bars and broken fields indicate the range of the representative value of 5% error, whereas the symbols indicate the values after the non-destructive analysis (see text for the data sources and sample abbreviations)

N means normalised value). Representative values for obsidian from the major sources in Table 3.1 were plotted in this diagram, which are shown by the fields with a 5% error margin. In addition, the results of the quantitative values after non-destructive analysis in Table 3.3 are shown by the symbols.

Although the profiles of obsidian from the Kirigamine area were quite indistinguishable in the multi-element diagram (see Figure 3.8), this variation diagram shows that obsidian from the Wada-toge and Omegura sources (see Figure 3.9) is clearly separated from the obsidian of the Higashimochiya, Takayama, and Hoshikuso-toge sources (see Figure 3.9). The composition of obsidian from the Takayama, Hoshikuso-toge, and Higashimochiya sources overlaps completely within a 5% error margin. This result may be explained by the petrogenetic homogeneity of the obsidian sources. The pyroclastic flow containing the HT-1 specimen is derived from the Higashimochiya volcanic vent (Yamasaki *et al.* 1976); and the TY-1 specimen collected from the Takayamagawa riverbed is derived from the pyroclastic flow. Obsidian from the Shirataki source has a lower Y/Ca ratio than obsidian from the Kirigamine area. Obsidian from the Tsumetayama source has lower Y/Ca and Rb/Mn ratios than obsidian from the Shirataki source and the Kirigamine area. Obsidian from the Quispisisa source is characterised by the lowest Y/Ca ratio and the highest Rb/Mn ratio among the types of obsidian analysed in the present study.

The results from the non-destructive analysis of flaked obsidian indicate that obsidian from the Wada-toge, Hoshikuso-toge, Shirataki, and Quispisisa sources is predominantly plotted in the field of the representative value of obsidian in the sources. This means that the quantitative value after the non-destructive method using the flaked obsidian is comparable to that from the destructive method (i.e., Fusion Bead Method) within a 5% error margin. Namely, the quantitative values after the non-destructive analysis presented in this paper can be plotted in this diagram, which gives appropriate guidance for the provenance identification of archaeological obsidian.

Acknowledgements

This paper is based on a presentation at the International Obsidian Workshop which took place at the Centre for Obsidian and Lithic Studies, Meiji University (Nagawa Town, Nagano Pref., Japan), on 5–6 November 2011. I would like to express my sincere thanks to all the participants, especially Prof. A. Ono, Mr K. Shimada and Mr J. Hashizume. Special thanks also go to Prof. K. Yajima for bringing the specimen from the Hoshikuso archaeological mining site, and to Prof. S. Aida, Mr N. Ohotake, Ms S. Ohotake, Mr A. Moriya, and Mr H. Kawata for organising and collaborating in the fieldwork conducted in the Kirigamine–Yatsugatake area. The original manuscript was improved by Drs M. D. Glascock, Y. V. Kuzmin, and S. G. Keates, Ms C. C. Lindsey, and Ms L. Dogiama (McMaster University, Hamilton, Canada). I am very grateful to Prof. J. Kawai and Dr A. M. De Francesco for constructive reviews of the first version of the manuscript. The present study was supported by Grants-in-Aid for Young Researchers from Meiji University in 2012, and by a grant from the Strategic Research Foundation Grant-aided Project for Private Universities from the MEXT, 2011–6 (S1101020).

References

BERTIN, E. P. 1978. *Introduction to X-Ray Spectrometric Analysis*. New York, Plenum Press.

BURGER, R. L., and M. D. GLASCOCK. 2000. Locating the Quispisisa Obsidian Source in the Department of Ayacucho, Peru. *Latin American Antiquity* 11, 258–268.

IKEYA, N. 2009. *Kokuyouseki Kokogaku* [Archaeology of Obsidian]. Tokyo, Shinsensha Publishers.

IMAI, N., S. TERASHIMA, S. ITOH, and A. ANDO. 1995. The 1994 Compilation Values for GSJ Reference Samples, "Igneous Rock Series". *Geochemical Journal* 29, 91–95.

KUZMIN, Y. V. 2011. The Patterns of Obsidian Exploitation in the Late Upper Pleistocene of the Russian Far East and Neighbouring Northeast Asia. In *Natural Resource, Environment and Humans* (Proceedings of the Meiji University Centre for Obsidian and Lithic Studies. No. 1), edited by A. Ono, 67–82. Tokyo, Centre for Obsidian and Lithic Studies, Meiji University.

NAKANO, S., K. TAKEUCHI, H. KATO, A. SAKAI, S. HAMASAKI, T. HIROSHIMA, and M. KOMAZAWA. 1998. *Geological Map of Japan 1:200,000. Nagano* (Cartographic Material NJ-85-36). Tokyo, Geological Society of Japan.

RIGAKU. 1982. *Keikou Ekkususen Bunseki no Tebiki* [Guidebook for the X-ray Fluorescent Analysis]. Tokyo, Rigaku Co., Ltd.

SHACKLEY, M. S. (ed.). 2011. *X-Ray Fluorescence Spectrometry (XRF) in Geoarchaeology*. New York, Springer.

SUDA, Y. 2012. Chemical Analysis of Obsidian by Wavelength-Dispersive X-Ray Fluorescence Spectrometry: Application to Non-Destructive Analysis of Archaeological Obsidian Artefacts. In *Natural Resource, Environment and Humans* (Proceedings of the Meiji University Centre for Obsidian and Lithic Studies. No. 2), edited by A. Ono, 1–14. Tokyo, Centre for Obsidian and Lithic Studies, Meiji University.

SUDA, Y. 2013. Quantitative Analytical Methodology for Major Elements in Silicate Rocks by Wavelength Dispersive X-ray Fluorescence Spectrometer. In *Natural Resource, Environment and Humans* (Proceedings of the Meiji University Centre for Obsidian and Lithic Studies. No. 3), edited by A. Ono, 31–45. Tokyo, Centre for Obsidian and Lithic Studies, Meiji University (in Japanese with English Abstract).

SUDA, Y., N. KOIZUMI, and T. OKUDAIRA. 2011. Keikou Ekkususen Bunseki Souchi wo Mochiita Keisanengan (Kaseigan, Taisekigan, Taisekibutsu, Dojyou) Chyu no Syuseibun, Biryou Seibun no Teiryou Bunseki [X-Ray Fluorescence Analysis of Major, Trace and Rare Earth Elements for Igneous Rocks, Sedimentary Rocks, Sediments, and Soil]. *Magma* 93, 19–32.

SUDA, Y., and Y. MOTOYOSHI. 2011. X-Ray Fluorescence (XRF) Analysis of Major, Trace, and Rare Earth Elements for Silicate Rocks by Low-Dilution Glass Bead Method. *Nankyoku Shiryou* 55, 93–108 (in Japanese with English Abstract).

SUDA, Y., T. OKUDAIRA, and K. FURUYAMA. 2010. Teikishyaku Galass Beed Hou ni Yoru Keikou Ekkususen Bunseki Souchi (RIX-2100) wo Mochiita Keisanen Gan Chuy no Syuseibun Biryou Seibun no Teiryou Bunseki [X-Ray Fluorescence (XRF; RIX-2100) Analysis of Major and Trace Elements for Silicate Rocks by Low-Dilution Glass Bead Method]. *Magma* 92, 21–39.

TAKAHASHI, K., and K. NISHIKI. 2006. Volcanostratigraphy of the Lower Pleistocene in the Northern Flank of the Northern Yatsugatake Volcanoes, Central Japan: A Voluminous Magmatism in the Northern Yatsugatake and Enrei areas. *Journal of Geological Society of Japan* 112, 549–567 (in Japanese with English Abstract).

YAMASAKI, T., T. KOBAYASHI, and S. KAWACHI. 1976. Geology and Petrography of the Wada Pass and Adjacent Area, Nagano Prefecture, Central Japan. *Journal of Geological Society of Japan* 82, 127–137 (in Japanese with English Abstract).

Chapter 4
THE EFFECTIVENESS OF ELEMENTAL INTENSITY RATIOS FOR SOURCING OBSIDIAN ARTEFACTS USING ENERGY DISPERSIVE X-RAY FLUORESCENCE SPECTROMETRY: A CASE STUDY FROM JAPAN

Tarou KANNARI, Masashi NAGAI, and Shigeo SUGIHARA

Abstract: *Non-destructive methods using elemental intensity ratios for analysing prehistoric obsidian artefacts in Japan have primarily been employed based on regulations protecting items of cultural heritage. The Energy Dispersive X-ray Fluorescence spectrometer is the mostly widely used type of equipment, and is a suitable appliance for this purpose. However, since results from this instrument in comparison to others have not been discussed in the literature, this paper examines its effectiveness compared to studies which employ a Wavelength Dispersive X-ray Fluorescence spectrometer.*

Keywords: *Obsidian Sourcing, Geochemistry, Energy Dispersive X-ray Fluorescence, Wavelength Dispersive X-ray Fluorescence, Non-Destructive Method, Intensity Ratios, Japan*

INTRODUCTION

Research on the geologic origins of lithic artefacts can shed light on important aspects of past human behaviour with regard to raw material procurement and use, and the distribution of valuable cultural items across different time periods. Obsidian is a volcanic glass with a specific chemical composition peculiar to the source volcano, and each eruption event corresponds to a variety of magma generations and evolutionary processes. In addition, obsidian has a homogeneous structure due to element partitioning between the crystalline and liquid phases. Various chemical methods are used to analyse obsidian artefacts based on its geochemical characteristics (e.g., Frahm 2012; Glascock *et al.* 2010; Gratuze 1999; Hughes 2010a; Shackley 2005; Kim *et al.* 2007; Le Bourdonnec *et al.* 2005).

In particular, the Energy Dispersive X-ray Fluorescence (hereafter – EDXRF) spectrometer is an advantageous tool because it is easy to operate, simple to prepare samples, non-destructive, and has a comparatively short measuring time (Shackley 2011). There have been many studies for sourcing obsidian artefacts by non-destructive EDXRF (e.g., Biró *et al.* 1986; Carter and Shackley 2007; Carter *et al.* 2013; Craig *et al.* 2007; Davis *et al.* 1998; Giauque *et al.* 1993; Hughes 1988, 1994; James *et al.* 1995; Poupeau *et al.* 2010; Shackley 1991, 1995; Weisler and Clague 1998). A semi-quantitative (net intensity) approach was used in the early work by XRF (Jack and Heizer 1968; Jack and Carmichael 1969), and some of the above cited studies in the 1980–90s also used this method. Giauque *et al.* (1993), Shackley (1995), and Davis *et al.* (1998) reported quantitative data, and Shackley (2005) criticised the intensity approach. Thus, the quantitative method for sourcing obsidian artefacts became more popular. However, as Neutron Activation Analysis (henceforth – NAA) was recommended for small specimens by Shackley (1998), the problems for small obsidian samples using commercially available EDXRF remained, except in a research by Giauque *et al.* (1993) that indicated a high accuracy using a customised instrument (Hughes 2010b). In the cited studies that employed EDXRF the elements heavier than Ti were chosen for sourcing obsidian artefacts. Many authors reported that trace elements such as Rb, Sr, Y, and Zr were effective in discriminating obsidian sources (e.g., Cann and Renfrew 1964; Jack and Heizer 1968; Nelson *et al.* 1975, Nelson *et al.* 1977; Smith *et al.* 1977; Hughes 1984, 1994; Newman and Nielsen 1985; Shackley 1988, 1995, 1998).

In Japan, previous studies of obsidian artefacts employed several methods, such as NAA (Osawa 1991; Osawa *et al.* 1977; Kuzmin *et al.* 2013); Electron Probe Microanalysis (EPMA) (e.g., Wada and Sano 2011); XRF (e.g., Higashimura 1980 and Warashina and Higashimura 1988, with a total of over 100 analytical reports by these authors; Mochizuki *et al.* 1994 and Ikeya and Mochizuki 1998, with a total of over 100 analytical reports by these authors; Ninomiya and Simadate 2001; Tateishi *et al.* 2004; Hall and Kimura 2002; Kakubuchi and Utsunomiya 2002, 2003; Obata *et al.* 2010; Inoue and Noto 2002; Ooya *et al.* 2006, 2007); and NAA and XRF (Yoshitani *et al.* 1999, 2001). Many of the results in Japan were reported based on net X-ray intensity.

Many studies for sourcing obsidian artefacts presented precise quantitative data by EDXRF. However, as Hughes (2010b) indicated the problem was not solved because small samples did not provide accurate quantitative values obtained by the non-destructive EDXRF analysis. As noted by Mochizuki *et al.* (1994), Eerkens *et al.* (2007), and Tsutsumi (2010), analysis of small specimens is necessary. We recognised that analysis for all stone tools found at a site (naturally, containing small samples) needed to use a simple method. In this paper, we study again the potential of the intensity approach for sourcing obsidian artefacts through non-destructive method using EDXRF. The second issue is – which elements are the most suitable? These matters are discussed from a geochemical point of view.

HISTORICAL BACKGROUND OF SOURCING OBSIDIAN ARTEFACTS IN JAPAN

Why are Non-Destructive Techniques and Intensity Ratios the Preferred Methods?

Non-destructive analysis has been applied to archaeological obsidian artefacts in Japan based on regulations protecting cultural properties and traditional Japanese cultural perceptions. Since flaked stone tools made by humans are unique, their destruction during analysis has many negative consequences for archaeological studies and cultural heritage. For this reason, it is problematic and often unethical to conduct destructive analysis in Japan.

Dr Takenobu Higashimura, a pioneer with regard to the method employing elemental intensity ratios in Japan, pointed out that "In the elemental ratio method, the content of each element is not obtained but merely the ratio of content between two elements is generated ... The important question is – can we obtain a useful parameter that can be used to 'fingerprint' the obsidian source? Even if accurate element content in the sample can be measured, it does not help directly with source determination. The most important is to obtain the same value for all samples from the source." (Higashimura 1986, 22; translation by T. Kannari).

The elemental intensity ratio mentioned above is the ratio of spectrum areas for two elements. The intensity ratio becomes an approximate value, since the peak areas will be different for samples with dissimilar shapes or thicknesses (Higashimura 1986). There are relevant works for the identification of the sources of sanukite (glassy andesite) using the same method (Higashimura and Warashina 1975; Warashina *et al.* 1978).

Mochizuki (1997a) examined element concentrations by the Fundamental Parameter (FP) method, and intensity ratios using EDXRF. As a result, the intensity ratio method was chosen for sourcing obsidian artefacts, and the origin of obsidian in Japan was determined using Mn, Fe, and K for major elements along with Rb, Sr, Y, and Zr for trace elements. Identification parameters were: Rb fraction as Rb x 100 / A (A is the sum of Rb, Sr, Y, and Zr contents); Sr fraction as Sr x 100 / A; Zr fraction as Zr x 100 / A; Mn fraction as Mn x 100 / Fe; and log (Fe/K). Those parameters involve a few elements and scatterplots with practical combinations of elements to determine the sources of obsidian artefacts. They are easy to understand, and the results of sourcing the obsidian used for artefacts are immediately clear. The validity of those parameters was confirmed by Shimano *et al.* (2004) using element concentrations measured by WDXRF. Numerous studies employing this method have been conducted in Japan (e.g., Tsutsumi and Mochizuki 2012) and in Anatolia (e.g., Mochizuki 1997b, Kannari *et al.* 2012), with excellent results using this approach.

We recognised that intensity ratios satisfy the aim "Do not destroy the artefacts" which is required by Japanese archaeologists. Intensity by direct irradiation is a qualitative value, and fluctuates on account of sample conditions and changes to the instruments over time. Although the researchers cited above examined carefully these potential problems, they selected intensity by irradiation as a method.

Obsidian Occurrence in Nature

Obsidian is produced when rhyolitic-to-andesitic magma is quenched. The content of SiO_2 (silica dioxide) falls within a range of 60–78 wt.%, mostly within 74–78 wt.%. It has low combined water content, generally less than 1 wt.%, but in some rare cases more than 1 wt.%. There are two main obsidian occurrences: 1) the primary sources derived from volcanoes; and 2) the secondary sources when obsidian was transported from primary sources by natural forces (Figure 4.1). In volcanic geology, the main emphasis is given to the primary sources. A sufficient amount of obsidian suitable for making stone artefacts is produced from the volcanic vent in the form of lavas and pyroclastic flow deposits. Obsidian from the primary sources is transported by natural agents such as erosion, and appears later in riverbeds or terraces' gravel as a secondary source. Hunter–gatherers selected raw materials such as obsidian that was suitable for the production of stone tools. Therefore, it is necessary to study obsidian from the perspective of both primary and secondary distributions. For example, riverbed gravel has impact scars on its surface due to the rolling action of fluvial transportation, which reduces its weight in proportion to the travel distance. Consequently, observing the weathered surface of obsidian artefacts may be one way of deducing the geographic point where the material was collected by hunter–gatherers.

Advantages of Energy Dispersive X-ray Fluorescence

X-ray Fluorescence spectrometers include two types: WDXRF, which measure the spectrum of an X-ray fluorescence-radiated sample using a dispersive crystal; and EDXRF, which detects multi-element X-ray Fluorescence on a solid-state detector (SSD). WDXRF is superior to EDXRF in spectral resolution and its capacity for measuring light and trace elements, while EDXRF containing SSD has high energy resolution (Frankel and Aitken 1970). In general, the chamber design is larger for EDXRF than for WDXRF. WDXRF's mechanical limitation is that the dispersive crystal and the detector are in a spectroscopic chamber. In contrast, the EDXRF has only a stationary detector in the chamber. The analytical time for EDXRF is shorter than WDXRF because the latter measures only one element at a time, while EDXRF measures all elements simultaneously. In summary, the advantages of EDXRF are: 1) few limitations regarding sample size or shape for a large

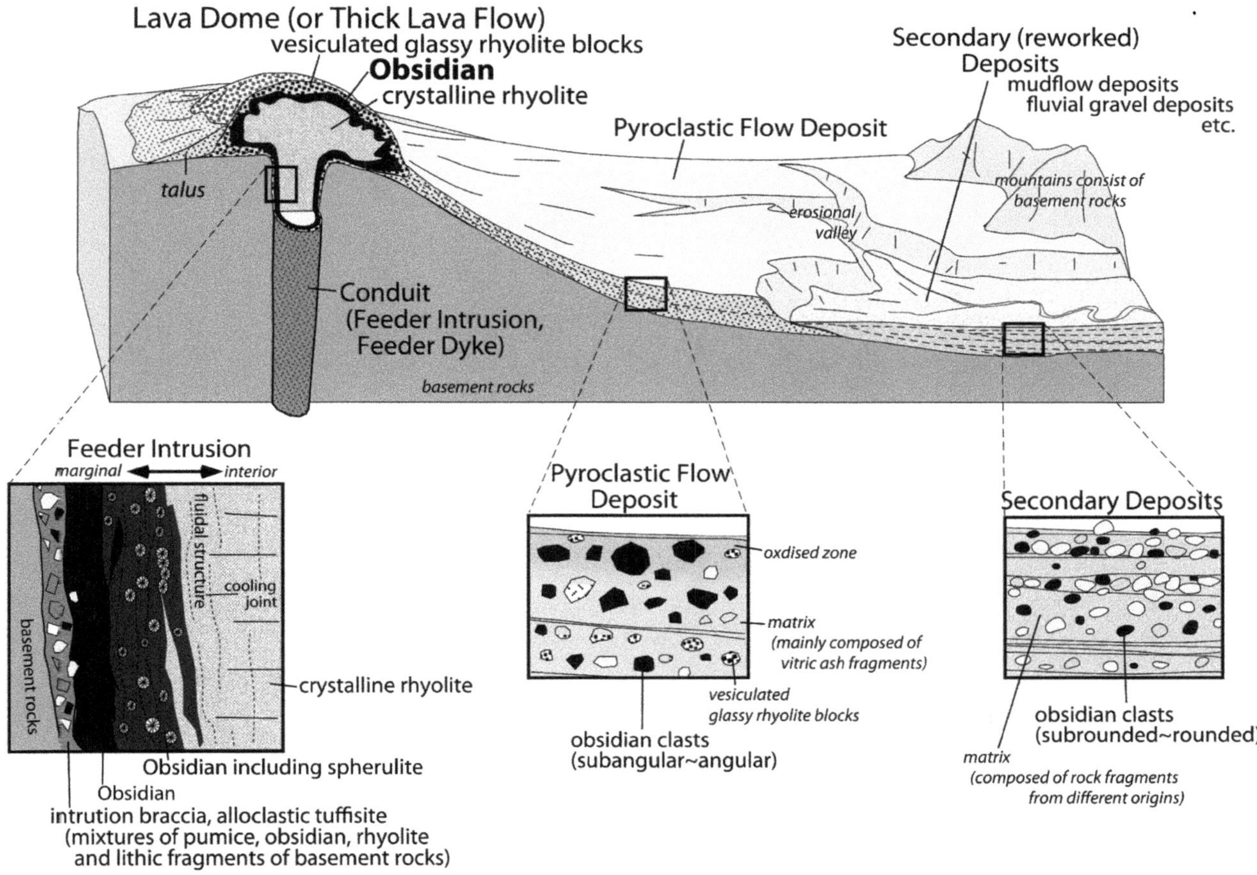

Figure 4.1. Schematic diagram of obsidian occurrences in nature

chamber, making it non-destructive and easy to use; 2) very short analytical times such that many samples can be analysed quickly; and 3) low costs of initial investment, operation, and maintenance. Thus, EDXRF is a more than suitable instrument for identifying the obsidian sources used for artefact manufacture than WDXRF.

METHODS AND MATERIALS

Sample Preparation

More than 100 geological sources of obsidian can be found in Japan. A Database describing them is almost complete (Sugihara and Kobayashi 2004, 2006). Obsidian sources in Japan were classified as the districts of Hokkaido–Tohoku (Kannari *et al.* 2010), Chubu–Kanto (Sugihara *et al.* 2008, 2009a, 2009b; Nagai *et al.* 2007), Hokuriku–Chubu (Sugihara *et al.* 2011), and Oki–Kyushu (Kannari *et al.* 2011), as shown in Figure 4.2.

For the current study, 12 sources are selected from the Database possessing a variety of chemical compositions, including: Akaishiyama, Tokachiishizawa, Ikutahara, Oga, Wada-toge (Higashimochiya), Nishikirigamine (Hoshigato), Takaharayama, Asinoyu, Oki, Himeshima, Ureshino, and Kamiushibana. Obsidian samples from each source were analysed under three conditions: 1) fused glass beads for WDXRF (Figure 4.3, a) where the glass bead has a flux-to-sample ratio of 5:1 (details on sample preparation can be found in Appendix 1); 2) mirror surface samples for EDXRF (Figure 4.3, b) in which mirror surface samples were removed from source nodules with a petrographic saw, and surfaces were polished using diamond paste; and 3) flake samples for EDXRF obtained with a rock hammer from the Akaishiyama source containing no spherulites, and samples from the Takaharayama source with many spherulites (Figure 4.3, c). For small flake samples a macromolecule film (Chemplex®, CAT.NO:100) was used to prevent specimens from falling into the instrument.

Instrument Conditions

The XRF analyses were conducted using WDXRF (Rigaku, RIX–1000) and EDXRF (JEOL, JSX–3100s) spectrometers at the Centre for Obsidian and Lithic Studies (COLS), Meiji University (Tokyo). Element concentrations for WDXRF were calculated using the calibration curve established for each element from international rock standards. Concentrations for EDXRF were established by the FP method and were not corrected. X-ray intensities from EDXRF were calculated using software developed in our laboratory only for automation. Further details of analytical instruments and operating conditions for EDXRF are provided in Appendix 2.

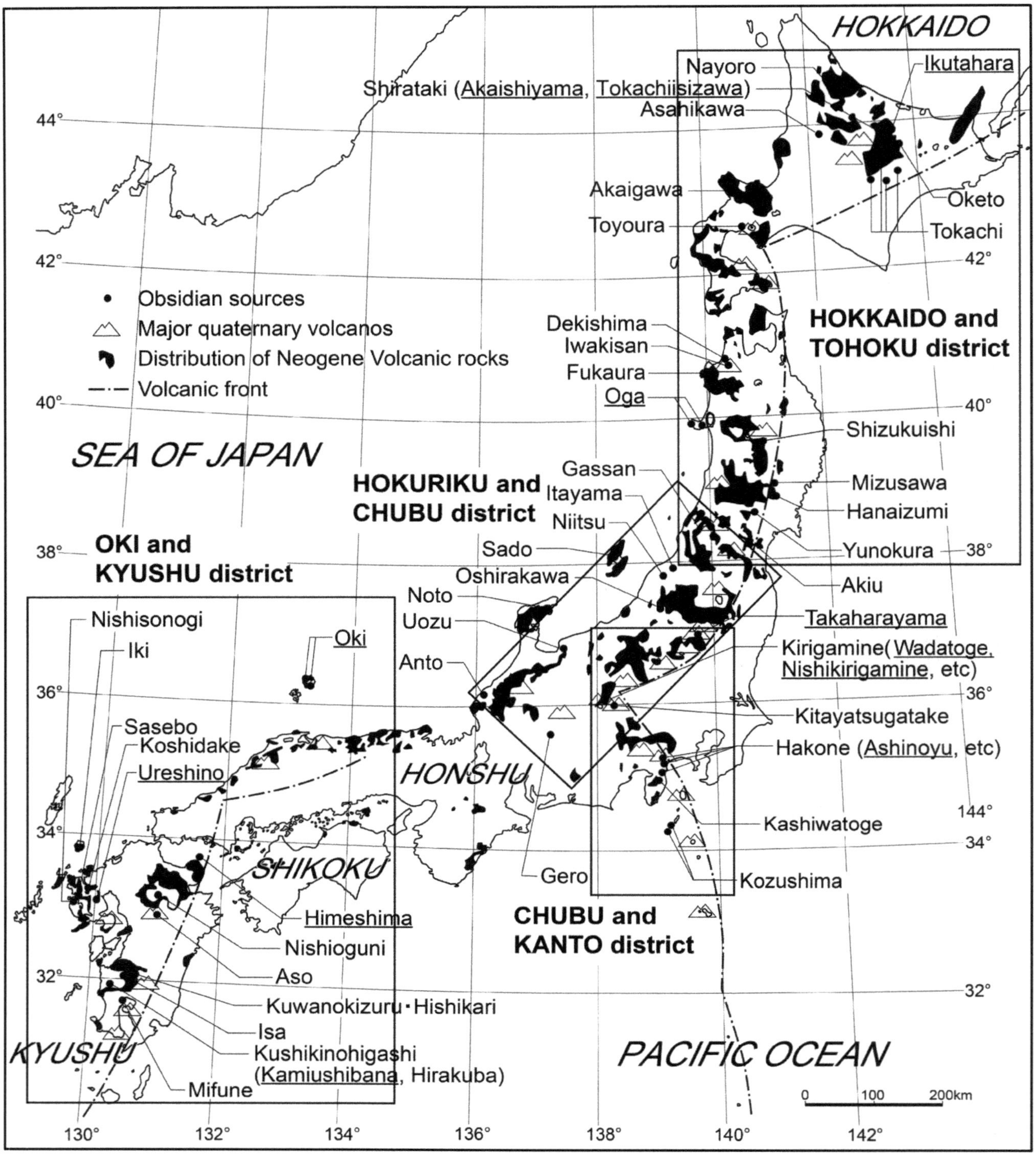

Figure 4.2. Major obsidian sources for prehistoric cultural complexes in Japan

RESULTS AND DISCUSSION

Element Concentrations Using Wavelength Dispersive X-ray Fluorescence and Energy Dispersive X-ray Fluorescence

Table 4.1 shows a comparison between element concentrations of glass beads using WDXRF with the calibration curve method, and mirror surface samples using EDXRF with the FP method. It is necessary to discuss the results for all major elements and selected trace elements (Rb, Sr, Y, Zr, Ba, and Zn). Differences relative to those concentrations (Diff.) range from -5% to 5% for SiO_2, TiO_2, Al_2O_3, CaO, Na_2O, and K_2O. Almost all Diff. for FeO* and MnO are less than 30%. Diff. for MgO and P_2O_5 by EDXRF are almost over 50%. Almost all Diff. for trace elements fall in the range of about 20%. Barium by EDXRF was detected when the obsidian of mirror surface samples contained more than 72 wt.% SiO_2, and more than 100 ppm Ba, while Ba was not

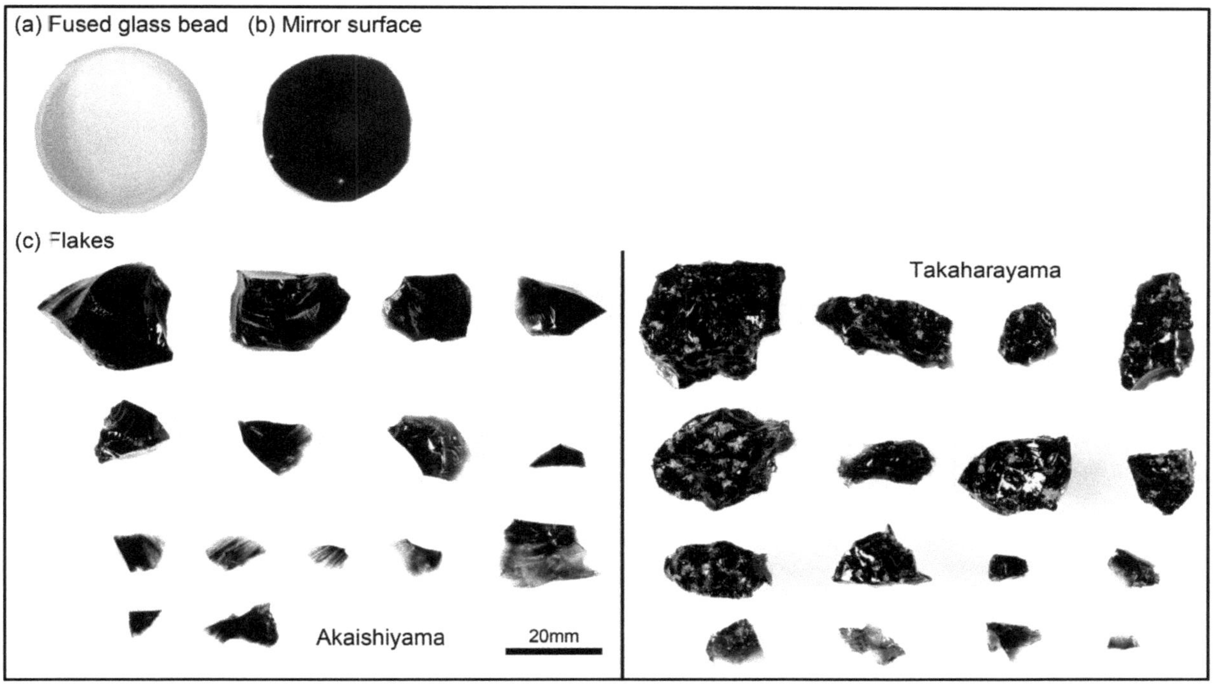

Figure 4.3. Obsidian samples from the Akaishiyama and Takaharayama sources in three conditions (a–c)

Table 4.1. Element concentrations for major and selected trace elements in obsidian samples from Japanese sources

Sample Name			SiO_2	TiO_2	Al_2O_3	FeO*	MnO	MgO	CaO	Na_2O	K_2O	P_2O_5
A. Glass Beads by WDXRF			in wt.%									
Akaishiyama			76.46	0.04	13.14	1.43	0.05	0.05	0.58	3.67	4.57	0.02
Tokachiishizawa			76.71	0.02	13.04	1.31	0.06	0.03	0.49	3.77	4.55	0.01
Ikutahara			74.48	0.13	13.69	1.99	0.02	0.08	0.84	3.98	4.78	0.01
Oga			76.49	0.08	13.24	0.87	0.16	0.09	0.89	3.71	4.47	0.01
Wada-toge			77.05	0.07	12.79	0.90	0.11	0.04	0.52	3.72	4.79	0.01
Nishikirigamine			77.62	0.09	12.43	0.80	0.08	0.08	0.52	3.76	4.62	0.01
Takaharayama			76.57	0.20	12.51	2.15	0.06	0.25	1.58	3.52	3.14	0.03
Asinoyu			68.26	0.74	14.77	5.12	0.19	1.31	4.13	4.49	0.82	0.17
Oki			75.46	0.12	12.56	2.12	0.06	0.01	0.59	3.98	5.10	0.004
Himeshima			75.30	0.00	14.62	1.31	0.12	0.09	0.55	4.15	3.75	0.11
Ureshino			75.39	0.06	14.11	1.30	0.06	0.14	1.14	3.69	4.07	0.04
Kamiushibana			68.02	0.87	15.12	4.17	0.09	1.12	3.26	4.31	2.85	0.19
B. Mirror Surface Samples by EDXRF												
Akaishiyama	n = 10	mean	75.84	0.04	13.31	0.96	0.05	0.11	0.53	3.89	4.51	0.61
		S.D.[#]	0.05	0.00	0.03	0.00	0.00	0.03	0.01	0.09	0.02	0.02
		Diff. (%)[#]	-1	6	1	-33	16	144	-9	6	-1	3657
Tokachiishizawa	n = 10	mean	75.88	0.03	13.35	0.97	0.06	0.10	0.45	3.94	4.57	0.59
		S.D.	0.09	0.00	0.03	0.00	0.00	0.03	0.01	0.07	0.02	0.06
		Diff. (%)	-1	17	2	-26	12	245	-8	5	1	3894
Ikutahara	n = 10	mean	73.63	0.13	13.92	1.51	0.04	0.20	0.82	4.21	4.71	0.68
		S.D.	0.07	0.00	0.02	0.00	0.00	0.02	0.01	0.07	0.01	0.02
		Diff. (%)	-1	4	2	-24	76	161	-3	6	-2	6475
Oga	n = 10	mean	75.36	0.08	13.60	0.62	0.15	0.17	0.88	3.91	4.46	0.63
		S.D.	0.05	0.00	0.02	0.00	0.00	0.03	0.01	0.06	0.02	0.03
		Diff. (%)	-1	6	3	-29	-4	89	-1	5	0	6274

Sample Name			SiO_2	TiO_2	Al_2O_3	FeO*	MnO	MgO	CaO	Na_2O	K_2O	P_2O_5
B. Mirror Surface Samples by EDXRF (continued)												
Wada-toge	n = 10	mean	76.21	0.07	13.06	0.60	0.11	0.13	0.49	3.93	4.70	0.63
		S.D.	0.06	0.00	0.03	0.00	0.00	0.02	0.00	0.07	0.01	0.03
		Diff. (%)	-1	4	2	-33	-3	195	-5	6	-2	10256
Nishikirigamine	n = 10	mean	76.22	0.10	13.08	0.57	0.08	0.15	0.49	3.99	4.61	0.63
		S.D.	0.06	0.00	0.02	0.00	0.00	0.02	0.00	0.06	0.01	0.02
		Diff. (%)	-2	5	5	-29	3	98	-6	6	0	6042
Takaharayama	n = 10	mean	75.59	0.20	12.85	1.63	0.07	0.31	1.57	3.79	3.10	0.77
		S.D.	0.07	0.00	0.03	0.01	0.00	0.02	0.01	0.06	0.02	0.03
		Diff. (%)	-1	3	3	-24	15	24	-1	8	-1	2796
Asinoyu	n = 10	mean	66.73	0.74	15.44	4.47	0.20	1.62	4.26	4.84	0.72	0.90
		S.D.	0.07	0.00	0.06	0.01	0.00	0.04	0.01	0.11	0.01	0.03
		Diff. (%)	-2	0	5	-13	6	24	3	8	-13	433
Oki	n = 10	mean	74.15	0.11	13.10	1.75	0.07	0.13	0.57	4.19	5.08	0.74
		S.D.	0.07	0.00	0.03	0.00	0.00	0.02	0.00	0.07	0.01	0.03
		Diff. (%)	-2	-3	4	-18	23	1386	-4	5	0	18007
Himeshima	n = 10	mean	74.18	0.01	15.16	0.97	0.11	0.16	0.51	4.38	3.73	0.66
		S.D.	0.05	0.00	0.03	0.00	0.00	0.02	0.01	0.05	0.01	0.03
		Diff. (%)	-1	595	4	-26	-5	85	-7	6	-1	499
Ureshino	n = 10	mean	74.29	0.06	14.53	1.00	0.07	0.24	1.14	3.84	4.08	0.63
		S.D.	0.05	0.00	0.02	0.00	0.00	0.03	0.01	0.05	0.01	0.03
		Diff. (%)	-1	6	3	-23	15	72	0	4	0	1515
Kamiushibana	n = 10	mean	67.25	0.85	15.42	3.56	0.11	1.35	3.14	4.60	2.71	0.89
		S.D.	0.09	0.00	0.03	0.01	0.00	0.04	0.01	0.11	0.01	0.03
		Diff. (%)	-1	-2	2	-15	20	21	-3	7	-5	371
C. Flake Samples by EDXRF												
Akaishiyama	n = 2	mean	73.65	0.05	14.29	1.45	0.07	n.d.***	0.72	3.85	5.71	n.d.
Akaishiyama**	n = 13	mean	77.18	0.07	12.17	1.88	0.09	n.d.	1.13	0.34	6.96	0.25
		S.D.	0.25	0.00	0.16	0.06	0.00		0.10	0.35	0.13	0.21
Takaharayama	n = 4	mean	73.37	0.26	14.07	2.40	0.08	n.d.	2.07	3.57	3.96	n.d.
		S.D.	1.33	0.05	0.51	0.55	0.01		0.19	0.67	0.69	
Takaharayama**	n = 12	mean	75.56	0.36	11.95	3.27	0.11	n.d.	3.18	0.40	4.78	0.58
		S.D.	1.56	0.08	0.51	0.48	0.01		0.69	0.22	0.22	0.56

Sample Name			Rb	Sr	Y	Zr	Ba	Zn
A. Glass Beads by WDXRF			in ppm					
Akaishiyama			158	25	29	67	923	41
Tokachiishizawa			183	< 3	32	54	189	37
Ikutahara			157	47	42	208	917	60
Oga			176	100	22	70	888	32
Wada-toge			272	< 3	47	83	38	31
Nishikirigamine			144	38	25	72	363	28
Takaharayama			111	126	42	157	758	49
Asinoyu			19	264	33	101	337	102
Oki			212	< 3	49	326	19	88
Himeshima			89	64	10	25	871	60
Ureshino			145	208	10	74	639	40
Kamiushibana			82	388	29	231	472	66

Sample Name			Rb	Sr	Y	Zr	Ba	Zn
B. Mirror Surface Samples by EDXRF								
Akaishiyama	n = 10	mean	154	27	27	69	895	34
		S.D.[#]	5	4	4	5	38	4
		Diff. (%)[#]	-2	9	-8	4	-3	0
Tokachiishizawa	n = 10	mean	180	9	35	59	157	33
		S.D.	6	4	5	5	23	4
		Diff. (%)	-1	740	9	11	-17	-21
Ikutahara	n = 10	mean	164	47	41	215	809	51
		S.D.	3	4	6	5	31	6
		Diff. (%)	5	0	-3	3	-12	-14
Oga	n = 10	mean	173	89	25	76	842	28
		S.D.	6	4	4	4	47	5
		Diff. (%)	-2	-10	17	9	-5	-13
Wada-toge	n = 10	mean	276	6	45	84	n.d.	31
		S.D.	4	2	3	5		4
		Diff. (%)	2	-	-5	1		0
Nishikirigamine	n = 10	mean	140	40	27	77	322	26
		S.D.	4	3	5	4	38	5
		Diff. (%)	-3	7	10	7	-11	-6
Takaharayama	n = 10	mean	109	114	39	154	616	42
		S.D.	6	4	3	3	32	7
		Diff. (%)	-1	-9	-6	-2	-19	-15
Asinoyu	n = 10	mean	12	249	34	102	n.d.	94
		S.D.	5	8	6	7		5
		Diff. (%)	-36	-6	2	1		-8
Oki	n = 10	mean	219	3	49	345	n.d.	80
		S.D.	4	3	5	5		5
		Diff. (%)	3	-	1	6		-9
Himeshima	n = 10	mean	81	59	10	27	875	55
		S.D.	4	3	4	5	29	2
		Diff. (%)	-8	-8	-6	8	0	-8
Ureshino	n = 10	mean	140	178	13	63	615	36
		S.D.	5	4	3	5	27	4
		Diff. (%)	-4	-14	24	-14	-4	-10
Kamiushibana	n = 10	mean	86	347	32	229	n.d.	58
		S.D.	3	5	5	6		5
		Diff. (%)	4	-10	11	-1		-12
C. Flake Samples by EDXRF								
Akaishiyama	n = 2	mean	240	49	51	115	1301	56
Akaishiyama**	n = 13	mean	285	52	54	124	1460	75
		S.D.	33	8	12	24	59	10
Takaharayama	n = 4	mean	180	194	71	282	1134	81
		S.D.	67	51	28	80	267	40
Takaharayama**	n = 12	mean	220	244	89	348	1334	96
		S.D.	27	42	14	58	113	14

* Total iron calculatied as FeO.
** Analysis with a macromolecule film.
*** Not detected.

[#] S.D. – standard deviation; Diff. – difference relative to other analytical results: Diff. = [mean (B or C) / mean A – 1] x 100.

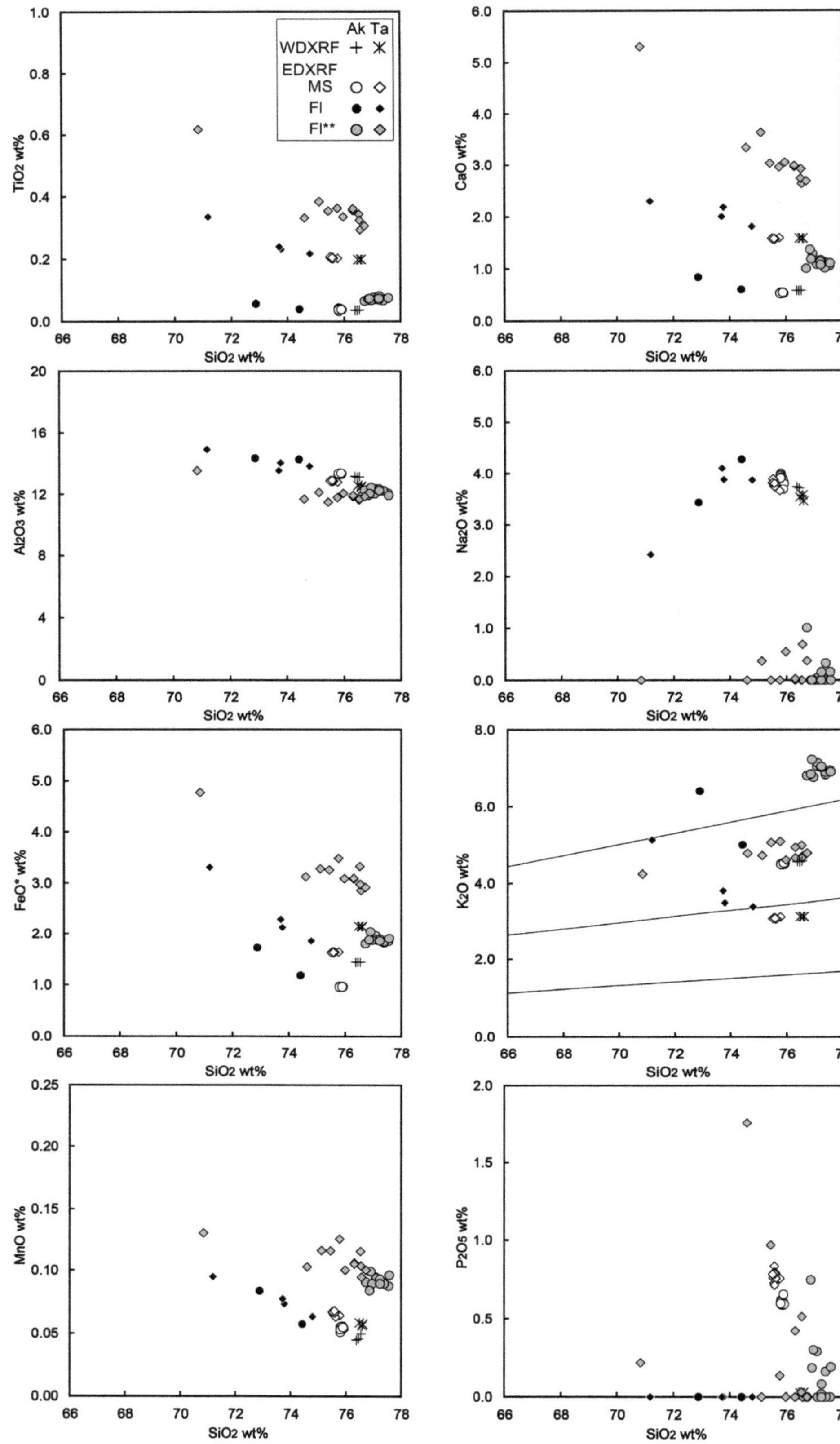

A

Figure 4.4. A. The SiO_2 variation diagrams for element concentrations of glass beads by WDXRF, and mirror surface (MS) and flake (Fl) samples by EDXRF, from Akaishiyama (Ak) and Takaharayama (Ta) sources. FeO is the total Fe as FeO; Fl** is for use of macromolecule film. The K_2O discrimination lines were based on Gill (1981)*

detected when the content was below that value. The FP data by EDXRF are in good agreement with the WDXRF data, except for MgO, P_2O_5, and Ba.

The element concentrations obtained for glass beads using WDXRF, and for mirror surface and flake samples using EDXRF, are presented in Figure 4.4. The SiO_2

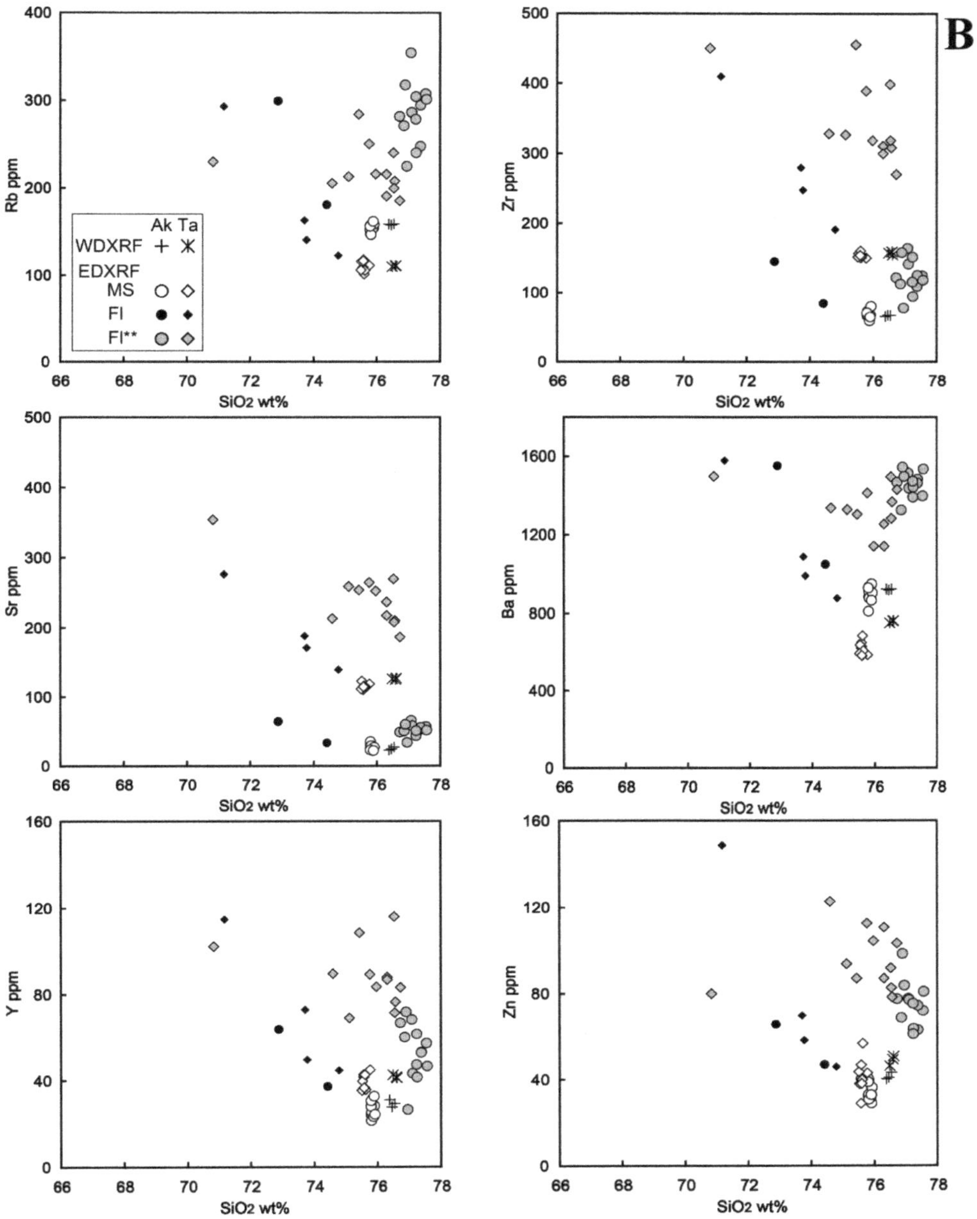

Figure 4.4. B. See captions on Figure 4.4. A

variation diagrams (usually call Harker diagrams) are typically used in petrology. The results from the analysis of flake samples, whether using a macromolecule film or not, are not in good agreement with the WDXRF data. MgO was not detected at all, and the content of Na_2O is too low in the case of using a macromolecule film because the film attenuated X-ray fluorescence (Honma 2005). Although Davis *et al.* (1998) have shown highly reliable quantitative values, we did not obtain good results for flake samples of various shapes and sizes with the FP method.

Two issues for sourcing obsidian artefacts by non-destructive methods now appear. First, the signal intensity from the sample must be constant if a repeatable element concentration is obtained, whether using the calibration curve or FP method. In other words, the measured surface must be smooth, and the irradiated area must be fixed. However, stone tools made by flaking have irregular surfaces and shapes such that their analytical conditions are inferior. Second, a macromolecule film is necessary for a small sample analysis when EDXRF is used. However, the film attenuates the X-ray fluorescence of light elements (Honma 2005). Therefore, the attenuation of intensity is influential in other element concentrations using the FP method because it calculates the intensities for each element based on standardising the chemical composition to 100%.

Comparison between Element Concentration and Intensity

Scatterplots of the element concentrations of the three conditions referred to above, using A. Mochizuki's parameters, are displayed in Figure 4.5. The centres of each distribution range do not overlap visually. In

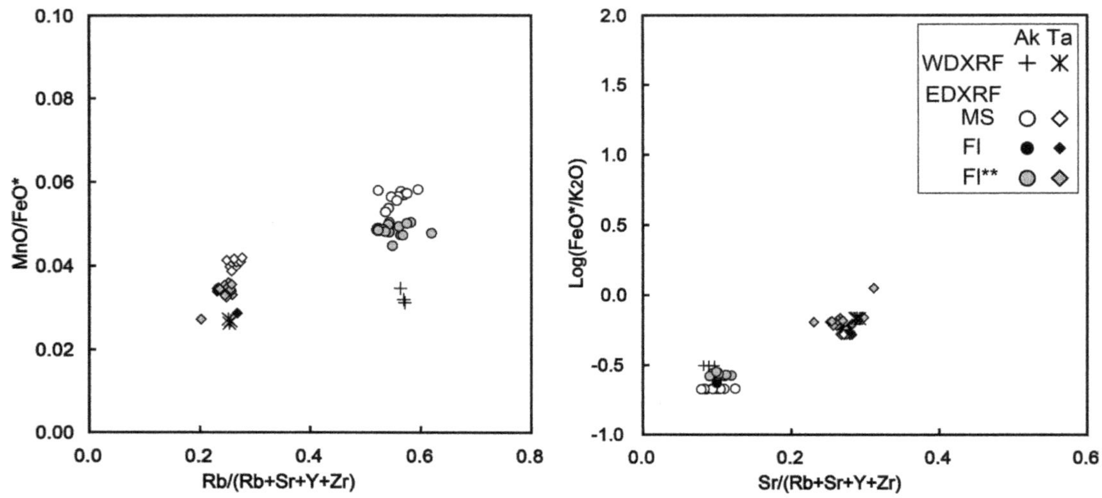

Figure 4.5. Scatterplots of element concentrations for glass beads by WDXRF, and for mirror surface (MS) and flake (Fl) samples by EDXRF, from Akaishiyama (Ak) and Takaharayama (Ta) sources. FeO is total Fe as FeO. Fl** is for use of macromolecule film*

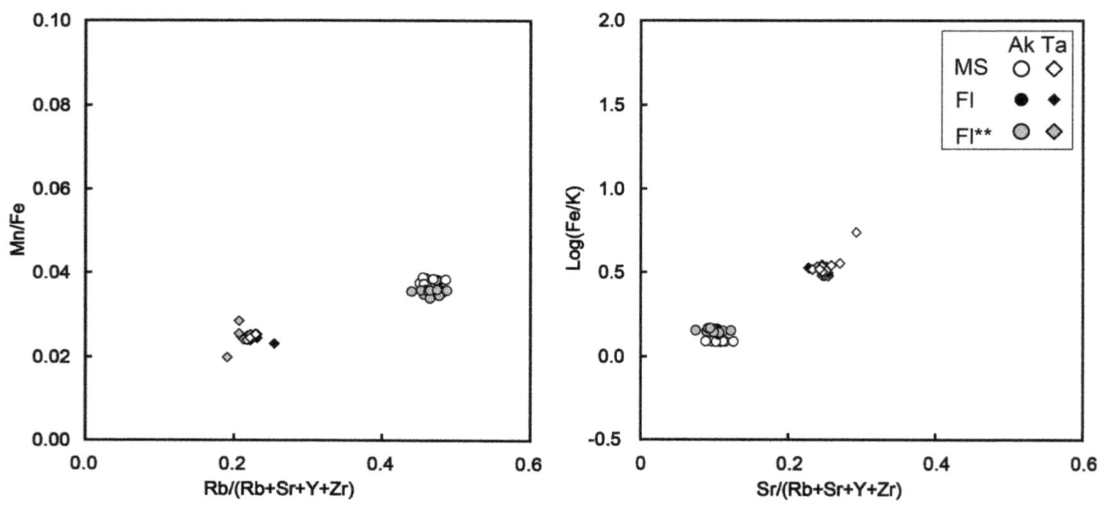

*Figure 4.6. Scatterplots of intensity ratios for mirror surface (MS) and flake (Fl) samples by EDXRF, from Akaishiyama (Ak) and Takaharayama (Ta) sources. Fl** is for use of macromolecule film*

particular, if the ranges of two conditions of the FP method are different, it will be difficult to identify the obsidian source used for artefact manufacture. Usually standard samples may be selected in the best condition (in this study, the mirror surface). However, surface conditions, whether it is a mirror or irregular, influence element concentrations obtained by the FP method. These results suggest that the FP method is not suitable for non-destructive analysis of obsidian artefacts. In addition, as the FP method depends on the coefficient of sensitivity-equipped instruments, it is unclear whether or not the result obtained by the FP method is a quantitative value.

Figure 4.6 shows scatterplots based on intensity ratios using EDXRF. Since the clusters of mirror surface and flake samples overlap, the obsidian source used for artefacts was identified. This coincides with results obtained by Mochizuki (1997a). Table 4.2 shows Mahalanobis distances for each cluster, which are calculated using element concentrations of three conditions and intensities of two conditions. Mahalanobis distance for intensities is equal to or more than element concentration employing the FP method. Whereas we recognise the importance of element concentrations, it seems that the elemental intensity ratios approach is effective to detect obsidian sources used for artefact manufacture. Assuming that the accuracy of element concentration using the FP method for non-destructive technique is low, this approach is sufficient for limited sourcing of artefact obsidian. Moreover, we recognise that this method requires a geochemical Database for obsidian, in order to prevent erroneous determinations. It should be noted that the distribution range of flake samples is large due to irregular surfaces. In order to avoid errors due to uneven surfaces, it is necessary to analyse a minimum of 20 samples from each source. In addition, a final decision for identifying the obsidian source should be made using discriminant analysis.

Table 4.2. The Mahalanobis distances of each cluster for the Akaishiyama and Takaharayama sources

Elemental Concentrations							
Akaishiyama				Takaharayama			
	GB	Ms	Fl		GB	Ms	Fl
GB	0			GB	0		
Ms	12,110	0		Ms	4318	0	
Fl	143	**52**	0	Fl	14	**13**	0
Intensities							
Akaishiyama				Takaharayama			
	Ms	Fl			Ms	Fl	
Ms	0			Ms	0		
Fl	**68**	0		Fl	**5**	0	

Abbreviations are: GB – glass beads; Ms – mirror surface sample; and Fl – flake sample.

Notes on Using Intensity Ratios

Although A. Mochizuki's parameters effectively deal with a limited number of elements for sourcing obsidian, it is possible that intensity ratios are accidentally approximate (Table 4.3). When obsidian sources are 300 km apart, it is unlikely they were created from the same magma chamber. Additional petrographic analysis is clearly required, although it is possible to classify these sources through discriminant analysis. Different amounts of crystallites are visible in these obsidian samples through the microscope (Table 4.4). This problem was discussed by Shackley (1988); it indicates that the raw materials from obsidian sources should be carefully analysed by both geochemical and petrographic methods.

The intensity ratios method is effective for discriminating obsidian artefacts from an unknown source, as shown in the blank area where artefacts from unknown sources are plotted; see, for example, the results for the Yadegawa site, Nagano Pref. (Figure 4.7). When the results based only on qualitative analysis were uncertain, we measured glass beads made from several artefacts using WDXRF. As a result, we obtained more reliable data (Figure 4.8). Thus, it is evident that some obsidian artefacts from the Yadegawa site were from an unknown source. Despite having investigated and collected obsidian from over 95% of the Japanese sources, some artefacts of unknown provenance were found.

The Advantage of Using EDXRF Compared to WDXRF

EDXRF can suppress the generation of X-rays due to its high energy resolution and simple optical system. Thus, an advantage of EDXRF is that damage to samples is minimal. The recent studies by De Francesco *et al.* (2008) and Suda (2012) employed a non-destructive method (i.e., WDXRF). Because WDXRF needs a strong X-ray current, it can change the colour of glass beads. We therefore expected obsidian colours to be changed, and conducted experiments with conditions similar to Suda (2012) – Cr anode X-ray tube; 50 kV; 50 mA; and ca. 50 minutes. Our results confirmed that the obsidian colour became yellow (Figure 4.9). In archaeology, there are cases when this kind of change due to chemical analysis is considered as destruction. Accordingly, if this occurs, the analyst must inform the client in advance.

Table 4.3. Element compositions of obsidian samples from the Tokachi and Nishiaomori areas (after Kannari *et al.* 2010)

Area	**Tokachi**	**Nishiaomori**
Locality	**Tokachi-Mitsumata**	**Amadanaigawa**
Sample Name	**HD040d**	**051010-1 Tentanai**
in wt. %		
SiO_2	75.91	75.15
TiO_2	0.04	0.04
Al_2O_3	13.12	13.00
FeO*	1.24	1.33
MnO	0.05	0.05
MgO	0.11	0.32
CaO	0.73	0.71
Na_2O	4.12	3.85
K_2O	4.12	4.28
P_2O_5	0.01	0.01
Major total	99.45	98.74
in ppm		
Rb	153	148
Sr	51	63
Ba	961	822
Y	32	11
Zr	77	86
Nb	4.8	6.1
Th	12.3	6.4
Zn	34	34
Rb / (Rb + Sr + Y + Zr)	0.49	0.48
Sr / (Rb + Sr + Y + Zr)	0.16	0.20
Mn/Fe	0.042	0.039
log (Fe/K)	- 0.52	- 0.51

* Total iron calculatied as FeO.

The analytical time of both EDXRF and WDXRF was compared. The time required for EDXRF is 240 seconds per sample which is much shorter than for WDXRF (e.g., 52 minutes in Suda 2012). However, the method using intensity ratios needs to construct a Database for comparison of the source and artefact samples. Although we need one month to build a Database, several thousand samples can be analysed in one year. We conclude that a shorter measurement time has more advantages for archaeology, since most studies involve numerous artefacts found during the excavations.

Table 4.4. Petrographic results of microscopic examination of obsidians from the Tokachi-Mitsumata and Amadanaigawa localities (after Nagai *et al.* 2012). Crystallite classification is based on Johannsen (1931)

Locality	Colour	Spherulites	Phenocrysts			Microlites		Crystallites				
			Pla	*Qu*	*Bi*	*Pla*	*Bi*	*Globulite*	*Clavalite*	*Margarite*	*Trichites*	*Scopulite*
Tokachi-Mitsumata	Black+	•	•	•	•	•	•	⊙	•	⊙	⊙	•
Amadanaigawa	Black#	•	□	□	•	⊙	○	•	•	•	•	•

Symbols and abbreviations are: + – strong, # – weak; Pla – plagioclase; Qu – quartz; Bi – biotite; ⊙ – high; ○ – medium; □ – low; • – free.

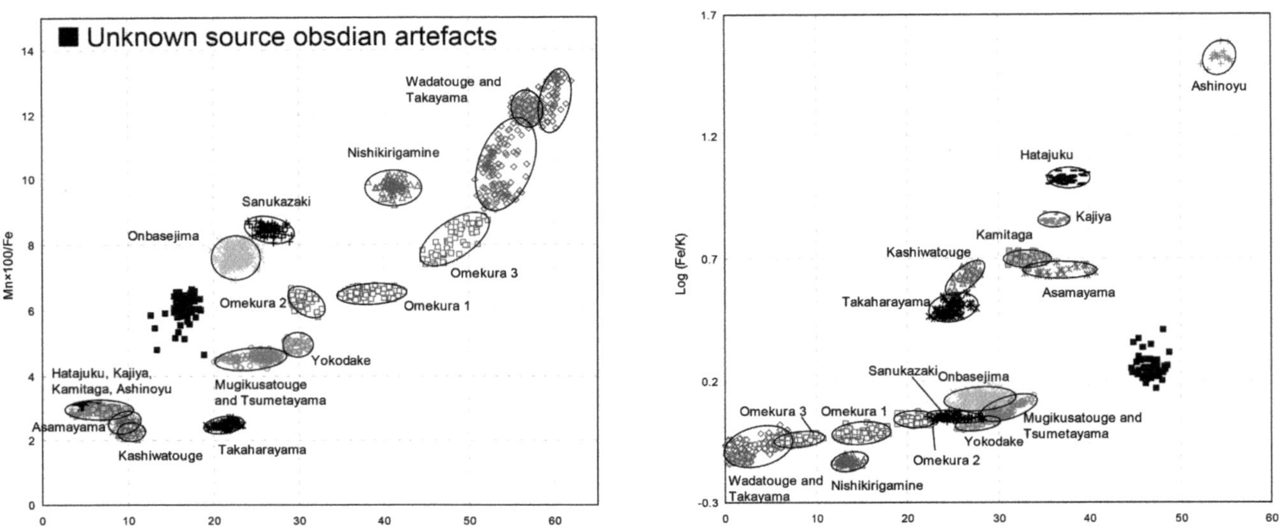

Figure 4.7. Scatterplots of intensity ratios for obsidian artefacts found at the Yadegawa site, including those from an unknown source (modified from Nagai et al. *2012)*

What are the Optimal Elements for Sourcing Obsidian Artefacts?

Mochizuki (1997a) used K, Mn, and Fe as the major elements, and Rb, Sr, Y, and Zr as trace elements. The method selected not only simple ratios of two elements, but also fractions of target trace element/summed four trace elements. Ternary diagrams (Jack and Carmichael 1969; Hughes 1984; Anderson *et al.* 1986) use three elements: Rb, Sr, and Zr. We also consider K and Y as important elements for distinguishing the origin of obsidian used for artefacts. Since these works indicated few interpretations, let us discuss the selected elements from a geochemical viewpoint.

It is well-known that despite the composition of the magma, the same source material can vary depending on the physical conditions. Magma composition varies due to processes such as crystallisation, differentiation, contamination, and mixing after generation. Since obsidian is produced by solidification of evolved felsic magma, its chemical composition is affected by these various processes. Magmas are formed by partial melting of silicate rocks, either in the Earth's mantle or crust that usually consists of several minerals. Elements such as Rb, Sr, Y, Zr, and K are representative incompatible elements which preferentially concentrate in the liquid phase of magma rather than in the crystalline phase. Therefore, when magma is generated by partial melting of the source materials, the degree of melting is smaller and the concentration of incompatible elements tends to increase. The distribution ratio between the liquid and crystalline phases varies depending on the elements. When the composition changes by separation of the newly formed crystals with the cooling of magma (i.e., crystallisation differentiation), the concentration of the incompatible elements also increases, inheriting the original compositional ratio in parental magma. Therefore, the dissimilarities in the concentrations of incompatible elements are easier to recognise in the case of differentiated magma.

There is zonal arrangement in the chemical composition of volcanic rocks in island arc zones. This is observed as differences in the concentration of total alkali (K and Na) in the transverse direction (e.g., Kuno 1966). The reason for this is differences in the magma source region (in this case, the degree of partial melting of mantle peridotite) that occurred underground, and thereby is reflected in the compositional differences in generated primary magmas.

The ratios of the incompatible elements should be kept constant by the crystallisation differentiation process in response to magmas made of the same source material and with the same degree of melting. In fact, the actual magma composition can acquire more diversity because

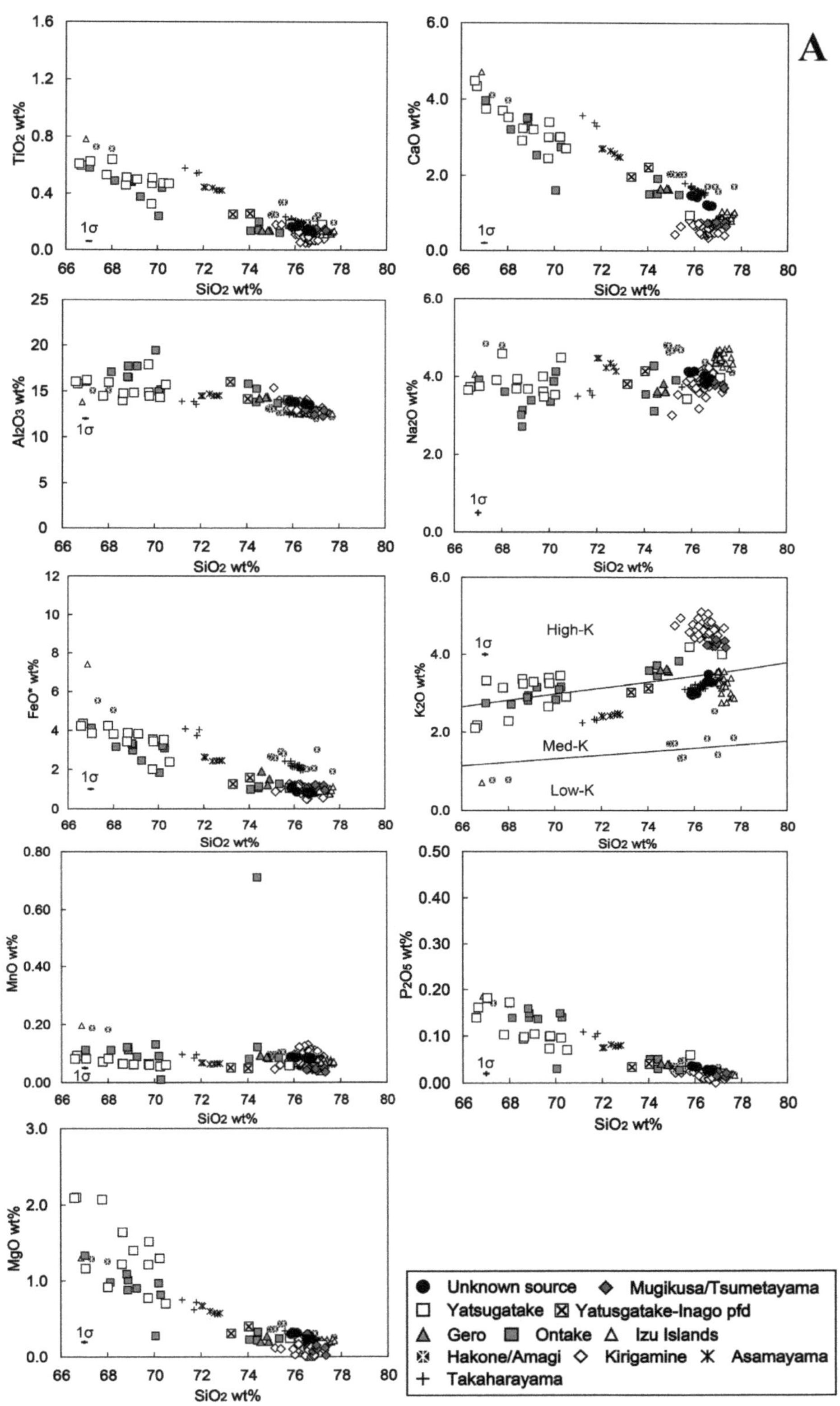

Figure 4.8. A. The SiO_2 variation diagrams for major and trace elements in obsidian artefacts from an unknown source found at the Yadegawa site (modified from Nagai et al. *2012). FeO* is total Fe as FeO. K_2O discrimination lines are based on Gill (1981). Most of the northern Yatugatake Volcano data are from Takahashi and Nishiki (2006), Nishiki* et al. *(2007), and Nakamura (1991). Most of the Ontake Volcano data are from Yamada and Kobayashi (1988), and Kimura and Yoshida (1999). Some of the Rb, Sr, Y, Zr, Ba, and Zn were not analysed*

each incompatible element does not show exactly the same behaviour. There is such a case in which a more incompatible element may be distributed in crystal, depending on the magma composition or type of crystalline phase. Rubidium exhibits nearly perfect incompatibility in the crystallisation differentiation

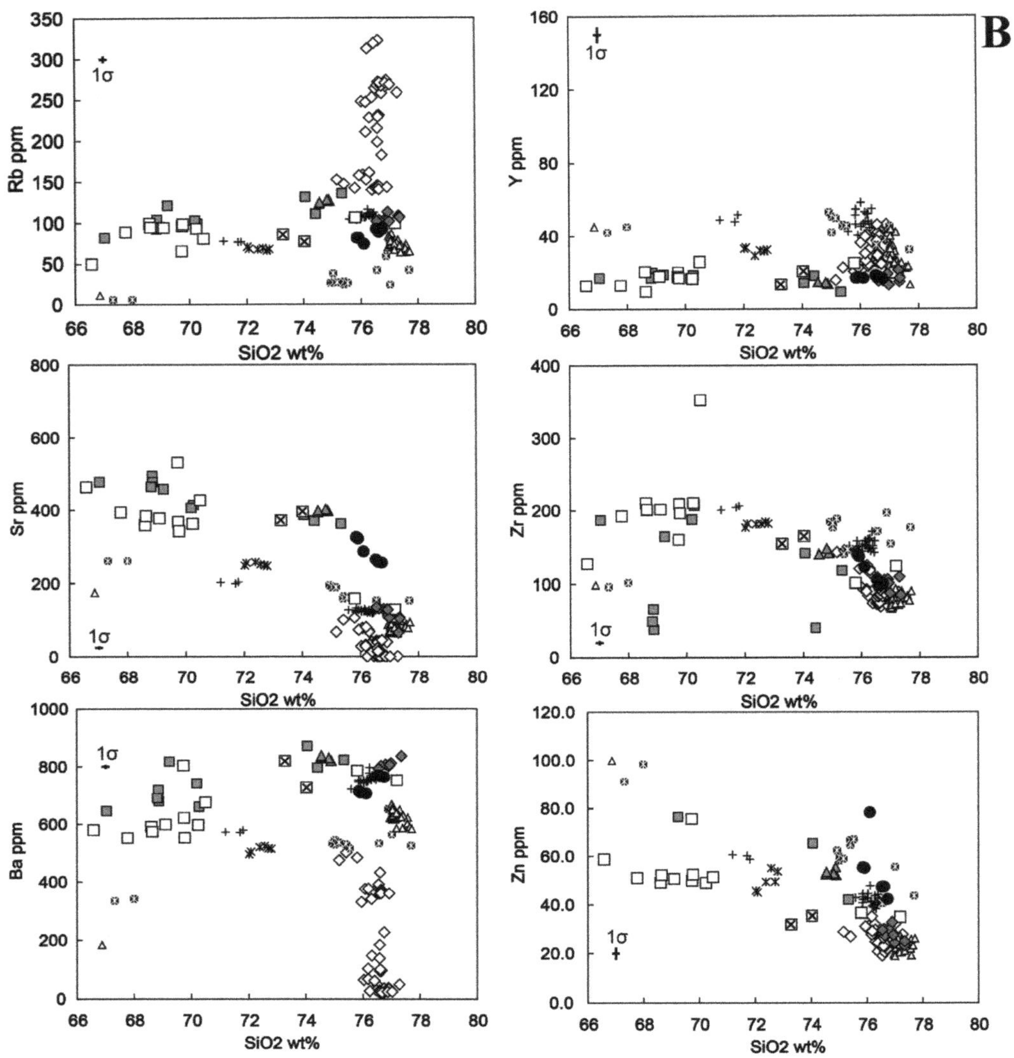

Figure 4.8. B. See captions on Figure 4.8. A

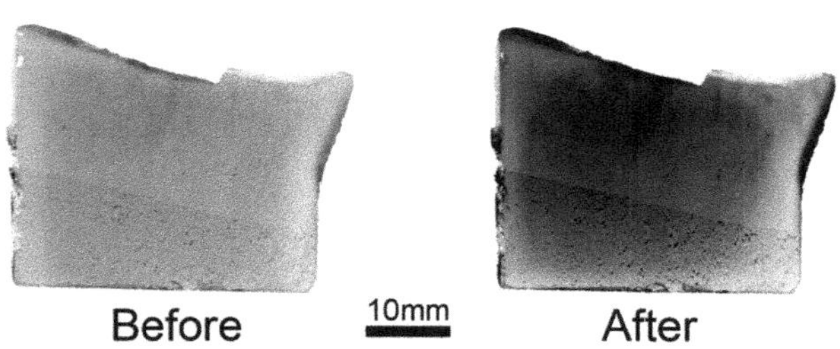

Figure 4.9. Obsidian sample which changed colour after X-ray irradiation by WDXRF

process. However, the concentration of Y may change due to a decrease when amphibole and/or garnet are abundantly generated in this process in felsic magmas because Y is taken up in these minerals. Similarly, Zr shows decreasing behaviour in zircon-saturated conditions. Strontium behaviour differs depending on the magmatic environment. In mantle-derived primitive basaltic magma, Sr demonstrates incompatibility. However, in the plagioclase-saturated condition, Sr is substituted for Ca which is the main component of plagioclase, because the charge and ionic radius of Sr are similar to Ca. Therefore, when magma evolved in shallow levels, Sr is no longer an incompatible element, and its concentration tends to be reduced in almost all composition ranges (from basaltic to felsic magma). On the other hand, in cases where felsic magma was generated by partial melting in the deep parts of thick continental crust or subducted oceanic crust (where garnet can exist instead of plagioclase), Sr may behave as an incompatible element showing high concentration (adakite magma; see Defant and Drummond 1990). Consequently, if there are differences in the timing of plagioclase crystallisation in the processes of generation and evolution of magmas, Sr will be useful in the identification of the rocks on a wide composition range.

Iron is a typical element that is distributed in the crystalline phase. However, Fe can form ferromagnesian silicate minerals and oxide minerals in different oxidation states. Hence, oxygen fugacity and/or the water content of magma control mineral assemblage and/or composition. A Fe-enriched magma series with the relatively reductive differentiation process is called tholeiitic, while the relatively oxidised Fe-poor series is calc-alkaline. Manganese behaviour is generally similar to Fe in felsic magma, but this is not well understood. According to recent studies, the crystallisation of distinctive minerals (e.g., cordierite, biotite, and tourmaline) in Al-rich felsic magma (S-type granite magma) may change the Mn content and Mn/Fe ratio in the magma at the time of the partial melting and differentiation processes (London *et al.* 2001). Alternatively, Mn-rich magmas may have been generated by melting of sediment-enriched Mn oxides or other compounds. Manganese behaves differently than Fe because it undergoes various oxidation states in the Earth's surface environment.

CONCLUSIONS

This study discussed the results of three conditions for differently prepared obsidian samples measured using WDXRF and EDXRF techniques. We did not obtain good results with the FP method using EDXRF because intensities were attenuated by the irregular surface and shape of samples, or by a macromolecule film. We indicated that the elements chosen by Mochizuki (1997a) are effective for geochemical sourcing of obsidian used for artefacts. Intensity ratios of Mochizuki's (1997a) elements (K, Mn, Fe, Rb, Sr, Y, and Zr) measured by non-destructive EDXRF are sufficient to determine the origin of obsidian, and for the classification of obsidian artefacts to both known and unknown sources. The EDXRF method is superior to WDXRF because it inflicts minimal damage to artefacts and has a short analytical time. We conclude that analysis of obsidian artefacts by EDXRF cannot indicate their chemical composition but can determine their geological source. Therefore, we suggest that intensity ratios are useful to determine the provenance of obsidian artefacts, despite the fact that challenges remain.

Acknowledgements

This study was greatly supported by late Saburo Kobayashi and Toshihiro Nakamura (Meiji University) who contributed to founding the organisation of the Cultural Properties Laboratory at Meiji University (integrated into COLS). We are grateful to Taketo Shimano (Fuji Tokoha University, Fuji, Shizuoka Pref.), Hirofumi Suzuki (formerly at Meiji University), Jiro Komori (Teikyo Heisei University, Ichihara, Chiba Pref.) and Toru Shibata (Tokai University, Tokyo) for developing the obsidian artefacts' sourcing system. We thank Prof. A. Ono for useful comments, and Drs M. D. Glascock, Y. V. Kuzmin and S. G. Keates for grammar corrections. We are grateful to all post–graduate students at Meiji University and Nihon University for preparation of samples and operating the instruments. We also thank Dr E. Frahm for helpful comments on an earlier version of this paper. This work was supported by the Research Project "The Environmental Development and Human Activity" (2006–10) from the MEXT.

References

ANDERSON, D. C., J. A. TIFFANY, and F. W. NELSON. 1986. Recent Research on Obsidian from Iowa Archaeological Sites. *American Antiquity* 51, 837–852.

BIRÓ, K. T., A. POZSGAI, and A. VLADÁR. 1986. Electron Beam Microanalyses of Obsidian Samples from Geological and Archaeological Sites. *Acta Archaeologica Academiae Scientiarum Hungaricae* 38, 257–278.

CANN, J. R., and C. RENFREW. 1964. The Characterization of Obsidian and Its Application to the Mediterranean Region. *Proceedings of the Prehistoric Society* 30, 111–133.

CARTER, T., and M. S. SHACKLEY. 2007. Sourcing Obsidian from Neolithic Çatalhöyük (Turkey) Using Energy Dispersive X-ray Fluorescence. *Archaeometry* 49, 437–454.

CARTER, T., S. GRANT, M. KARTAL, A. COŞKUN, and V. ÖZKAYA. 2013. Networks and Neolithisation: Sourcing Obsidian from Körtik Tepe (SE Anatolia). *Journal of Archaeological Science* 40, 556–569.

CRAIG, N., R. J. SPEAKMAN, R. S. POPELKA-FILCOFF, M. D. GLASCOCK, J. D. ROBERTSON, M. S. SHACKLEY, and M. S. ALDENDERFER. 2007. Comparison of XRF and PXRF for Analysis of Archaeological Obsidian from Southern Peru. *Journal of Archaeological Science* 34, 2012–2024.

DAVIS, M. S., T. L. JACKSON, M. S. SHACKLEY, T. TEAGUE, and J. H. HAMPEL. 1998. Factors Affecting the Energy-Dispersive X-ray Fluorescence (EDXRF) Analysis of Archaeological Obsidian. In *Archaeological Obsidian Studies: Method and Theory*, edited by M. S. Shackley, 159–180. New York & London, Plenum Press.

DEFANT, M. J., and M. S. DRUMMOND. 1990. Derivation of Some Modern Arc Magmas by Melting of Young Subducted Lithosphere. *Nature* 347, 662–665.

DE FRANCESCO, A. M., G. M. CRISCI, and M. BOCCI. 2008. Non-Destructive Analytic Method Using XRF for Determination of Provenance of Archaeological Obsidians from the Mediterranean Area: a Comparison with Traditional XRF Methods. *Archaeometry* 50, 337–350.

EERKENS, J. W., J. R. FERGUSON, M. D. GLASCOCK, C. E. SKINNER, and A. WAECHTER. 2007. Reduction Strategies and Geochemical Characterization of Lithic Assemblages: a Comparison of Three Case Studies from Western North America. *American Antiquity* 72, 585–597.

FRAHM, E. 2012. Distinguishing Nemrut Dağ and Bingöl A Obsidians: Geochemical and Landscape Differences and the Archaeological Implications. *Journal of Archaeological Science* 39, 1436–1444.

FRANKEL, R. S., and D. W. AITKEN. 1970. Energy-Dispersive X-ray Emission Spectroscopy. *Applied Spectroscopy* 24, 557–566.

GIAUQUE, R. D., F. ASARO, F. H. STROSS, and T. R. HESTER. 1993. High-Precision Non-Destructive X-ray Fluorescence Method Applicable to Establishing the Provenance of Obsidian Artifacts. *X-Ray Spectrometry* 22, 44–53.

GILL, J. B. 1981. *Orogenic Andesites and Plate Tectonics.* Berlin–Heidelberg–New York, Springer-Verlag.

GLASCOCK, M. D., P. C. WEIGAND, R. ESPARZA LÓPEZ, M. A. OHNERSORGEN, M. GARDUÑO AMBRIZ, J. B. MOUNTJOY, and J. A. DARLING. 2010. Geochemical Characterisation of Obsidian in Western Mexico: The Sources in Jalisco, Nayarit, and Zacatecas. In *Crossing the Straits: Prehistoric Obsidian Source Exploitation in the North Pacific Rim* (B.A.R. International Series 2152), edited by Y. V. Kuzmin and M. D. Glascock, 201–217. Oxford, BAR Publishing.

GRATUZE, B. 1999. Obsidian Characterization by Laser Ablation ICP–MS and Its Application to Prehistoric Trade in the Mediterranean and the Near East: Sources and Distribution of Obsidian within the Aegean and Anatolia. *Journal of Archaeological Science* 26, 869–881.

HALL, M., and H. KIMURA. 2002. Quantitative EDXRF Studies of Obsidian Sources in Northern Hokkaido. *Journal of Archaeological Science* 29, 259–266.

HIGASHIMURA, T. 1980. *Kokogaku to Butsurikagaku* [Archaeology and Physical Chemistry]. Tokyo, Gakuseisha Publishers.

HIGASHIMURA, T. 1986. *Sekki Sanchi Suiteihou* [The Method of Sourcing Stone Implements]. Tokyo, Nyusaiencesha Publishers.

HIGASHIMURA, T., and T. WARASHINA. 1975. Sourcing of Sanukite Stone Implements by X-ray Fluorescence Analysis. *Journal of Archaeological Science* 2, 169–178.

HONMA, H. 2005. Shiryo Chouseiho [Method of Sample Preparation]. In *Keiko Xsen Bunseki no Jissai*, edited by I. Nakai, 62–77. Tokyo, Asakura Shoten.

HUGHES, R. E. 1984. Obsidian Sourcing Studies in the Great Basin: Problems and Prospects. In *Obsidian Studies in the Great Basin* (Contributions of the University of California Archaeological Research Facility 45), edited by R. E. Hughes, 1–19. Berkeley, University of California.

HUGHES, R. E. 1988. The Coso Volcanic Field Reexamined: Implications for Obsidian Sourcing and Hydration Dating Research. *Geoarchaeology* 3, 253–265.

HUGHES, R. E. 1994. Intrasource Chemical Variability of Artefact–Quality Obsidians from the Casa Diablo Area, California. *Journal of Archaeological Science* 21, 263–271.

HUGHES, R. E. 2010a. Trace Element Characterisation of Archaeologically Significant Volcanic Glasses from the Southern Great Basin of North America. In *Crossing the Straits: Prehistoric Obsidian Source Exploitation in the North Pacific Rim* (B.A.R. International Series 2152), edited by Y. V. Kuzmin and M. D. Glascock, 165–181. Oxford, BAR Publishing.

HUGHES, R. E. 2010b. Determining the Geologic Provenance of Tiny Obsidian Flakes in Archaeology Using Non-Destructive EDXRF. Electronic document, available online at: http://www.americanlaboratory.com/492–Determining–the–Geologic–Provenance–of–Tiny–Obsidian–Flakes–in–Archaeology–Using–Nondestructive–EDXRF/. Accessed 9 October 2013.

IKEYA, N., and A. MOCHIZUKI. 1998. Ashitaka Sanroku ni okeru Sekizaisosei no Hensen [Composition and Transition of Raw Material at Mount Ashitaka during the Upper Palaeolithic]. *Shizuoka-ken Kokogaku Kenkyu* 36, 21–44.

INOUE, I., and N. NOTO. 2002. The Classification of Obsidian and Glassy Andesite by the Petrogical Classification. *Kokogaku to Shizen Kagaku* 45, 55–77 (in Japanese with English Abstract).

JACK, R. N., and R. F. HEIZER. 1968. "Fingerprinting" of Some Mesoamerican Obsidian Artifacts. In *Papers on Mesoamerican Archaeology* (Contributions of the University of California Archaeological Research Facility 5), 81–100. Berkeley, University of California.

JACK, R. N., and I. S. E. CARMICHAEL. 1969. The Chemical "Fingerprinting" of Acid Volcanic Rocks. In *Short Contributions to California Geology* (California Division of Mines and Geology Special Report 100), 17–32. Sacramento, CA, California Division of Mines and Geology.

JAMES, M. A., J. BAILEY, and J. M. D'AURIA. 1996. A Volcanic Glass Library for the Pacific Northwest: Problems and Prospects. *Canadian Journal of Archaeology* 20, 93–122.

JOHANNSEN, A. 1931. *A Descriptive Petrography of the Igneous Rocks. Volume 1.* Chicago, University of Chicago Press.

KAKUBUCHI, S., and M. UTSUNOMIYA. 2002. Keiko Xsen Bunsekiho ni yoru Kokuyoseki no Sanchi Dotei (1) [Identification of Source of Obsidian by X-ray Fluorescence Analysis, Part 1]. *Saga Daigaku Bunka Kyoiku Gakubu Kenkyu Ronbun Shu* 7(1), 49–52.

KAKUBUCHI, S., and M. UTSUNOMIYA. 2003. Keiko Xsen Bunsekiho ni yoru Kokuyoseki no Sanchi Dotei (2) [Identification of Source of Obsidian by X-ray Fluorescence Analysis, Part 2]. *Saga Daigaku Bunka Kyoiku Gakubu Kenkyu Ronbun Shu* 7(2), 47–58.

KANNARI, T., M. NAGAI, T. SHIBATA, S. NAGAOKA, and S. SUGIHARA. 2011. Quantitative and Qualitative Analysis of Obsidian Sources in Oki and Kyushu Districts: Method of Sourcing Obsidian Artefacts. *Kankyosi to Jinrui* 4, 3–40 (in Japanese with English Abstract).

KANNARI, T., S. SUGIHARA, M. NAGAI, and T. SHIBATA. 2010. Quantitative and Qualitative Analysis of Obsidian Sources in Hokkaido and Tohoku Districts. *Kokogaku to Shizen Kagaku* 60, 57–81 (in Japanese with English Abstract).

KANNARI, T., Y. NISHIAKI, M. NAGAI, T. SHIBATA, F. GULIYEV, and S. SUGIHARA. 2012. Azerubaijan, Gyoitepe Iseki Shutsudo Kokuyosekisei Ibutsu no Gensanchi Suitei – Teisei Teiryo Bunseki ni Motozuite [Sourcing Obsidian Artefacts from Göytepe Site, Azerbaijan, Based on Qualitative and Quantitative Analyses]. In *Nihon Bunkazai Kagaku Kai Dai 29 Kai Taikai*, 262–263. Kyoto, Nihon Bunkazai Kagakukai.

KIM, J.-C., D. K. KIM, M. YOON, C. C. YUN, G. PARK, H. J. WOO, M.-Y. HONG, and G. K. LEE. 2007. PIXE Provenancing of Obsidian Artefacts from Paleolithic Sites in Korea. *Bulletin of the Indo-Pacific Prehistory Association* 27, 122–128.

KIMURA, J., and T. YOSHIDA. 1999. Magma Plumbing System beneath Ontake Volcano, Central Japan. *Island Arc* 8, 1–29.

KUNO, H. 1966. Lateral Variation of Basalt Magma Type across Continental Margins and Island Arcs. *Bulletin of Volcanology* 29, 195–222.

KUZMIN, Y. V., M. D. GLASCOCK, and M. IZUHO. 2013. The Geochemistry of the Major Sources of Archaeological Obsidian on Hokkaido Island (Japan): Shirataki and Oketo. *Archaeometry* 55, 355–369.

LE BOURDONNEC, F.-X., S. DELERUE, S. DUBERNET, P. MORETTO, T. CALLIGARO, J.-C. DRAN, and G. POUPEAU. 2005. PIXE Characterization of Western Mediterranean and Anatolian Obsidians and Neolithic Provenance Studies. *Nuclear Instruments and Methods in Physics Research B* 240, 595–599.

LONDON, D., J. M. EVENSEN, E. FRITZ, J. P. ICENHOWER, G. B. MORGAN, VI, and M. B. WOLF. 2001. Enrichment and Accommodation of Manganese in Granite–Pegmatite Systems. In *Eleventh Annual V. M. Goldschmidt Conference. Abstract No. 3369.* Houston, TX, Lunar and Planetary Institute.

MOCHIZUKI, A. 1997a. Identification of Sources of Obsidian Found in Chubu and Kanto Districts by X-ray Fluorescence Analysis. *Xsen Bunseki no Shinpo* 28, 157–168 (in Japanese with English Abstract).

MOCHIZUKI, A. 1997b. Keiko Xsen Bunseki ni yoru Toruko no Kokuyoseki Sanchi no Bunrui (1): Chuo Anatoria no Kokuyoseki Sanchi no Bunrui to Kaman-Karehoyukku Shutsudo no Kokuyosekisei Sekki no Sanchi Suitei [Classification of Obsidian Sources in Turkey (I): Classification of Obsidian Sources in Central Anatolia and Source Identification of Obsidian Artefacts from Kaman-Kalehöyük]. *Anatoria Kokugaku Kenkyu* 6, 169–185.

MOCHIZUKI, A., N. IKEYA, K. KOBAYASHI, and Y. MUTO. 1994. Isekinai ni okeru Kokuyosekisei Sekki no Gensanchibestu Bunpu ni tsuite: Numazu-shi Doteue Iseki BBVsou no Gensanchi Suitei kara [Intra-Site Distribution Patterns of Obsidian Tools by Different Sources: A Case of Provenance Analysis of the BBV Layer, Doteue Site, Numazu City, Shizuoka Prefecture, Central Japan]. *Shizuoka-ken Kokogaku Kenkyu* 26, 1–24.

NAGAI, M., T. KANNARI, T. SHIBATA, K. SHIMADA, and S. SUGIHARA. 2012. Petrographical and Geochemical Characteristics of Obsidian Artefacts of Unknown Source in Yadegawa Site: Fundamental Research for the Identification of the 'NK Obsidian Source'. *Meiji Daigaku Hakubutsukan Kenkyu Hokoku* 17, 57–66 (in Japanese with English Abstract).

NAGAI, M., T. SHIMANO, and S. SUGIHARA. 2008. Implementation and Evaluation of Routine Measurements of Major and Trace Element Contents in Igneous Rock Samples by X-ray Fluorescence Spectrometry: the Fundamental Study for Stone Sources of Lithic Tools. *Meiji Daigaku Hakubutsukan Kenkyu Hokoku* 13, 69–80 (in Japanese with English Abstract).

NAGAI, M., S. SUGIHARA, and T. DANHARA. 2007. Enrei Ruiso, Wadatoge–Kirigamine Chiiki no Kazan Sojyo to Fuisshon-Torakku Nendai [Volcanostratigraphy and Fission-Track Ages of the Enrei Formation in Wadatoge–Kirigamine area, Central Japan]. In *Nihon Chishitsu Gakkai Dai 114 Nen Gakujutsu Taikai*, 144. Sapporo, Nihon Chishitsu Gakkai.

NAKAMURA, M. 1991. A Petrologic Model for the Younger Period of Northern Yatsugatake Volcanoes, Central Japan. *Bulletin of the Volcanological Society of Japan* 36, 93–112 (in Japanese with English Abstract) (in Japanese with English Abstract).

NELSON, D. E., J. M. D'AURIA, and R. B. BENNETT. 1975. Characterization of Pacific Northwest Coast Obsidian by X-ray Fluorescence Analysis. *Archaeometry* 17, 85–97.

NELSON, F. W., K. K. NIELSON, N. F. MANGELSON, M. W. HILL, and R. T. MATHENY. 1977. Preliminary Studies of the Trace Element Composition of Obsidian Artifacts from Northern Campeche, Mexico. *American Antiquity* 42, 209–225.

NEWMAN, J. R., and R. L. NIELSEN. 1985. Initial Notes on the X-ray Fluorescence Sourcing of Northern New Mexico Obsidians. *Journal of Field Archaeology* 12, 377–383.

NISHIKI, K., A. MATSUMOTO, K. UTO, K. TAKAHASHI, and Y. MIYAKE. 2007. Reexamination of Volcanic Activity of Yatsugatake Area, Central Japan. *Journal of the Geological Society of Japan* 113, 193–211 (in Japanese with English Abstract).

NINOMIYA, S., and K. SIMADATE. 2001. Shizen Kagaku teki Syuhou ni yoru Bunseki – Keiko Xsen ni yoru Boso Hanto Syutsudo Sentoki Sekkigun no Kokuyouseki Gensanchi Suitei [Analysis Using the Physical Science – Sourcing of Obsidian Points from Boso Peninsula by X-ray Fluorescence]. *Chiba-ken Bunkazai Senta Kenkyu Kiyo* 22, 65–100.

OBATA, H., I. MORIMOTO, and S. KAKUBUCHI. 2010. Obsidian Trade between Sources on Northwestern Kyushu Island and the Ryukyu Archipelago (Japan) during the Jomon Period. In *Crossing the Straits: Prehistoric Obsidian Source Exploitation in the North Pacific Rim* (B.A.R. International Series 2152), edited by Y. V. Kuzmin and M. D. Glascock, 57–71. Oxford, BAR Publishing.

OOYA, M., Y. NISHII, M. UENO, N. KAMEDA, S. KUNITAKE, K. SHIMADATE, T. TAMURA, and A. MOCHIZUKI. 2006. Sekki Zairyo oyobi Sekki no Rikagakuteki Bunsekichi (1) – XRF ni yoru Kokuyogan Bunsekichi (2005 Nendo) [Chemical Analytical Values of Raw Materials and Stone Tools (1) – Analytical Values of Obsidian Using XRF (2005)]. *Saitama-ken Maizo Bunkazai Chosa Jigyodan Kenkyu Kiyo* 21, 199–210.

OOYA, M., M. UENO, M. ARAYA, K. MURAHATA, K. SASAMORI, S. KUNITAKE, M. MATSUMOTO, T. TAMURA, and H. KATO. 2007. Sekki Zairyo oyobi Sekki no Rikagakuteki Bunsekichi (2) – XRF ni yoru Kokuyogan Bunsekichi (2006 Nendo) [Chemical Analytical Values of Raw Materials and Stone Tools (2) – Analytical Values of Obsidian Using XRF (2006)]. *Saitama-ken Maizo Bunkazai Chosa Jigyodan Kenkyu Kiyo* 22, 91–98.

OSAWA, M. (ed.). 1991. Kokuyoseki no Kagakusosei – Iseki Syutsudo Kokuyoseki Sekki no Gensanch Suitei no Kiso toshite [Chemical Compositions of Obsidian in Japan – An Approach to Archaeological Provenience Studies]. Grants-in-Aid for Scientific Research Report. Manuscript on File. Tokyo, Tokyo Gakugei Daigaku.

OSAWA, M., H. KASUYA, and Y. SAKAKIBARA. 1977. Trace Element Abundances in Stone Artifacts and Related Materials from Japan by Neutron Activation Analysis. An Approach to Archaeological Provenience. *Journal of Radioanalytical Chemistry* 39, 137–152.

POUPEAN, G., F.-X. LE BOURDONNEC, T. CARTER, S. DELERUE, M. S. SHACKLEY, J.-A. BARRAT, S. DUBERNET, P. MORETTO, T. CALLIGARO, M. MILÆ, and K. KOBAYASHI. 2010. The Use of SEM–EDS, PIXE, and EDXRF for Obsidian Provenance Studies in the Near East: a Case Study from Neolithic Çatalhöyük (Central Anatolia). *Journal of Archaeological Science* 37, 2705–2720.

SHACKLEY, M. S. 1988. Sources of Archaeological Obsidian in the Southwest: an Archaeological, Petrological, and Geochemical Study. *American Antiquity* 53, 753–772.

SHACKLEY, M. S. 1991. An Energy Dispersive X-ray Fluorescence (EDXRF) Analysis of 57 Obsidian Artifacts from Seligman to Sunset Crater, Northern Arizona (Archaeological Research Facility Laboratory Reports). Electronic document, available at: http://www.escholarship.org/uc/item/6kj3r95c. Accessed 9 October 2013.

SHACKLEY, M. S. 1995. Sources of Archaeological Obsidian in the Greater American Southwest: an Update and Quantitative Analysis. *American Antiquity* 60, 531–551.

SHACKLEY, M. S. 1998. Gamma Rays, X-rays and Stone Tools: Some Recent Advances in Archaeological Geochemistry. *Journal of Archaeological Science* 25, 259–270.

SHACKLEY, M. S. 2005. *Obsidian: Geology and Archaeology in the North American Southwest.* Tucson, University of Arizona Press.

SHACKLEY, M. S. 2011. An Introduction to X-ray Fluorescence (XRF) Analysis in Archaeology. In *X-ray Fluorescence Spectrometry (XRF) in Geoarchaeology*, edited by M. S. Shackley, 7–44. New York, Springer.

SHIMANO, T., S. ISHIHARA, M. NAGAI, H. SUZUKI, and S. SUGIHARA. 2004. Hachobunsangata Keiko Xsen Bunseki Souchi ni yoru Nihon Zenkoku no Kokuyoseki Zengan Teiryo Bunseki [Whole Rock Components of Obsidians in Japan using WDXRF]. In *Nihon Bunkazai Kagaku Kai Dai 21 Kai Taikai*, 140–141. Kyoto, Nihon Bunkazai Kagakukai.

SMITH, I. E., G. K. WARD, and W. R. AMBROSE. 1977. Geographic Distribution and the Characterization of Volcanic Glasses in Oceania. *Archaeology and Physical Anthropology in Oceania* 12, 173–201.

SUDA, Y. 2012. Chemical Analysis of Obsidian by Wavelength-Dispersive X-Ray Fluorescence Spectrometry: Application to Non-Destructive Analysis of Archaeological Obsidian Artefacts. In *Natural Resource, Environment and Humans* (Proceedings of the Meiji University Centre for Obsidian and Lithic Studies. No. 2), edited by A. Ono, 1–14. Tokyo, Centre for Obsidian and Lithic Studies, Meiji University.

SUGIHARA, S., and S. KOBAYASHI. 2004. The Study of Archaeological Remains by Analysis of Natural Science: Database of Obsidian Producing Points. *Meiji Daigaku Jinbun Kagaku Kenkyujyo Kiyo* 55, 1–83 (in Japanese with English Abstract).

SUGIHARA, S., and S. KOBAYASHI. 2006. The Study of Cultural Region of Cultural Properties by Analysis of Natural Science. *Meiji Daigaku Jinbun Kagaku Kenkyujyo Kiyo* 59, 43–94 (in Japanese with English Abstract).

SUGIHARA, S., M. NAGAI, T. KANNARI, and T. SHIBATA. 2011. Quantitative and Qualitative Analysis of Obsidian Sources in Hokuriku District: Method of Sourcing Obsidian Artefacts. *Kankyosi to Jinrui* 4, 41–66 (in Japanese with English Abstract).

SUGIHARA, S., M. NAGAI, and T. SHIBATA. 2008. Petrography and Geochemistry of Obsidian from the Izu Islands off the Pacific Coast of Japan: Toward Sourcing Obsidian Artefacts in Japanese Prehistory. *Sundai Shigaku* 133, 45–76 (in Japanese with English Abstract).

SUGIHARA, S., M. NAGAI, T. SHIBATA, T. DANHARA, and H. IWANO. 2009b. Petrographic and Geochemical Study, and Fission-Track Dating of Obsidian from the Kirigamine and Kitayatsugatake Areas in the Central Part of Japan: the Fundamental Research for Sourcing of Obsidian Artefacts. *Sundai Shigaku* 136, 57–109 (in Japanese with English Abstract).

SUGIHARA, S., K. TSURUMAKI, T. SHIBATA, M. NAGAI, T. DANHARA, and H. IWANO. 2009a. Fission-Track Dating and Petrological Study of Obsidian from

Takahara Volcano Located in the Northern Part of Tochigi Prefecture, Japan: the Fundamental Research for Sources of Obsidian Artefacts. *Meiji Daigaku Hakubutsukan Kenkyu Hokoku*, 14, 43–70 (in Japanese with English Abstract).

TAKAHASHI, K., and K. NISHIKI. 2006. Volcanostratigraphy of the Lower Pleistocene in the Northern Flank of the Northern Yatsugatake Volcanoes, Central Japan: a Voluminous Magmatism in the Northern Yatsugatake and Enrei Area. *Journal of the Geological Society of Japan* 112, 549–567 (in Japanese with English Abstract).

TATEISHI, T., H. TSUMURA, and S. NINOMIYA. 2004. Kasori Minami Kaizuka Syutsudo Kokuyoseki no Gensanchi Suitei [Sourcing of Obsidian Artefacts Found at South Kasori Shellmound]. *Kaizuka Hakubutsukan Kiyo* 31, 1–15.

TSUTSUMI, T. 2010. Prehistoric Procurement of Obsidian from Sources on Honshu Island (Japan). In *Crossing the Straits: Prehistoric Obsidian Source Exploitation in the North Pacific Rim* (B.A.R. International Series 2152), edited by Y. V. Kuzmin and M. D. Glascock, 27–55. Oxford, BAR Publishing.

TSUTSUMI, T., and A. MOCHIZUKI. 2012. Lithic Artefacts of Microblade Industry from the Yadegawa Site and Their Provenances: the Sixth Analysis. In *Natural Resource, Environment and Humans* (Proceedings of the Meiji University Centre for Obsidian and Lithic Studies. No. 2), edited by A. Ono, 73–82. Tokyo, Centre for Obsidian and Lithic Studies, Meiji University (in Japanese with English Abstract).

WADA, K., and K. SANO. 2011. Chemical Composition and Microstructure of Shirataki Obsidian, Northern Hokkaido: Geological and Petrological Source Data for the Precise Obsidian Source Identification. *Kyusekki Kenkyu* 7, 57–73 (in Japanese with English Abstract).

WARASHINA, T., and T. HIGASHIMURA. 1988. Sekki Genzai no Sanchi Bunseki [Sourcing of Raw Material of Stone Implements]. In *Kokogaku to Kanren Kagaku*, edited by C. Serizawa, 447–491. Okayama, Okayama Daigaku.

WARASHINA, T., Y. KAMAKI, and T. HIGASHIMURA. 1978. Sourcing of Sanukite Stone Implements by X-ray Fluorescence Analysis II. *Journal of Archaeological Science* 5, 283–291.

WEISLER, M. I., and D. A. CLAGUE. 1998. Characterization of Archaeological Volcanic Glass from Oceania. In *Archaeological Obsidian Studies: Method and Theory*, edited by M. S. Shackley, 103–128. New York & London, Plenum Press.

YAMADA, N., and T. KOBAYASHI. 1988. *Ontakesan* [Geology of the Ontakesan District, with Geological Sheet Map of 1:50,000]. Tsukuba, Chishitsu Chosajyo.

YOSHIDA, H., and N. TAKAHASHI. 1997. Chemical Behaviour of Major and Trace Elements in the Horoman Mantle Diaper, Hidaka Belt, Hokkaido, Japan. *Journal of Mineralogy, Petrology and Economic Geology* 92, 391–409 (in Japanese with English Abstract).

YOSHITANI, A., M. KAWABE, O. SUDA, I. MIZUFUNE, and K. NAGAHATA. 2001. On the Primary Occurrence Places of Obsidian Rocks Found at the Oribe 17 Site Distributed in Kamishihoro-cho, Tokachi Region in Central Hokkaido. *Higashi Taisetsu Hakubutsukan Kenkyu Hokoku* 23, 1–19 (in Japanese with English Abstract).

YOSHITANI, A., O. SUDA, M. KAWABE, T. SUEMORI, H. KATAYAMA, M. WAKUSHIMA, and S. UEMURA. 1999. On the Chemical Composition of Microelements in Obsidian Found in Tokachi District, Central Hokkaido. *Higashi Taisetsu Hakubutsukan Kenkyu Hokoku* 21, 1–11 (in Japanese with English Abstract).

Appendix 1
DESCRIPTION OF THE WDXRF METHOD

Preparation of glass bead samples for the WDXRF involved the following procedures. A portion of the weathered surface of specimens was removed with a rock trimmer and hammer, in order to obtain a homogeneous sample. Samples were further divided into chips weighing approximately 25–100 g and a maximum 2 cm lenght. Using a petrographic saw and sandblasting machine as needed, we were careful to avoid contamination from iron powder, paint from the saw, or sandblasting propellants. The rock chips were cleaned with distilled water in an ultrasonic cleaning device, and dried in an oven at 110° C for eight hours afterwards. The chips were then ground with an iron mortar until they were 2 mm long or shorter. They were mixed using an Ishikawa's Raikai agate mortar (Ishikawa Kojo, AGB) until the specimens were 20 μm long or shorter. Mixing time was eight minutes. If the crushing was incomplete, the time was extended. Each powder sample (weigh approximately 1.2 g) was placed in an aluminum crucible and combusted in a muffle furnace for one hour at 600° C. When the samples had been cooled in a desiccator to a normal temperature, 1 g (± 0.0004 g) of powder from each sample and 5 g (± 0.0004 g) of the flux (Merck, $Li_2B_4O_7$, Spectromelt A10) were precisely measured using an electronic balance, and then mixed using an agate mortar. Specimens were put into a platinum crucible (alloy of 95% Pt and 5% Au), combined with a mould of glass beads 36 mm in diameter, and placed on an automated high frequency bead sampler (Nippon Thermonics Co., Ltd., NT–2100). Each specimen was heated to 800° C for one minute, and to 1100° C for two minutes in atmosphere. The fusion was stirred and swirled automatically for an additional five minutes at 1100° C. After heating, the specimen was moved to a fan cooler unit facilitated by the bead sampler. LiI solution was unnecessary, and was not used in this study. The flux was combusted in a muffle furnace and atmosphere at 700° C for five hours, and then retained at 110° C in a drying oven.

The WDXRF machine used is a Rigaku RIX–1000 at COLS. This instrument contains a 3 kW Cr anode X-ray tube. The acceleration voltage and tube current were set at 50 kV and 50 mA, respectively. The spectrum chamber was a vacuum (under 3.0 Pa), and the sample mask's diameter was 30 mm. The optimal combination was selected from four analysing crystals (LiF200, Ge, PET, and TAP) and two types of detector (F-PC and SC) for each element. The selected analysis periods were selected from 40 to 200 live times for an X-ray peak, and from 20 to 100 live times for background of the optimal value. Ten major elements were shown in wt.%; total iron was converted to total FeO; and trace elements were shown at parts-per-million (ppm). Calibration curves were produced by Nagai *et al.* (2008). Standard samples were provided by the Geological Survey of Japan (GSJ): JA-1, JA-2, JA-3, JB-1b, JB-2, JB-3, JG-1, JG-1a, JG-2, JG-3, JGb-1, JGb-2, JR-1, JR-2, JR-3; the United States Geological Survey (USGS): RGM-1; National Institute of Standards and Technology (NIST): SRM-278; and volcanic rock samples from Iki Island (Yoshida and Takahashi 1997): IK507, IK565, IK625, IK630, IK686, IK731, and IK790. For accuracy of calibration, we refer to Nagai *et al.* (2008).

Appendix 2
DESCRIPTION OF THE EDXRF METHOD

For EDXRF analysis, a JEOL JSX–3100s spectrometer (same performance as JSX–3100) at COLS was used. The X-ray tube was an end-window type Rh X-ray target. The accelerating voltage was set at 30 kV, and the tube current was on an automatic setting from 1 to 0.1 mA for the optimum counting rate. X-ray was detected using a Si/Li SSD. The atmosphere of the sample chamber was a vacuum. The collimator had a diameter of 7 mm, and the diameter of an irradiated X-ray was 15 mm. Counting time was 240 live seconds. The major elements (Si, Ti, Al, Fe, Mn, Mg, Ca, Na, K, and P) were measured, as well as trace elements (Rb, Sr, Ba, Y, Zr, Nb, Th, V, Zn, Cr, Ni, and Co). Some of these elements were not detected. The net intensity was calculated for each element in the peak area of the Kα line, with the exception of Ba which was in the Lα line.

Chapter 5
CHEMICAL COMPOSITION OF OBSIDIANS IN HOKKAIDO ISLAND, NORTHERN JAPAN: THE IMPORTANCE OF GEOLOGICAL AND PETROLOGICAL DATA FOR SOURCE STUDIES

Keiji WADA, Masayuki MUKAI, Kyohei SANO, Masami IZUHO, and Hiroyuki SATO

Abstract: *The chemical compositions of whole-rock and volcanic glass from obsidian sources in Hokkaido Island, northern Japan, were determined by X-ray Fluorescence and Electron Probe Microanalysis. A diagram of CaO/Al_2O_3 vs. TiO_2/K_2O was used to identify unknown obsidian sources in Hokkaido. We distinguished whole-rock and glass compositional groups from 16 geological obsidian sources in Hokkaido: Okushiri-Katsumayama, Toyoura, Akaigawa, Ishikarigawa-Takikawa, Ishikarigawa-Asahikawa, Oumu, Nayoro, Monbetsu-Kamimobetsu, Shirataki, Engaru, Ikutahara, Rubeshibe, Oketo, Tokachi-Mitsumata, Shikaribetsu, and Kushiro-Akan. Six obsidian sources (Ishikarigawa-Asahikawa, Shirataki, Ikutahara, Rubeshibe, Oketo, and Kushiro-Akan) have two geochemical groups for both whole-rock and glass. The obsidians from Shirataki, the most important source of artefact raw material in Hokkaido, are further sub-divided into four groups of glass composition based on the CaO/FeO* ratio. For source identification, it is effective to use the polished thin sections made from tiny portions of artefacts because reliable data from both the glass composition and microscopic examination can be obtained. Obsidian sources in Hokkaido are concentrated in or near the Monbetsu-Kamishihoro graben in the northeastern part of the island. The formation zone of these obsidians gradually shifted towards the south 12–1 Ma ago, which is almost consistent with volcanic activity associated with the tensional stress field in the upper crust of Hokkaido. The formation of obsidian in Hokkaido is almost always related to the rhyolite lava.*

Keywords: *Obsidian, Source Analysis, Whole-Rock, Glass, Hokkaido Island, Japan*

INTRODUCTION

Obsidian is one of the most important rocks in the study of prehistoric human activity (e.g., Shackley 2005). A plethora of information can be gleaned from composition, shape, and quality of obsidian artefacts at ancient sites. For example, studies of obsidian distribution networks in prehistoric societies, namely routes taken to carry obsidian from its place of origin to utilisation sites, have been conducted on the basis of obsidian chemical composition (e.g., Tykot 1997; Glascock *et al.* 1999; Gratuze 1999; Kuzmin *et al.* 2002; Sato *et al.* 2002; Phillips and Speakman 2009). However, these studies were not focused on geological and petrological aspects such as why there is more than one chemical group of obsidian at a single source, and what it means for the geology. If the origin of obsidian and the occurrence and distribution of each source are not understood, scholars will not be able to clarify the processes of prehistoric obsidian use and population movements quantitatively.

Wada and Sano (2011) obtained basic data for the geology and petrology of obsidian from Shirataki in Hokkaido Island, one of the most important obsidian sources in Japan. The Shirataki obsidian–rhyolite lava complex was formed by the quenching of aphyric rhyolite magmas (with SiO_2 content of 76.7–77.4 wt.%), forming the monogenetic volcano consisting of ten lavas at approximately 2.2 Ma ago. According to the chemical composition of the whole-rock, Shirataki obsidians consist of two groups, and by using the volcanic glass composition it can be further sub-divided into four groups.

Reliable data on the chemical composition of obsidian can be obtained by conducting geological and petrological studies similar to research on the Shirataki source. This approach can be used for every obsidian locality. Therefore, if artefacts are carefully sampled and analysed, and the obsidian source is accurately identified, specific issues such as temporal variations in obsidian use by prehistoric humans or geographical spread and distribution routes of obsidian, can be studied.

The purpose of this study is to clarify the geology, formation process, and geochemical composition of obsidian from the Hokkaido sources, using geological and petrological techniques. The results can provide identification of the sources of obsidian used for making artefacts.

GEOLOGICAL OBSIDIAN SOURCES IN HOKKAIDO

General Characteristics

A list of 21 obsidian sources in Hokkaido was compiled by Izuho and Sato (2007). They divided the sources into 11 primary and ten secondary sources. The latter are derived independently from geologically primary obsidian sources. For six secondary sources, we can establish their primary obsidian localities based on the existence of rhyolite lava or pyroclastic products on the geological map of the neighbouring area. However, we cannot specify the primary locations for four secondary sources: Chikabumidai, Ubundai, Hokuryu, and Chippubetsu. They contain obsidian from several primary sources because obsidian fragments occur in alluvial deposits of the large river systems. On the geochemical basis, Mukai (2010) identified 22 compositional groups

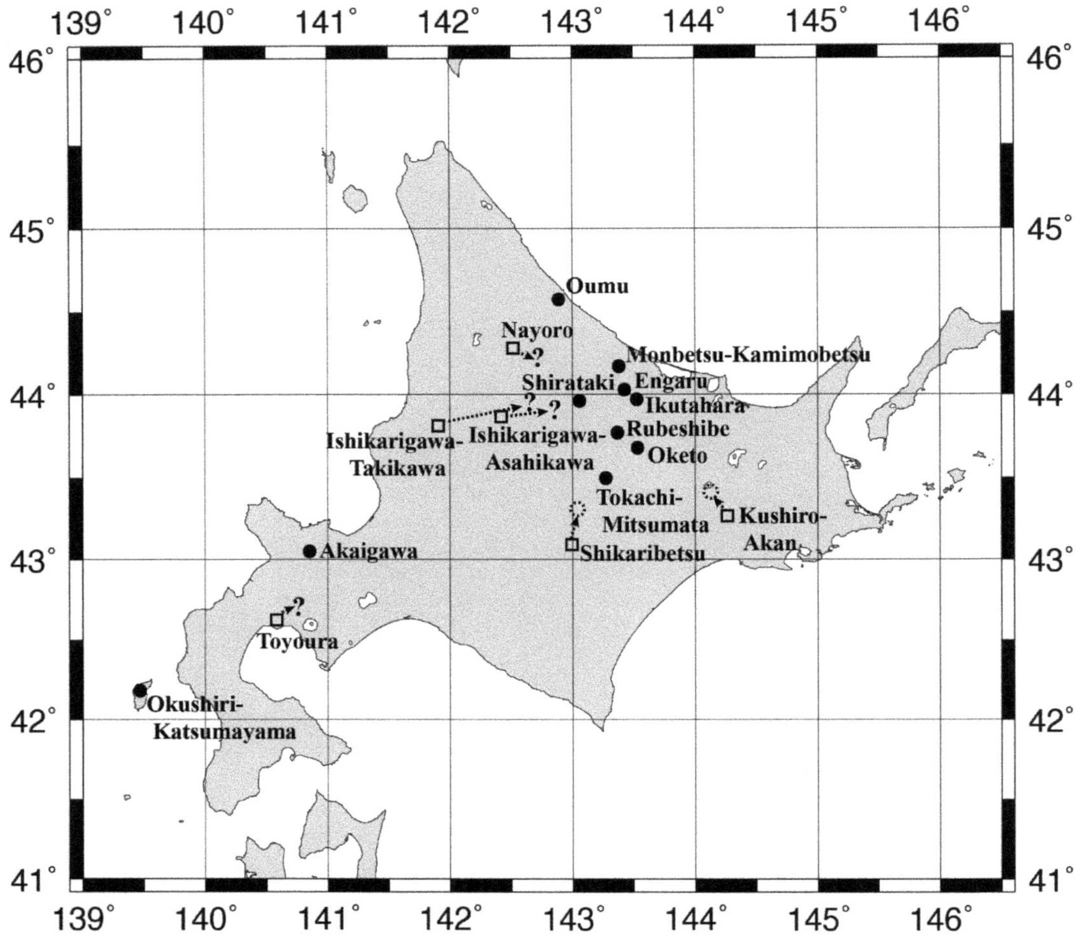

Figure 5.1. Map of obsidian sources in Hokkaido. Open squares are secondary sources, and their suggested geological sources are shown as dotted circles or dotted arrows

for 15 obsidian sources in Hokkaido; and Ferguson *et al.* (this volume) recognise 21 source groups in Hokkaido obsidian using Mukai's (2010) samples.

In addition to these classifications (Izuho and Sato 2007; Mukai 2010; see also Ferguson *et al.*, this volume), data on the obsidian sources in Hokkaido have clearly been enhanced according to the geological and petrochemical data presented in this chapter. The Hokkaido sources are defined as being geologically of the same origin and area. The meaning of difference between geological and chemical sources of obsidian must be considered. For source identification, we must know if there are multiple chemical groups in a single geological source. It is fundamental that the name of an obsidian source in Hokkaido is a geographical name showing the place of origin. However, there are examples of geographical name other than the place of origin which have been traditionally used, such as 'Asahikawa', 'Takikawa', and 'Kushiro'.

In this paper, the geographical names that have been used previously were employed as much as possible. We confirm the following 16 localities in Hokkaido: Okushiri-Katsumayama, Toyoura, Akaigawa, Ishikarigawa-Takikawa (new name), Ishikarigawa-Asahikawa (new name), Oumu, Nayoro, Monbetsu-Kamimobetsu, Shirataki, Engaru, Ikutahara, Rubeshibe, Oketo, Tokachi-Mitsumata, Shikaribetsu, and Kushiro-Akan (new name), as geological obsidian sources (Table 5.1, Figure 5.1). These sources originated from the geological localities that were formed by the rhyolite magma activity in the respective volcanic centres. In total, the 21 geochemical groups from these sources (Table 5.1) are the same as Ferguson *et al.*'s (this volume) sources.

K–Ar Age Determinations

This analysis was carried out by the Hiruzen Institute for Geology and Chronology Co., Ltd. The obsidians dated are fresh and unaltered samples from the Akaigawa, Nayoro, Shirataki, Ikutahara, Rubeshibe, Oketo, Tokachi-Mitsumata, and Shikaribetsu sources. Rocks were crushed and sieved using a 60- to 80-mesh size fraction. Iron oxides and plagioclase crystals were removed using an isodynamic separator to segregate the glass for dating. The sieved fraction was ground with an agate mortar, and analysed for potassium content.

Analyses of K and Ar in obsidian and age calculations were carried out using methods described by Nagao *et al.*

Table 5.1. List of Hokkaido obsidian sources

No.	Name of Source	Source Type	Name of Chemical Composition for Source	Sub-Type by Glass Composition	Source Name by Ferguson *et al.* (this volume)
1	Okushiri-Katsumayama	primary	Okushiri-Katsumayama		Okushiri
2	Toyoura	secondary	Toyoura		Toyoura
3	Akaigawa	primary	Akaigawa		Akaigawa
4	Ishikarigawa-Takikawa	secondary	Ishikarigawa-Takikawa		Takikawa
5	Ishikarigawa-Asahikawa	secondary	Ishikarigawa-Asahikawa 1		Asahikawa (Syunkodai)
			Ishikarigawa-Asahikawa 2		Asahikawa (Higashitakasu)
6	Oumu	primary	Oumu		Oumu
7	Nayoro	secondary	Nayoro		Nayoro
8	Monbetsu-Kamimobetsu	primary	Monbetsu-Kamimobetsu		Monbetsu
9	Shirataki	primary	Shirataki-Akaishiyama	Akaishiyama-A, -B	Shirataki (Akaishiyama)
			Shirataki-Tokachiishizawa	Tokachiishizawa-A, -B	Shirataki (Tokachiishizawa)
10	Engaru	primary	Engaru		Engaru
11	Ikutahara	primary	Ikutahara	Ikutahara-A, -B	Ikutahara
12	Rubeshibe	primary	Rubeshibe-Iwayamanosawa		Rubeshibe (Iwayama)
			Rubeshibe-Kayokozawa		Rubeshibe (Kayokozawa)
13	Oketo	primary	Oketo-Tokoroyama	Tokoroyama-A, -B	Oketo (Tokoroyama)
			Oketo-Oketoyama		Oketo (Oketoyama)
14	Tokachi-Mitsumata	primary	Tokachi-Mitsumata		Tokachi (Mitsumata)
15	Shikaribetsu	secondary	Shikaribetsu		Tokachi (Shikaribetsu)
16	Kushiro-Akan	secondary	Kushiro-Akan 1		Kushiro (Shitakara)
			Kushiro-Akan 2		Kushiro (Kutyorogawa)

No.	Name of Source	Utilisation as Artefacts[∀]	Geological Occurrence	K–Ar Age (Ma)
1	Okushiri-Katsumayama	NE	Katsumayama lava (Kano *et al.* 2006)	0.2–0.7*
2	Toyoura	LOC	pyroclastic deposits, originally of lava origin	
3	Akaigawa	EXT	lava	3.1**
4	Ishikarigawa-Takikawa	LOC(P)	?	
5	Ishikarigawa-Asahikawa	LOC	?	3.9***
		LOC(P)	?	
6	Oumu	NE	Motoineppu lava (Suzuki *et al.* 1966)	
7	Nayoro	LOC	pyroclastic deposits, originally of lava origin?	12.4**; 5.9****
8	Monbetsu-Kamimobetsu	NE	Kamimobetsu lava (Yahata *et al.* 1988)	11.8*****
9	Shirataki	EXT	Akaishiyama Summit lava; Akaishiyama upper lava, Kitayunosawa lava; and Shikatoride upper lava (Wada and Sano 2011)	2.2[#]
		EXT	Tokachiishizawa lava; Horokayubetsu lava; 850 m lava; Akaishiyama lower lava; Shikatoride lower lava; and Tokachiishizawa 830 m lava (Wada and Sano 2011)	
10	Engaru	NE	pyroclastic deposits, originally of lava origin ?	
11	Ikutahara	LOC(P)	lava (Yamada *et al.* 1963; Mukai 2010)	7.2**
12	Rubeshibe	LOC	lava (Mukai 2010)	3.9**
		LOC	lava (Mukai 2010)	
13	Oketo	EXT	Tokoroyama lava; Kita-Tokoroyama lava (this paper)	3.8**
		EXT	Oketoyama lava (Sawamura and Minato 1965; this paper)	4.5**
14	Tokachi-Mitsumata	EXT	lava (Mukai 2010)	2.1**

No.	Name of Source	Utilisation as Artefacts[∀]	Geological Occurrence	K–Ar Age (Ma)
15	Shikaribetsu	LOC(P)	lava? (Mukai 2010)	0.75**
16	Kushiro-Akan	LOC(P)	Akan welded pyroclastic flow deposits	1.0–1.5[##]

[∀] EXT – sources with extensive geographic distribution; LOC – sources with utilisation in the vicinities; LOC(P) – sources with possible utilisation in the vicinities; and NE – sources with non-expected utilisation in the vicinities.

[#] After Wada and Sano (2011).

[##] After Hasegawa *et al.* (2011).

* After Kano *et al.* (2006).

** This paper.

*** After Mukai and Wada (2002).

**** After Oka *et al.* (1991).

***** After Yahata and Nishido (1995).

(1984) and Itaya *et al.* (1991). Potassium content was analysed by flame photometry using a 2000 ppm Cs buffer, with analytical error within 2% at the 2σ confidence level. The argon content was analysed on a 15 cm radius sector type mass spectrometer with a single collector system and with an isotopic dilution method using a ^{38}Ar spike (Itaya *et al.* 1991). Multiple runs using a standard (JG-1 biotite, 91 Ma old) indicated that the error for Ar analysis was within 1% at the 2σ confidence level. The decay constants for ^{40}K to ^{40}Ar, ^{40}Ca, and the ^{40}K content in potassium used in the age calculation are from Steiger and Jager (1977): 0.581 x 10^{-10}/year, 4.962 x 10^{-10}/year, and 0.0001167, respectively. Table 5.2 shows the results of K–Ar age determination of obsidians from eight sources.

Geological Description of Obsidian Sources

Okushiri-Katsumayama

Okushiri-Katsumayama obsidian occurs in the rhyolite lava of the Katsumayama Volcano on Okushiri Island 15 km west of the Oshima Peninsula, southwest Hokkaido (Kano *et al.* 2006) (Figure 5.1). The eruption of the rhyolite lava is dated by the fission-track method to 0.2–0.7 Ma ago (Kano *et al.* 2006). As the obsidian does not produce a successful conchoidal fracture because of the existence of several percent of phenocrysts, it is not suitable as a raw material for the production of stone tools.

Toyoura

Toyoura obsidian is found as fragments in the Toya pyroclastic flow deposits and in secondary deposits around Toyoura Town, west of the Toya Caldera (Mukai 2005). The geological source of Toyoura obsidian is still unknown (Figure 5.1); therefore, the source name is provisional. This obsidian, of a black colour and high transparency, is suitable for artefact manufacture.

Akaigawa

Akaigawa obsidian occurs as fragments in talus deposits on the northwest flank of an unnamed summit at an elevation of 613.5 m above sea level (a.s.l.), along the Dobokugawa and Magarigawa rivers, south of the Akaigawa Caldera (Yokoyama *et al.* 2003) (Figure 5.1). The geological source of this obsidian is probably rhyolitic lava on the plateau around this summit (Mukai *et al.* 2002). The K–Ar age is determined as 3.1 Ma (Table 5.2). Most of the obsidian contains small laminar spherulites, but some of the pieces are jet-black in colour and without spherulites, thus of high quality. This obsidian has conchoidal fractures and is suitable for the production of artefacts. Tools made of the Akaigawa obsidian are abundant in the southern part of Hokkaido.

Ishikarigawa-Takikawa (new name)

The Ishikarigawa-Takikawa obsidian occurs as fragments in secondary deposits in the Kita-Sorachi area around Takikawa, Fukagawa, and Asahikawa cities. Since obsidian gravel is found in the Ishikarigawa River system, for example, near Aibetsu Town east of Asahikawa City, it is assumed that the primary source is situated somewhere in the headwaters of the Ishikarigawa River (Figure 5.1). Therefore, this source name is provisional. This obsidian is characterised by an exceptionally high degree of transparency among the Hokkaido sources.

Ishikarigawa-Asahikawa (new name)

Ishikarigawa-Asahikawa obsidian occurs as fragments in secondary deposits along the Ishikarigawa River, especially in the Asahikawa City area. An unspecified primary source is located somewhere in the headwaters of the Ishikarigawa River (Figure 5.1). This source name is therefore provisional. Ishikarigawa-Asahikawa obsidian is divided into two types based on morphology. The first one has wrinkled surface and is jet-black in colour. It was probably used for tool-making because it was within easy access. The second one is grey-coloured because of the high content of microlites. The K–Ar age for the former type is determined as 3.9 Ma (Mukai and Wada 2002).

Oumu

Oumu obsidian was sampled only in small quantities from the gravel in the Otoineppu River, west of Oumu Town (Figure 5.1). It is highly probable that the primary source is part of the Motoineppu rhyolite lava (Mukai and Wada 2003), which is widely distributed south of the Otoineppu River (Suzuki *et al.* 1966). This obsidian would not have been used as a material for the production of tools because of its poor quality and small quantity.

Table 5.2. The K–Ar ages of obsidian from the Akaigawa, Nayoro, Shirataki, Ikutahara, Rubeshibe, Oketo, Tokachi-Mitsumata, and Shikaribetsu sources. The data for Shirataki obsidians (Nos. 9-1 and 9-2) are after Wada and Sano (2011)

Source No.	Sample No.	Obsidian Source	K (wt.%)	Radioactive ^{40}Ar (10^{-8} cc STP/g)	K –Ar Age (Ma)	Non-radioactive ^{40}Ar (%)
3	AK-1	Akaigawa	3.822 ± 0.076	46.55 ± 0.50	3.14 ± 0.07	9.6
7	NAY-1	Nayoro	3.030 ± 0.061	145.8 ± 1.4	12.36 ± 0.27	3.5
9-1	07917R2-02	Shirataki (Akaishiyama)	3.848 ± 0.077	33.44 ± 0.43	2.24 ± 0.05	13.2
9-2	08IK-2	Shirataki (Tokachiishizawa)	3.817± 0.076	33.45 ± 0.71	2.26 ± 0.07	47.1
9-2	08101301-2	Shirataki (Tokachiishizawa)	3.749 ± 0.075	31.90 ± 1.40	2.20 ± 0.11	71.3
9-2	0961702E	Shirataki (Tokachiishizawa)	3.848± 0.077	31.55 ± 0.37	2.11 ± 0.05	15.4
11	IKU-1	Ikutahara	3.943 ± 0.079	111.2 ± 1.1	7.25 ± 0.16	4.8
12	RB-102301-B3	Rubeshibe (Kayokozawa)	2.940 ± 0.059	44.58 ± 1.10	3.90 ± 0.12	53.3
13-1	TO-60502-D	Oketo (Tokoroyama)	3.336 ± 0.067	49.40 ± 0.56	3.81 ± 0.09	13.4
13-2	OK-1	Oketo (Oketoyama)	2.830 ± 0.057	49.70 ± 1.39	4.52 ± 0.15	57.6
14	TK-92901-A	Tokachi-Mitsumata	3.457 ± 0.069	28.86 ± 3.37	2.15 ± 0.25	87.7
14	TK-92903-E	Tokachi-Mitsumata	3.434 ± 0.069	28.32 ± 0.36	2.12 ± 0.05	21.8
15	TSHI-1	Shikaribetsu	3.016 ± 0.060	8.76 ± 0.19	0.75 ± 0.02	45.9

Nayoro

Nayoro obsidian occurs as fragments in secondary deposits derived from the late Miocene Kawanishi Formation (Imanishi 1956). The primary source is still unknown, so the name is provisional. The fission-track age of this obsidian is 5.9 Ma (Oka *et al.* 1991), but the K–Ar age is 12.4 Ma (Table 5.2) which is concordant with the stratigraphy of this area. Most of the obsidian samples were collected from the gravel along the Chureppu River southeast of Nayoro City. The Nayoro obsidian is characterised by a pockmarked surface, and is suitable as a raw material for artefact production.

Monbetsu-Kamimobetsu

Monbetsu-Kamimobetsu obsidian derives from the Kamimobetsu rhyolite lava (Yahata *et al.* 1988; Matsunami *et al.* 2002). The K–Ar age is 11.8 Ma (Yahata and Nishido 1995). In the lava outcrop, the obsidian shows a perlite structure (Mukai 2005). One can also find this obsidian in the gravel along the Kamimobetsu River located southeast of Kamimobetsu, south of Monbetsu City. Since the obsidian is rich in water and changes into pitchstone, it is not suitable for tool-making.

Shirataki

The Shirataki obsidian comes from the vicinity of Shirataki Village, west of Engaru Town, and it was the most important source of raw material used for manufacture obsidian artefacts in Hokkaido (Kimura 2005). In the Shirataki obsidian–rhyolite field, there are many outcrops with densely compact obsidian layers of both excellent quality and abundant quantity. The Shirataki obsidian lavas erupted approximately 2.2 Ma ago (four K–Ar dates, see Table 5.2; see also Wada and Sano 2011), and formed a monogenetic volcano composed of ten obsidian flow units (Wada and Sano 2011). These obsidians are divided into two types. The first type has a jet-black lustre, and consists almost entirely of glass with trace amounts of tiny magnetite crystals (smaller than 70 microns). The second type has a rough surface, and consists almost entirely of glass with trace amounts of plagioclase crystals (size of 20–200 microns) with tiny magnetite microlites (Wada and Sano 2011). Tools manufactured from Shirataki obsidian spread throughout not only Hokkaido but also Sakhalin Island, the southern Kurile Islands, and the northern part of Honshu Island, more than 500 km from the source in the Palaeolithic and the following Jomon period (e.g., Kimura 2005; Kuzmin *et al.* 2002, 2013; Phillips and Speakman 2009).

Engaru

Engaru obsidian occurs as fragments in pyroclastic rocks of the late Miocene Shanafuchi Formation (Tajika and Yahata 1991; Mukai 2003). Samples were collected from the gravel of the Sanabuchi River, west of Engaru Town. Some of the obsidians are suitable as raw material for artefact manufacture but they are found in small quantities only.

Ikutahara

Ikutahara obsidian is found in the Nitappu River area southwest of Yasukuni, Ikutahara District, Engaru Town. The primary source is probably the rhyolite lava from the Ikutahara Formation (Mukai *et al.* 2004) distributed in the southern part of the Setani-Ushiyama andesite lava field. The K–Ar age of the Ikutahara obsidian is 7.2 Ma (Table 5.2) which agrees well with the age of the Setani-Ushiyama lava (7.2 Ma; see Yahata and Nishido 1995). This obsidian is suitable for the production of artefacts.

Rubeshibe

Rubeshibe obsidian comes from the obsidian–rhyolite lava distributed in the upper course of the Keshomappu River, northwest of Rubeshibe Town (Figure 5.1). The K–Ar age of this obsidian is 3.9 Ma (Table 5.2). Samples were collected at the Kayokozawa Outcrop with a dense obsidian layer, and from the talus deposits at the Iwayamanosawa Slope (Mukai 2005). It is possible to obtain large pieces (> 10 cm). This obsidian is suitable as a raw material for artefact-making.

Oketo

Oketo obsidian derives from two rhyolitic lavas — Tokoroyama and Oketoyama — located west of Oketo Town. Tokoroyama lava can be further sub-divided into two lavas, Tokoroyama and Kita-Tokoroyama. These three lavas are distributed along a NE–SW direction, and each was formed independently. The K–Ar age of obsidian from the Tokoroyama is 3.8 Ma, and from Oketoyama 4.5 Ma (Table 5.2). The Oketo obsidian was frequently used for the manufacture of artefacts in prehistory.

Tokachi-Mitsumata

Tokachi-Mitsumata obsidian is widely distributed in the gravel deposits of the Tokachi Plain. The primary source is probably located west of Mount Minami-Kumaneshiridake (1560 m a.s.l.), in the headwaters of Jyusannosawa, a branch of the Otofuke River, in the Tokachi-Mitsumata Caldera (Ishii *et al.* 2008) of the same name county. The new K–Ar age determination of two obsidian samples is 2.1 Ma (Table 5.2), showing the same age as the Shirataki source. The Tokachi-Mitsumata obsidian was widely used for artefact-making in prehistory.

Shikaribetsu

Shikaribetsu obsidian is found in gravel deposits of the Tokachi Plain in the western part of the Otofuke River basin, which is the southern piedmont of the Shikaribetsu Volcano. The K–Ar age of the obsidian is 0.75 Ma (Table 5.2). This shows that it was formed directly before the early activity of the Shikaribetsu Volcano group. Therefore, the source of this obsidian comes from rhyolitic lava in the Shikaribetsu Volcano area. The obsidian is jet-black in colour but has a dull lustre.

Kushiro-Akan (new name)

Kushiro-Akan obsidian occurs as rounded fragments in the Akan pyroclastic flow deposits or as gravel from secondary deposits in the Kuchoro River (Mukai 2009). Since it is related to the welded glass from the Akan pyroclastic flow deposits belonging to the Kushiro Formation dated to 1.0–1.5 Ma ago (Hasegawa *et al.* 2011), the primary source of this obsidian is possibly derived from pyroclastic activity of the Akan Volcano (Figure 5.1).

FORMATION OF OBSIDIAN IN HOKKAIDO

In Hokkaido, the obsidian sources are concentrated in the northeastern part of the island (Figure 5.1). Most of them are distributed in or near the Monbetsu-Kamishihoro graben (Yahata 1997) (Figure 5.2). This graben was developed after 11.5 Ma ago, and it spread in a N–S direction (Yahata 1997). Since the graben was created under a tensional stress field of the Earth's crust, it is considered that the obsidian sources in the graben were formed in association with volcanic activity under tensile stress (Wada 2009). In the tensional field, normal faults are likely to be formed, and magma would easily ascend from the deep crust. As a result, even the viscous rhyolite magma could come through the volcanic conduit. Bimodal volcanism with basaltic and rhyolitic magmas and the formation of monogenetic volcanoes would also be restricted to the tensional stress field of the crust. The obsidian–rhyolite lava tends to be found in these volcanic fields.

The volcanic activity which formed the obsidians of the Oumu, Nayoro, and Monbetsu-Kamimobetsu sources started in or near the northern part of the Monbetsu-Kamishihoro graben around 11 or 12 Ma ago (Figure 5.2). Both the Engaru and Ikutahara obsidian sources, located on the east side of the central part of the graben, were formed at around 7 Ma ago. South of the Ikutahara obsidian source, the Rubeshibe and Oketo sources were created by rhyolite lava activity at around 4 Ma ago. Obsidians from the Ishikarigawa-Asahikawa and Ishikarigawa-Takikawa sources must have been formed outside the western side of the Monbetsu-Kamishihoro graben during that period (Figures 5.1–5.2). On the other hand, Akaigawa obsidian was created approximately 3 Ma ago in western Hokkaido due to the bimodal volcanic activity of rhyolitic and basaltic magmas at the back-arc side.

Shirataki is the largest obsidian source in all of Japan. In the Shirataki area, aphyric rhyolite magmas were effused 2.2 Ma ago, resulting in the Shirataki obsidian–rhyolite lava complex (Wada and Sano 2011). At the same time, the obsidian from the Tokachi-Mitsumata source (age of 2.1 Ma) was created in the southern part of the Monbetsu-Kamishihoro graben (Figure 5.2). The main group of the obsidian sources in northeastern Hokkaido was formed at around 4.5–2 Ma ago (Figure 5.2), and this is consistent with the age of the tensile stress field in Hokkaido which was very high at around 5–2 Ma ago (Hirose and Nakagawa 1999).

Obsidian of the Shikaribetsu source located in the southernmost part of the Monbetsu-Kamishihoro graben

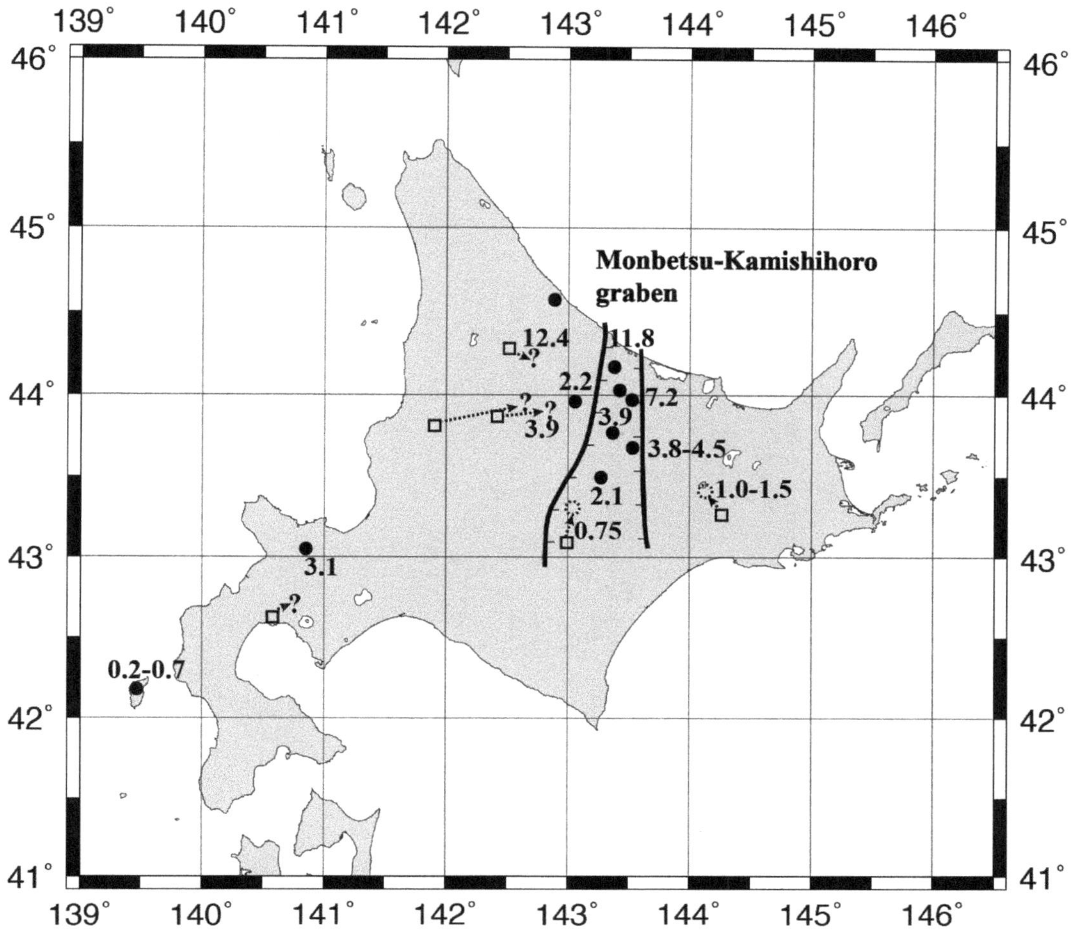

Figure 5.2. The K–Ar age (in Ma) of obsidian sources in Hokkaido and the location of the Monbetsu-Kamishihoro graben (after Yahata 1997)

shows the young age of 0.75 Ma. It must have been related to the activity of the Shikaribetsu volcanic group which is characterised by felsic magma. At the back-arc side, even younger Okushiri-Katsumayama obsidian was formed at 0.3–0.2 Ma ago under volcanism with a tensional field.

WHOLE-ROCK CHEMICAL COMPOSITION OF OBSIDIAN

Obsidian is macroscopically the most uniform volcanic rock. Although most of obsidian consists of glass, it also contains small amounts of minerals, so it is not completely homogeneous (Figure 5.3). Because the chemical composition of glass is different from that of minerals, if we consider obsidian composition on a scale on the order of several tens to several hundreds of microns, it is not necessarily homogeneous. Therefore, when the chemical composition of obsidian is stidied, the homogeneity of the analysed portion should be examined carefully.

Determining the standard chemical composition of obsidian is important for the identification of its sources. Volcanic rock is heterogeneous on a macro-scale because it is comprised of both minerals and glass. Therefore, before conducting chemical analysis, a rock must be homogenised by crushing it into a powder. In petrology, this kind of analysis examines the chemical content of whole-rock, and it shows the composition of the magma just before eruption (excluding the volatile component) (e.g., McBirney 1984). As most of obsidians consist of homogeneous glass with small amounts of minerals (Figure 5.3), whole-rock analysis should be performed. The whole-rock composition in this study represents the content of chemical elements in original obsidian–rhyolite magma, and it also shows the geochemical 'signature' of the source obsidian.

Analytical Procedure for Whole-Rock Chemical Analysis

Pretreatment of Samples

Fresh obsidian is crushed using a rock hammer into coarse chips 3–10 mm in diameter. The chips are rinsed with deionised water in an ultrasonic bath at least three times for one hour each, and afterwards dried at 110° C for six hours. The next step is to pulverise the obsidian using alloy mortar and then automatic agate mortar for about one hour to make a powder. This powder is heated up to 700–800° C. This calcined (i.e., total iron becomes

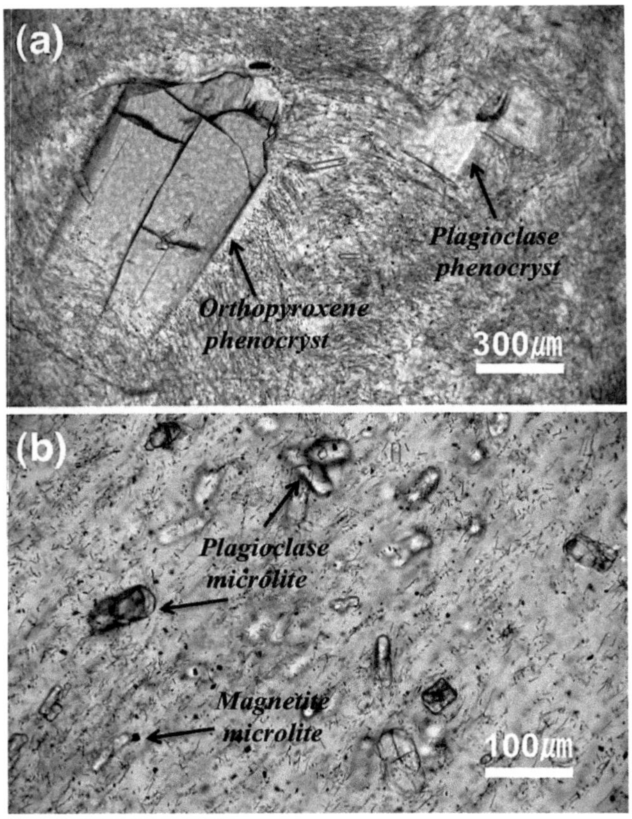

Figure 5.3. Microphotograhs of obsidian taken under open-polarised light: (a) from the Rubeshibe source; and (b) from the Shirataki source (Tokachiishizawa Locality)

ferric iron) powder (1.4 g) is added to 2.8 g of lithium borate flux ($Li_2B_4O_7$ and $LiBO_2$). It is mixed well, and fused in an automatic bead sampler at 1100° C. The melt is quenched in a Pt–Au crucible to make the glass bead disk. The glass beads are used for the analysis of ten major and 18 trace elements by X-ray Fluorescence (XRF) analysis.

XRF Analysis

Obsidian composition was determined using the Spectris MagiX PRO system with an Rh tube at the Department of Natural History Sciences, Graduate School of Science, Hokkaido University. Analysis is performed with the calibration curve method using the standard samples of the National Institute of Advanced Industrial Science and Technology (AIST): JB-1a, JB-2, JB-3, JA-2, JA-3, JR-1, JR-2, JR-3, JGb-1, JGb-2, JG-1a, JG-2, JG-3, JF-1, JF-2, JSy-1, and JP-1. The net count of differences between peak and background was determined. Matrix revisions with all elements were made, including the line overlap revisions for trace elements. The measurement time for trace elements is approximately one hour, and for major elements it is about ten minutes. Table 5.3 shows the content and standard deviation of each major and trace element on the reference rhyolite sample (HR-1A; aphyric pumice from the Shikotsu pyroclastic flow deposits in the Shikotsu Caldera, Hokkaido) designated by Hokkaido University.

Table 5.3. Standard deviation of the chemical composition for reference sample HR-1A (rhyolite standard; aphyric-type pumice from the Shikotsu pyroclastic flow deposits, used at Hokkaido University). The data are after A. Matsumoto

	N = 15	**Standard Deviation (1σ)**
wt.%		
SiO_2	76.51	0.034
TiO_2	0.18	0.004
Al_2O_3	13.46	0.009
Fe_2O_3*	1.67	0.013
MnO	0.07	0.001
MgO	0.19	0.001
CaO	1.60	0.007
Na_2O	3.87	0.017
K_2O	2.43	0.019
P_2O_5	0.02	0.000
Total	100.00	
ppm		
Sc	8.5	1.03
V	3.8	0.99
Cr	2.8	0.50
Co	11.5	0.42
Ni	5.1	0.33
Cu	20.1	0.92
Rb	68.9	1.16
Zn	54.5	0.84
Sr	141.5	1.13
Y	42.3	0.59
Zr	173.7	0.63
Nb	7.8	0.21
Ba	789.2	4.53
Pb	18.3	0.39
Th	7.0	0.39
Ga	12.9	1.06
La	17.3	1.31
Ce	45.0	2.44

Whole-Rock Composition of Hokkaido Obsidian Sources

Table 5.4 shows 19 whole-rock compositional groups from 16 obsidian sources in Hokkaido. The SiO_2 contents of these obsidians are 74.5 to 78.9 wt.%, showing rhyolitic magma composition. These data represent a reference whole-rock signature of the Hokkaido obsidian sources. In addition to it, whole-rock chemical analysis of 29 obsidian samples from the Shirataki source was conducted by Wada and Sano (2011) using the same analytical method.

Table 5.4. Whole-rock chemical compositions of the Hokkaido obsidians by XRF

Source No.	1	2	3	4	5-1	5-2	7	8
Name of Bulk Composition	Okushiri-Katsumayama	Toyoura	Akaigawa	Ishikarigawa-Takikawa	Ishikarigawa-Asahikawa 1	Ishikarigawa-Asahikawa 2	Nayoro	Monbetsu-Kamimobetsu
SiO_2	78.13	78.07	77.55	78.90	77.30	74.52	77.82	77.39
TiO_2	0.08	0.13	0.06	0.11	0.08	0.15	0.11	0.05
Al_2O_3	13.21	12.48	12.62	12.53	13.66	14.41	12.97	12.93
Fe_2O_3*	0.65	1.20	1.04	0.85	1.42	1.93	1.33	1.18
MnO	0.10	0.06	0.06	0.06	0.05	0.07	0.03	0.03
MgO	0.12	0.19	0.08	0.13	0.15	0.37	0.15	0.07
CaO	0.80	1.14	0.78	0.71	1.26	1.86	1.17	1.03
Na_2O	3.70	3.86	3.74	3.71	3.94	3.97	3.89	3.99
K_2O	4.69	3.56	4.56	4.37	3.69	3.32	3.68	3.69
P_2O_5	0.02	0.02	0.01	0.02	0.04	0.07	0.02	0.01
Total (wt.%)	101.50	100.71	100.51	101.39	101.58	100.67	101.18	100.37
ppm								
Sc	4.4	2.3	5.1	3.1	2.5	5.4	4.1	5.7
V	7.8	8.2	7.6	9.2	7.0	18.6	9.1	6.6
Cr	2.4	2.6	2.1	1.4	2.8	3.0	1.4	1.2
Co	0.0	1.6	0.4	1.4	3.2	3.9	1.8	0.8
Ni	0.8	0.3	1.0	0.9	0.7	0.9	0.7	1.1
Cu	3.7	3.3	5.3	4.8	5.8	2.7	3.5	1.8
Rb	209	93	137	155	133	120	128	194
Zn	20.5	28.1	26.1	25.6	41.0	49.2	34.5	37.6
Sr	118	97	46	58	109	150	95	56
Y	18.4	25.4	25.1	27.0	24.1	22.6	18.8	42.3
Zr	58	110	85	89	87	104	111	91
Nb	6.4	4.9	5.2	6.0	4.7	5.8	4.1	5.8
Ba	926	851	764	1090	772	718	674	1046
Pb	6.5	31.2	27.1	23.8	24.9	22.7	21.2	20.8
Th	19.4	11.4	20.5	19.2	10.4	9.2	13.1	11.7
Ga	12.0	10.9	11.5	13.9	16.1	16.7	15.6	16.6
La	22.3	17.0	27.3	24.7	21.5	17.1	20.2	19.0
Ce	40.8	43.5	53.4	52.3	44.2	33.9	45.9	40.0

Source No.	9-1*	9-2*	10	11	12-1	12-2
Name of Bulk Composition	Shirataki-Akaishiyama	Shirataki-Tokachiishizawa	Engaru	Ikutahara	Rubeshibe-Iwayamanosawa	Rubeshibe-Kayokozawa
SiO_2	77.55	77.93	78.27	75.88	76.39	75.90
TiO_2	0.04	0.02	0.09	0.12	0.08	0.10
Al_2O_3	13.05	13.20	13.10	13.72	13.61	13.84
Fe_2O_3*	1.17	1.10	1.37	1.62	1.69	1.87
MnO	0.05	0.06	0.05	0.03	0.05	0.06
MgO	0.04	0.03	0.05	0.08	0.15	0.17
CaO	0.55	0.47	0.72	0.82	1.35	1.53
Na_2O	3.90	4.12	4.45	4.31	3.88	3.97
K_2O	4.77	4.63	3.87	4.74	3.73	3.54
P_2O_5	0.02	0.02	0.01	0.02	0.03	0.03
Total (wt.%)	101.13	101.58	101.98	101.34	100.97	101.02

Source No.	9-1*	9-2*	10	11	12-1	12-2
Name of Bulk Composition	Shirataki-Akaishiyama	Shirataki-Tokachiishizawa	Engaru	Ikutahara	Rubeshibe-Iwayamanosawa	Rubeshibe-Kayokozawa
ppm						
Sc	2.5	4.3	2.0	8.9	5.7	5.3
V	4.5	4.0	0.4	9.3	7.3	8.5
Cr	1.4	3.1	2.5	1.2	3.1	2.4
Co	1.5	0.7	1.9	1.6	1.6	3.6
Ni	0.4	1.2	1.4	0.9	1.2	1.0
Cu	7.3	4.8	0.4	7.5	2.3	3.4
Rb	170	196	122	168	129	124
Zn	42.3	37.0	52.0	55.7	46.6	50.1
Sr	33	16	53	49	114	129
Y	31.0	35.7	42.6	40.7	27.7	26.4
Zr	74	66	135	215	110	125
Nb	6.7	8.6	8.1	7.3	6.2	5.9
Ba	926	209	825	927	837	827
Pb	19.1	18.5	18.1	22.0	20.7	22.0
Th	11.5	9.7	9.9	14.2	9.7	11.1
Ga	17.3	17.2	16.8	19.2	16.6	17.3
La	17.3	14.9	25.5	27.5	26.9	17.1
Ce	38.1	30.1	52.0	64.8	51.1	48.4

Source No.	13-1	13-2	14	15	16
Name of Bulk Composition	Oketo-Tokoroyama	Oketo-Oketoyama	Tokachi-Mitsumata	Shikaribetsu	Kushiro-Akan 1
SiO_2	78.13	78.02	77.58	77.30	78.30
TiO_2	0.10	0.11	0.04	0.11	0.04
Al_2O_3	12.71	12.95	13.02	13.00	13.16
Fe_2O_3*	1.06	1.32	1.05	1.43	1.05
MnO	0.04	0.05	0.05	0.04	0.05
MgO	0.13	0.13	0.06	0.18	0.06
CaO	0.87	1.01	0.70	1.23	0.71
Na_2O	3.82	4.32	4.11	3.79	4.08
K_2O	4.06	3.35	4.21	3.66	4.24
P_2O_5	0.02	0.02	0.02	0.03	0.02
Total (wt.%)	100.94	101.27	100.83	100.79	101.71
ppm					
Sc	5.7	3.9	6.0	5.8	6.3
V	9.7	7.5	5.3	10.5	6.9
Cr	1.9	2.2	4.3	2.2	1.3
Co	0.7	1.9	1.3	2.2	0.4
Ni	0.6	0.7	0.9	0.8	0.7
Cu	3.0	2.8	1.6	2.5	3.0
Rb	145	107	154	135	154
Zn	27.1	37.3	35.6	37.1	34.9
Sr	67	81	52	94	53
Y	24.9	24.4	33.6	24.6	33.8
Zr	97	127	80	92	82

Source No.	13-1	13-2	14	15	16
Name of Bulk Composition	Oketo-Tokoroyama	Oketo-Oketoyama	Tokachi-Mitsumata	Shikaribetsu	Kushiro-Akan 1
Nb	5.6	6.0	6.7	5.6	7.0
Ba	1078	791	967	829	970
Pb	17.8	15.8	24.8	20.1	23.7
Th	13.1	10.4	13.8	10.8	14.0
Ga	15.4	15.1	16.2	15.5	15.6
La	19.6	20.5	19.5	20.0	20.2
Ce	41.2	44.7	44.9	43.4	48.0

* The Sample 9-1 is No. JOSH-1 from Hachigosawa locality, Akaishiyama upper lava; and Sample 9-2 is No. JOSA-1 from Ajisainotaki locality, Horokayubetsu lava.

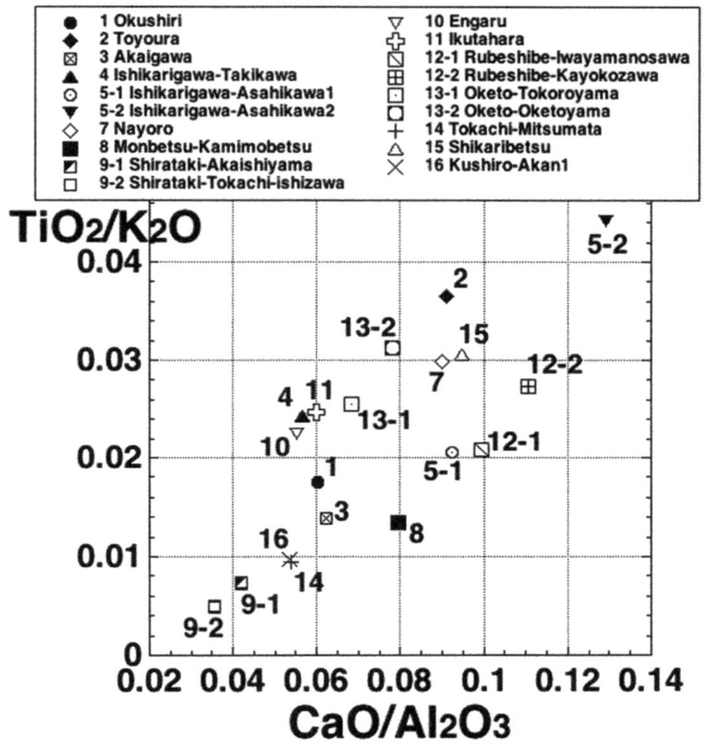

Figure 5.4. The CaO/Al_2O_3 vs. TiO_2/K_2O diagram of whole-rock composition for the Hokkaido obsidian sources; for the Shirataki-Akaishiyama (No. 9-1) and Shirataki-Tokachiishizawa (No. 9-2) sources, the average values are shown

Each percentage by total weight of a measured value of a major element is converted into 100 wt.% to compare the composition of the sources. According to this, various scatter diagrams based on these data were generated. As a result, each of the Hokkaido obsidian sources has a specific chemical composition.

From the content of TiO_2, CaO, and K_2O, the differences in the chemical composition are clear. The diagram of CaO/Al_2O_3 vs. TiO_2/K_2O is effective for the source identification of obsidians from Hokkaido (Figure 5.4). When the source obsidian compositions are similar to each other, the distinction is possible based on comparison to other diagrams, such as MgO vs. CaO; TiO_2 vs. Fe_2O_3*; Ba/Zr vs. Rb/Zr; and Ba/Zr vs. Sr/Zr (Figures 5.5–5.6).

Shirataki obsidians show the lowest values of CaO/Al_2O_3 and TiO_2/K_2O ratios (Figure 5.4) among the Hokkaido sources, and are clearly divided into two groups: Shirataki-Akaishiyama and Shirataki-Tokachiishizawa (Wada and Sano 2011). The Shirataki-Akaishiyama obsidian has a higher content of Fe_2O_3*, MgO, and CaO, and a lower content of MnO compared to the Shirataki-Tokachiishizawa group (Wada and Sano 2011). In terms of trace elements, the differences in the content of Ba, Rb, Zr, Sr, La, and Ce are remarkable: the Shirataki-Akaishiyama obsidian has a higher content of Ba, Sr, Zr, La, and Ce, and a lower concentration of Rb compared to the Shirataki-Tokachiishizawa group (Wada and Sano 2011). In the Ba/Zr vs. Rb/Zr plot, the Shirataki-Tokachiishizawa obsidian is distinctly different from the Shirataki-Akaishiyama one (Figure 5.6).

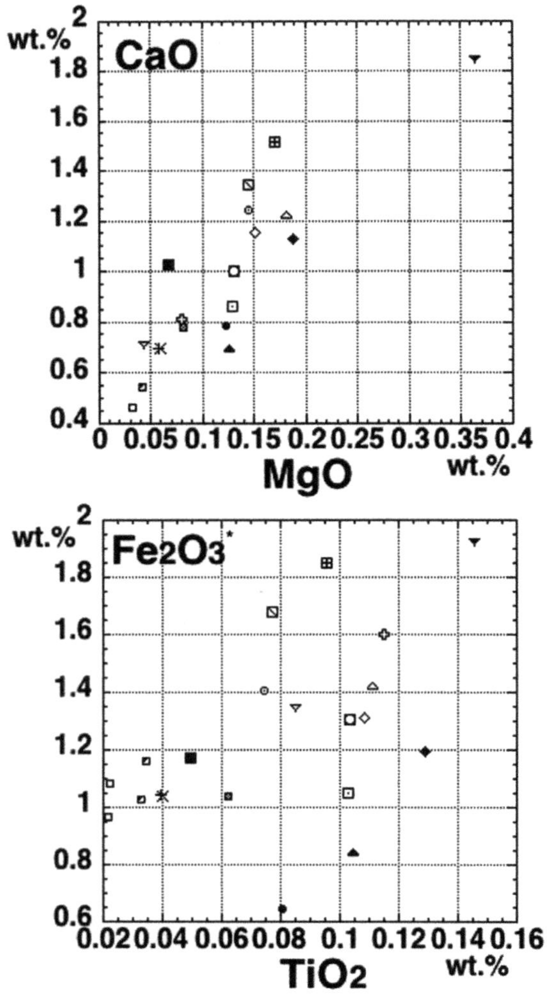

Figure 5.5. The MgO vs. CaO (upper) and TiO_2 vs. Fe_2O_3 (lower) diagrams. Fe_2O_3* shows the total iron content as Fe_2O_3. Symbols are the same as in Figure 5.4*

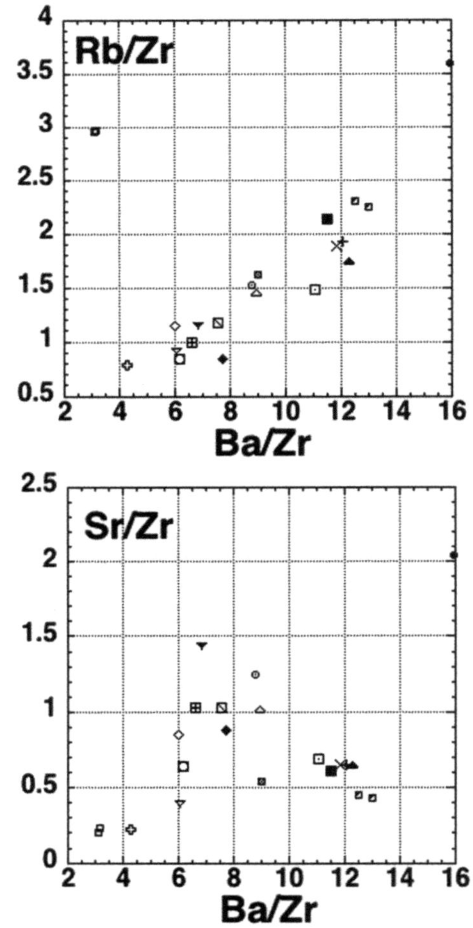

Figure 5.6. The Ba/Zr vs. Rb/Zr (upper) and Ba/Zr vs. Sr/Zr (lower) diagrams. Symbols are the same as in Figure 5.4

The Ishikarigawa-Asahikawa, Rubeshibe, and Oketo obsidian sources are also divided into two groups each: Ishikarigawa-Asahikawa 1 and Ishikarigawa-Asahikawa 2; Rubeshibe-Iwayamanosawa and Rubeshibe-Kayokozawa; and Oketo-Tokoroyama and Oketo-Oketoyama, respectively, in the CaO/Al_2O_3 vs. TiO_2/K_2O diagram (Figure 5.4).

The Ishikaraigawa-Takikawa, Engaru, and Ikutahara obsidian sources are located very close to each other in the CaO/Al_2O_3 vs. TiO_2/K_2O diagram (Figure 5.4). Two pairs (Nayoro and Shikaribetsu obsidians; and Ishikarigawa-Asahikawa 1 and Rubeshibe-Iwayamanosawa obsidians) also resemble each other (Figure 5.4). However, these sources can be clearly distinguished from each other on different scatter diagrams (Figures 5.5–5.6).

The Kushiro-Akan 1 obsidian resembles closely the Tokachi-Mitsumata obsidian, and can hardly be separated on any scatter diagram (Figures 5.4–5.6), although the Kushiro-Akan 1 source is slightly lower in Ba/Zr and Rb/Zr ratios than the Tokachi-Mitsumata obsidian.

CHEMICAL COMPOSITION OF VOLCANIC GLASS IN OBSIDIAN

Obsidian consists of more than 90 vol.% volcanic glass that is compositionally homogeneous. Therefore, the chemical composition of glass is similar to that of whole-rock, and it represents the geochemical content of liquid magma just prior to its being solidified at the glass transition temperature. However, as obsidian contains microlites (< 100 microns in size), such as magnetite and plagioclase (Figure 5.3), it is necessary to analyse carefully only certain spots in the glass, avoiding fine crystals (Figure 5.7). The analysed spot is a square with ten micron sides (i.e., an area of 100 μm^2; see Figure 5.7), and more than ten spots were analysed in each specimen. As a result, we used the average value for these spots.

Analytical Procedure for Glass Analysis

The content of major elements in obsidian glass was determined by Electron Probe Microanalyser, using the JEOL 8600 system of Hokkaido University of Education at Asahikawa. Analysis was conducted by simultaneous counting on three wavelength dispersive spectrometers equipped with TAP, PET, and LiF crystals with an

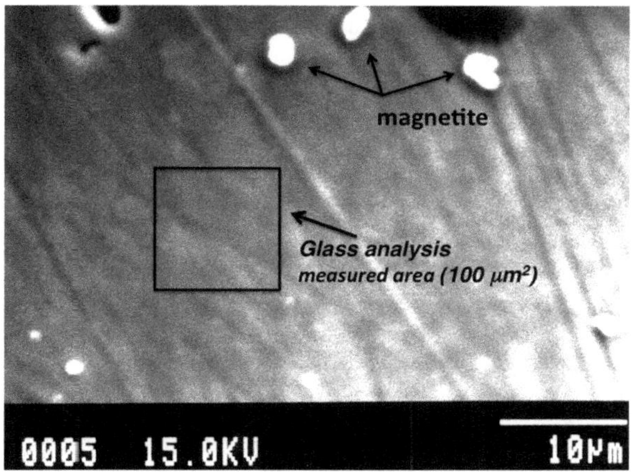

Figure 5.7. An example of glass analysis displayed in the Secondary Electron Images for the Shirataki obsidian

accelerating voltage of 15 keV, beam currents of 0.8 x 10^{-8} A on the Faraday cup, and counting times of 15 seconds (peak) and five seconds (background) for each element. The standards of the ten elements (see Table 5.5) are SiO_2, TiO_2, Al_2O_3, Fe_2O_3, MnO, MgO, $CaSiO_3$, $NaAlSi_2O_6$, $KTiOPO_4$, and $Na_4BeAlSi_4O_{12}Cl$. All analyses were adapted by oxide ZAF [atomic number (Z), absorption (A), and fluorescence (F)] corrections. Low probe currents and beam scanning areas of ten microns were used (Figure 5.7) to avoid the Na counting loss by beam irradiation. The accuracy and precision of analysis were simultaneously checked by reference samples of both obsidian glass (Astimex Scientific Ltd.) and dacite glass (Spfa-1 pumice erupted from Shikotsu Caldera, Hokkaido); both yielded values in good agreement with the recommended values (Wada *et al.* 2003).

Chemical Composition of Glass from Hokkaido Obsidian Sources

Table 5.5 shows the average chemical composition of glass from the Hokkaido obsidian sources. The reason why the total is less than 100 wt.% ("total average" in Table 5.5) is due to small quantities of water (almost < 1 wt.%) in the glass. It is difficult to analyse the content of water by Electron Probe Microanalysis (EPMA). The total weight of oxide is standardised to 100 wt.% for accurate comparison of each source (Table 5.5). Since the composition of glass shows an intrinsic value, it can be used for identification of obsidian sources in Hokkaido. A total of 24 groups are recognised from 16 Hokkaido obsidian sources.

The CaO/Al_2O_3 vs. TiO_2/K_2O diagram (Figure 5.8) is the most suitable for identification of obsidian (Mukai *et al.* 2000; Wada *et al.* 2003). The amount of CaO reflects most sensitively the specific source composition. Therefore, the values of CaO on the horizontal axis versus other oxides and Cl on the vertical axis (Figure 5.9) effectively show the chemical composition of each source.

Akaigawa and Toyoura obsidians are characterised by a high Cl content. Kushiro-Akan 2 obsidian has the highest CaO/Al_2O_3 and TiO_2/K_2O ratios among the Hokkaido sources. Glass of Shirataki obsidian is clearly divided into two groups. Both groups can additionally be subdivided into two groups in the CaO vs. FeO* diagram, and finally four groups can be distinguished: Akaishiyama-A and -B; and Tokachiishizawa-A and -B (Figure 5.9; see also Wada and Sano 2011). In Figure 5.9, the boundaries between the four groups are not clear because of an overlap. Therefore, care should be taken in

Table 5.5. Summary of chemical compositions of obsidian glass from the Hokkaido sources (in wt.%)

Source No.	1	2	3	4	5-1	5-2	6	7	8
Glass Composition Name	Okushiri-Katsumayama	Toyoura	Akaigawa	Ishikarigawa-Takikawa	Ishikarigawa-Asahikawa1	Ishikarigawa-Asahikawa2	Oumu	Nayoro	Monbetsu-Kamimobetsu
No. of Samples	23	14	19	22	16	15	13	14	10
SiO_2	77.44	78.35	77.84	78.39	77.09	75.42	77.95	77.62	78.31
TiO_2	0.08	0.13	0.07	0.12	0.08	0.15	0.09	0.12	0.07
Al_2O_3	12.87	12.24	12.29	12.10	13.13	14.02	12.49	12.60	12.61
FeO*	0.52	0.88	0.83	0.71	1.05	1.33	1.02	1.04	0.72
MnO	0.09	0.06	0.06	0.07	0.06	0.06	0.04	0.05	0.04
MgO	0.07	0.13	0.05	0.07	0.10	0.20	0.04	0.12	0.03
CaO	0.76	1.07	0.78	0.71	1.23	1.71	0.86	1.18	0.84
Na_2O	3.24	3.37	3.19	3.42	3.47	3.44	3.12	3.46	3.12
K_2O	4.85	3.64	4.72	4.30	3.73	3.62	4.31	3.72	4.22
Cl	0.08	0.17	0.18	0.11	0.06	0.05	0.08	0.09	0.05
Total Average	96.59	97.81	96.98	97.47	98.34	97.22	97.35	97.60	91.80
TiO_2/K_2O	0.016	0.036	0.015	0.027	0.021	0.042	0.022	0.032	0.016
CaO/Al_2O_3	0.059	0.087	0.063	0.059	0.093	0.122	0.069	0.094	0.066
FeO*/MgO	7.16	6.80	18.39	9.54	10.57	6.83	23.55	8.66	20.85
Na_2O/K_2O	0.667	0.924	0.676	0.796	0.930	0.950	0.725	0.928	0.740

Source No.	9-1A	9-1B	9-2A	9-2B	10	11-1	11-2
Glass Composition Name	Shirataki-Akaishiyama-A	Shirataki-Akaishiyama-B	Shirataki-Tokachiishizawa-A	Shirataki-Tokachiishizawa-B	Engaru	Ikutahara-A	Ikutahara-B
No. of Samples	45	42	17	45	18	10	10
SiO_2	77.92	77.64	77.75	77.71	77.03	75.32	75.73
TiO_2	0.04	0.04	0.03	0.03	0.07	0.13	0.14
Al_2O_3	12.63	12.61	12.77	12.57	12.56	13.43	13.42
FeO*	0.41	0.71	0.51	0.76	1.12	1.39	1.00
MnO	0.05	0.05	0.06	0.06	0.06	0.04	0.05
MgO	0.02	0.02	0.01	0.01	0.03	0.06	0.03
CaO	0.54	0.54	0.47	0.43	0.72	0.80	0.67
Na_2O	3.72	3.74	3.83	3.79	4.26	4.01	3.93
K_2O	4.57	4.55	4.50	4.55	4.07	4.76	4.97
Cl	0.11	0.11	0.08	0.09	0.07	0.08	0.09
Total Average	99.59	99.74	99.93	99.62	96.12	97.86	99.41
TiO_2/K_2O	0.009	0.009	0.007	0.006	0.017	0.027	0.027
CaO/Al_2O_3	0.043	0.043	0.036	0.034	0.057	0.059	0.050
FeO*/MgO	21.65	46.84	42.48	54.40	43.67	23.01	32.62
Na_2O/K_2O	0.813	0.823	0.850	0.832	1.047	0.843	0.791

Source No.	12-1	12-2	13-1	13-2	14	15	16-1	16-2
Glass Composition Name	Rubeshibe-Iwayamanosawa	Rubeshibe-Kayokozawa	Oketo-Tokoroyama	Oketo-Oketoyama	Tokachi-Mitsumata	Shikaribetsu	Kushiro-Akan1	Kushiro-Akan2
No. of Samples	10	22	34	20	40	13	10	10
SiO_2	76.84	76.11	78.25	78.01	77.83	78.01	77.88	77.88
TiO_2	0.08	0.12	0.11	0.11	0.05	0.11	0.04	0.22
Al_2O_3	13.20	13.65	12.28	12.64	12.68	12.55	12.65	12.06
FeO*	1.11	1.12	0.75	0.74	0.71	0.85	0.62	1.83
MnO	0.06	0.05	0.05	0.05	0.05	0.04	0.05	0.09
MgO	0.08	0.10	0.08	0.06	0.04	0.07	0.03	0.16
CaO	1.25	1.51	0.86	0.97	0.71	1.08	0.63	1.32
Na_2O	3.50	3.63	3.42	3.99	3.67	3.42	3.94	4.35
K_2O	3.81	3.63	4.13	3.37	4.20	3.80	4.11	1.99
Cl	0.08	0.08	0.07	0.07	0.07	0.06	0.07	0.12
Total Average	98.03	97.65	98.22	99.11	97.96	98.58	99.02	98.78
TiO_2/K_2O	0.022	0.032	0.028	0.033	0.011	0.030	0.010	0.112
CaO/Al_2O_3	0.095	0.111	0.070	0.077	0.056	0.086	0.049	0.109
FeO*/MgO	14.37	11.12	9.44	13.00	19.55	12.85	18.32	11.67
Na_2O/K_2O	0.920	1.001	0.829	1.184	0.875	0.899	0.961	2.184

identifying samples plotted on the compositional boundary (Wada and Sano 2011).

The obsidian sources with more than one chemical group are also discriminated by the glass composition. Ikutahara obsidian is sub-divided into two groups: Ikutahara-A and -B. The Ikutahara-A group is higher in CaO/Al_2O_3 ratio and FeO* content than the Ikutahara-B group (Figures 5.8–5.9). Rubeshibe-Kayokozawa obsidian is higher in CaO or CaO/Al_2O_3 ratio than Rubeshibe-Iwayamanosawa obsidian (Figures 5.8–5.9). Oketo-Tokoroyama obsidian is separated from the Oketo-Oketoyama obsidian by its lower Na_2O/K_2O, TiO_2/K_2O, and CaO/Al_2O_3 ratios (Table 5.5, Figure 5.8).

As described above, it is impossible to separate Tokachi-Mitsumata source from the Kushiro-Akan 1 one using the whole-rock composition (Figure 5.4). They can, however, be distinguished on the basis of their glass composition. Tokachi-Mitsumata obsidian has a slightly higher CaO/Al_2O_3 ratio than Kusho-Akan 1 obsidian (Figure 5.8).

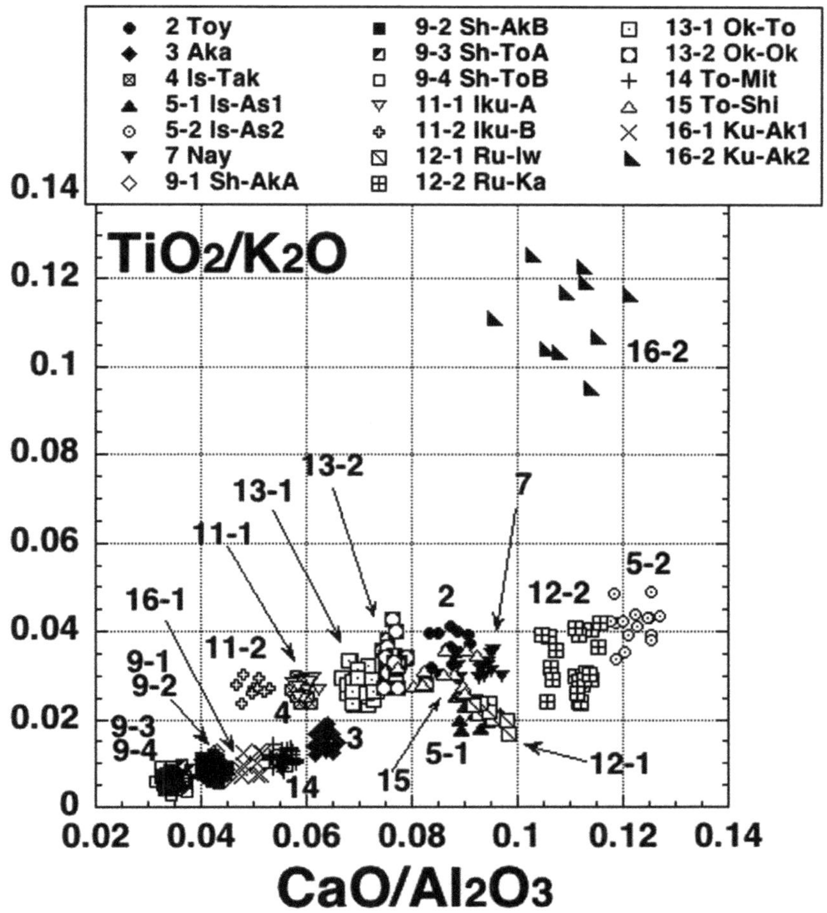

Figure 5.8. The CaO/Al_2O_3 vs. TiO_2/K_2O diagram of glass composition for the Hokkaido obsidian sources. Abbreviations are: Toy: Toyoura; Aka: Akaigawa; Is-Tak: Ishikarigawa-Takikawa; Is-As1: Ishikarigawa-Asahikawa 1; Is-As2: Ishikarigawa-Asahikawa 2; Nay: Nayoro; Sh-AkA: Shirataki-Akaishiyama A; Sh-AkB: Shirataki-Akaishiyama B; Sh-ToA: Shirataki-Tokachiishizawa A; Sh-ToB: Shirataki-Tokachiishizawa B; Iku-A: Ikutahara A; Iku-B: Ikutahara B; Ru-Iw: Rybeshibe-Iwayamanosawa; Ru-Ka: Rybeshibe-Kayokozawa; Ok-To: Oketo-Tokoroyama; Ok-Ok: Oketo-Oketoyama; To-Mit: Tokachi-Mitsumata; Shi: Shikaribetsu; Ku-Ak1: Kushiro-Akan 1; and Ku-Ak2: Kushiro-Akan 2. The Okushiri-Katsumayama, Oumu, Monbetsu-Kamimobetsu, and Engaru sources are not plotted because of space limitations

Application of Glass Composition to Identification of the Hokkaido Obsidian Sources

The source identification method for Hokkaido obsidian based on the chemical composition of glass is effective especially in the case of the destructive analysis for artefacts. The advantage is that we can analyse multiple samples at one time because more than 50 artefact slices can be included in one thin section for EPMA which can measure a micro-domain (Figures 5.7 and 5.10). We can set three thin sections into the sample holder of the EPMA system at one time.

In addition, the merits of the destructive analysis are: 1) we can remove the effects of compositional change caused by weathering, hydration, and burning of the artefacts on most surfaces by cutting and abrasion at the time of making the thin section; and 2) we can get clues from the characteristics of the microtexture and crystallinity of obsidian in a thin section by using the polarising microscope in transmission light mode. Furthermore, it is also effective for the non-destructive analysis, because we can analyse even very small portions of obsidian artefacts (< 10 microns) because of the use of a micron-scale electron beam.

CONCLUSIONS

We have revealed the geological and petrological features of obsidian sources in Hokkaido, along with new data for the K–Ar age of eight sources. For this study, the chemical compositions of obsidian — whole-rock and glass — were used.

In summary, the results are as follows: 1) We have established exact geological and petrological patterns for 16 Hokkaido obsidian sources; 2) The obsidian sources in

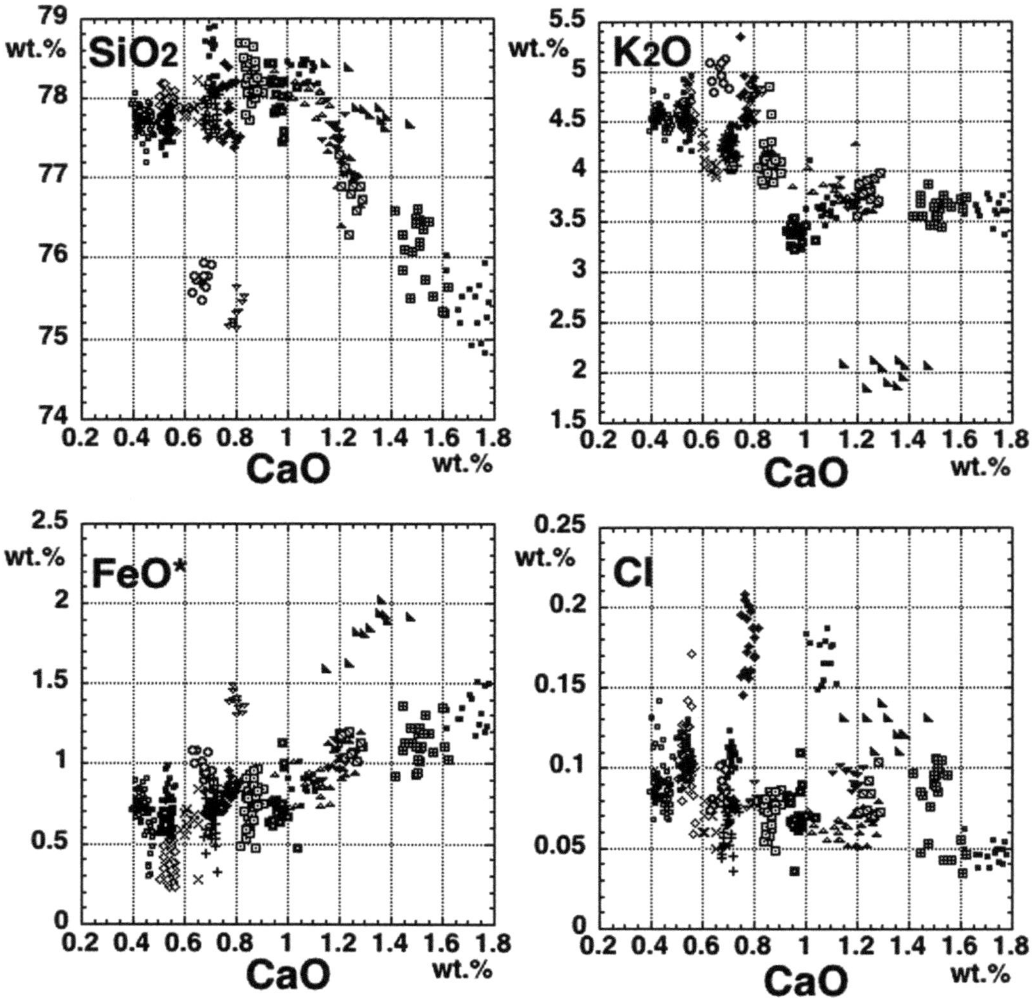

Figure 5.9. The CaO vs. SiO_2, K_2O, FeO, and Cl diagrams. FeO* shows the total iron content as FeO. Symbols are the same as in Figure 5.8. The Okushiri-Katsumayama, Oumu, Monbetsu-Kamimobetsu, and Engaru sources are not plotted because of space limitations*

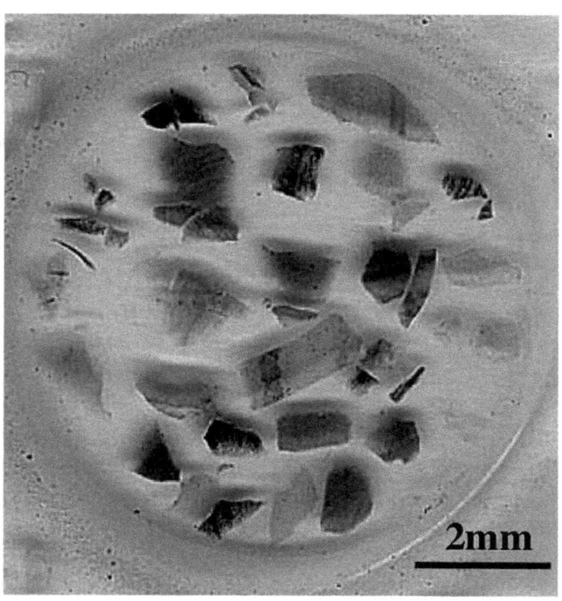

Figure 5.10. Obsidian artefacts with polished mirror surfaces embedded in epoxy resin

Hokkaido are concentrated in or near the Monbetsu-Kamishihoro graben in the northeastern part of the island; their formation zone moved from the north (at around 12 Ma ago) to the south (at around 1 Ma ago), which is almost consistent with volcanic activity under the tensional stress field in the upper crust of Hokkaido; 3) We have determined the 19 reference compositions (major and trace elements) of whole-rock for the 16 Hokkaido obsidian sources by XRF, using homogeneous glass beads; 4) We have established the standard composition for 24 volcanic glass groups in Hokkaido obsidian sources using EPMA. Combinations of scatter diagrams such as CaO/Al_2O_3 vs. TiO_2/K_2O are useful for identifying the source of obsidian artefacts. Shirataki obsidians are divided into four groups of glass composition in the CaO vs. FeO* diagram, corresponding to the respective rhyolite lava units; and 5) It is effective to use a polished thin section of obsidian in resin for source identification, because we can get reliable data of both glass composition and microscopic characteristics (such as crystallinity and magnetite morphology); only tiny portions of artefacts need to be destroyed for the thin sections.

Acknowledgements

This study was supported by Grants-in-Aid from the MEXT (No. 21242026 for H. Sato; and No. 19500866 for K. Wada). We are grateful to Prof. Mitsuhiro Nakagawa and Dr Akiko Matsumoto (Hokkaido University, Japan) for permission to use the XRF system and for technical support. We wish to thank Prof. A. Ono for giving us an opportunity to publish this article. We would like to thank Drs Y. V. Kuzmin, M. D. Glascock, and S. G. Keates, and Ms Candace C. Lindsey for editing and polishing the earlier version of this paper, and Dr La Fay Michelle (Hokkaido University of Education at Asahikawa) who corrected grammatical and other errors in the manuscript. Finally, many comments and suggestions from Drs E. Sato and Y. Suda helped to revise the original manuscript.

References

GLASCOCK, M., R. KUNSELMAN, and D. WOLFMAN. 1999. Intrasource Chemical Differentiation of Obsidian in the Jemez Mountains and Taos Plateau, New Mexico. *Journal of Archaeological Science* 26, 861–868.

GRATUZE, B. 1999. Obsidian Characterization by Laser Ablation ICP–MS and Its Application to Prehistoric Trade in the Mediterranean and the Near East: Sources and Distribution of Obsidian within the Aegean and Anatolia. *Journal of Archaeological Science* 26, 869-881.

HASEGAWA, K., M. NAKAGAWA, J. ITO, and T. YAMAMOTO. 2011. Depositional Ages of Quaternary Sequences in the Kushiro Region, Eastern Hokkaido, Japan: Correlations and Chronology on the Basis of High-Resolution Tephro-Stratigraphy. *Journal of the Geological Society of Japan* 117, 686–699 (in Japanese with English Abstract).

HIROSE, W., and M. NAKAGAWA. 1999. Neogene Volcanism in Central-Eastern Hokkaido: Beginning and Evolution of Arc Volcanism Inferred from Volcanological Parameters and Geochemistry. *Journal of the Geological Society of Japan* 105, 247–265 (in Japanese with English Abstract).

IMANISHI, S. 1956. *"Nayoro (Asahikawa-32)". Explanatory Text of the Geological Map of Japan (Scale 1:50,000).* Sapporo, Geological Survey of Hokkaido (in Japanese with English Abstract).

ISHII, E., M. NAKAGAWA, H. SAITO, and A. YAMAMOTO. 2008. The Pleistocene Tokachimitsumata Caldera and Associated Pyroclastic Flow Deposits in Central Hokkaido, Japan: Correlation of Large-Scale Pyroclastic Flow Deposits with Source Calderas. *Journal of the Geological Society of Japan* 114, 348–365 (in Japanese with English Abstract).

ITAYA, T., K. NAGAO, K. INOUE, Y. HONJOU, T. OKADA, and A. OGATA. 1991. Argon Isotope Analysis by a Newly Developed Mass Spectrometric System for K–Ar Dating. *Mineralogical Journal* 15, 203–221.

IZUHO, M., and H. SATO. 2007. Archaeological Obsidian Studies in Hokkaido, Japan: Retrospect and Prospects. *Bulletin of the Indo-Pacific Prehistory Association* 27, 114–121.

KANO, K., Y. YOSHIMURA, D. ISHIYAMA, G. ORTON, and T. OHGUCHI. 2006. Eruption Products and Structure of Katsuma-Yama Volcano, Okushiri Island, Hokkaido, Northern Japan. *Bulletin of the Volcanological Society of Japan* 51, 211–229 (in Japanese with English Abstract).

KIMURA, H. 2005. *Kita no Kokuyoseki no Michi: Shirataki Isekigun* [The Obsidian Road of the North: Shirataki Site Cluster]. Tokyo, Shinsensha Publishers.

KUZMIN, Y. V., M. D. GLASCOCK, and M. IZUHO. 2013. The Geochemistry of the Major Sources of Archaeological Obsidian on Hokkaido Island (Japan): Shirataki and Oketo. *Archaeometry* 55, 355–369.

KUZMIN, Y. V., M. D. GLASCOCK, and H. SATO. 2002. Sources of Archaeological Obsidian on Sakhalin Island (Russian Far East). *Journal of Archaeological Science* 29, 741–749.

MCBIRNEY, A. R. 1984. *Igneous Petrology.* San Francisco, Freeman.

MATSUNAMI, T., M. YAHATA, and K. MATSUSHITA. 2002. *"Kamishokotsu (Abashiri-15)". Explanatory Text of the Geological Map of Japan (Scale 1:50,000).* Sapporo, Geological Survey of Hokkaido (in Japanese with English Abstract).

MUKAI, M. 2003. Engaru-cho Shanabuchi Chiiki ni Bunpu suru Kokuyoseki no Tokucho [Character of the Obsidians Distribute to Shanabuchi Area, Engaru-cho, in Hokkaido]. *Asahikawa-shi Hakubutsukan Kenkyu Hokoku* 9, 27–30.

MUKAI, M. 2005. Monbetsu Chiiki Rubeshibe Chiiki Toyoura Chiiki kara Sanshutu suru Kokuyoseki Garasu no Kagaku Sosei [Chemical Compositions of the Obsidian Glasses from Monbetsu District, Rubeshibe District, and Toyoura District, Hokkaido, Japan]. *Asahikawa-shi Hakubutsukan Kenkyu Hokoku* 11, 9–20.

MUKAI, M. 2009. Kushiro Chiiki kara Sanshutu suru Kokuyoseki Garasu no Kagaku Sosei [Chemical Compositions of the Obsidian Glasses Produced from Kushiro Area, Hokkaido, Japan]. *Asahikawa-shi Hakubutsu Kagakukan Kenkyu Hokoku* 1, 27–38.

MUKAI, M. 2010. Hokkaido kara Sanshutu suru Kokuyoseki Garasu no Kagaku Sosei [Chemical Compositions of the Obsidian Glasses from Hokkaido, Japan]. *Asahikawa-shi Hakubutsu Kagakukan Kenkyu Hokoku* 2, 1–34.

MUKAI, M., H. HASEGAWA, and K. WADA. 2000. Asahikawa Shuhen Chiiki ni okeru Kokuyoseki Garasu no Kagaku Sosei: Kokuyoseki no Sanchi Tokutei eno Tekiyo [Chemical Compositions of the Obsidian Glasses from Asahikawa and Its Surrounding Area: an Application to the Specified Locality of the Obsidians]. *Asahikawa-shi Hakubutsukan Kenkyu Hokoku* 6, 51–64.

MUKAI, M., R. SHIBUYA, and K. WADA. 2004. Ikutahara Chiiki kara Sanshutu suru Kokuyoseki Garasu no Kagaku Sosei [Chemical Compositions of the Obsidian Glasses from Ikutahara District, Hokkaido, Japan]. *Asahikawa-shi Hakubutsukan Kenkyu Hokoku* 10, 35–40.

MUKAI, M., and K. WADA. 2002. Hokkaido Chuoubu Asahikawa Chiiki ni Sanshutu suru Kokuyoseki no K–Ar Nendai [K–Ar Age of the Obsidian from Asahikawa District in Central Hokkaido]. *Asahikawa-shi Hakubutsukan Kenkyu Hokoku* 8, 59–61.

MUKAI, M., and K. WADA. 2003. Engaru Chiiki Oumu Chiiki kara Sanshutu suru Kokuyoseki Garasu no Kagaku Sosei [Chemical Compositions of the Obsidian Glasses from Engaru District and Oumu District, Hokkaido, Japan]. *Asahikawa-shi Hakubutsukan Kenkyu Hokoku* 9, 19–26.

MUKAI, M., K. WADA, and C. OKURA. 2002. Oketo Chiiki Akaigawa Chiiki kara Sanshutu suru Kokuyoseki Garasu no Kagaku Sosei [Chemical Compositions of the Obsidian Glasses from Oketo District and Akaigawa District, Hokkaido, Japan]. *Asahikawa-shi Hakubutsukan Kenkyu Hokoku* 8, 47–58.

NAGAO, K., H. NISHIDO, T. ITAYA, and T. OGATA. 1984. K–Ar Age Determination Method. *Hiruzen Kenkyujo Kenkyu Hokoku* 9, 19–38 (in Japanese with English Abstract).

OKA, T., T. KOSHIMIZU, and T. FUKUOKA. 1991. Hokkaido Hokubu Nayoro Bonchi no Bifuka Sogun to Kawanishi So no Nendai ni Tsuite [Geological Age of the Bifuka Group and Kawanishi Formation in the Nayoro Basin, Northern Hokkaido]. *Nihon Chishitsu Gakkai Gakujutu Taikai Koen Yoshi* 98, 172.

PHILLIPS, S. C., and R. J. SPEAKMAN. 2009. Initial Source Evaluation of Archaeological Obsidian from the Kuril Islands of the Russian Far East Using Portable XRF. *Journal of Archaeological Science* 36, 1256–1263.

SATO, H., Y. V. KUZMIN, and M. D. GLASCOCK. 2002. Source Analysis of Obsidian in Prehistoric Sakhalin and an Assessment of Its Distribution in Northeast Asia. *Hokkaido Kokogaku*, 38, 1–13 (in Japanese with English Abstract).

SAWAMURA, K., and M. MINATO. 1965. *"Rubeshibe (Abashiri-46)". Explanatory Text of the Geological Map of Japan (Scale 1:50,000).* Sapporo, Geological Survey of Hokkaido (in Japanese with English Abstract).

SHACKLEY, M. S. 2005. *Obsidian: Geology and Archaeology in the North American Southwest.* Tucson, University of Arizona Press.

STEIGER, R., and E. JAGER. 1977. Subcommission on Geochronology, Convention on the Use of Decay Constants in Geo- and Cosmo-Chronology. *Earth and Planetary Science Letters* 36, 359–362.

SUZUKI, M., M. KONOYA, and T. FUJIWARA. 1966. *"Omu (Abashiri-6)". Explanatory Text of the Geological Map of Japan (Scale 1:50,000).* Sapporo, Geological Survey of Hokkaido (in Japanese with English Abstract).

TAJIKA, J., and M. YAHATA. 1991. *"Engaru (Abashiri-24)". Explanatory Text of the Geological Map of Japan (Scale 1:50,000).* Sapporo, Geological Survey of Hokkaido (in Japanese with English Abstract).

TYKOT, R. 1997. Characterization of the Monte Arci (Sardinia) Obsidian Sources. *Journal of Archaeological Science* 24, 467–479.

WADA, K. 2009. Petrologic Model of Shirataki Obsidian, Northern Hokkaido, Japan; Its Structure, Composition, and the Origin. In *2009 Portland GSA Annual Meeting. Abstracts with Programs. Vol. 41, No. 7,* 553. Boulder, CO, Geological Society of America.

WADA, K., M. MUKAI, and O. TAKEDA. 2003. Chemical Compositions of Obsidian Glasses Determined by EPMA: the Origin of Obsidians from the Remains at the Mouth of Tokoro River, Eastern Hokkaido. *Hokkaido Kyoiku Daigaku Taisetsusan Shizen Kyoiku Kenkyu Shisetsu Hokoku* 37, 59–70 (in Japanese with English Abstract).

WADA, K., and K. SANO. 2011. Chemical Composition and Microstructure of Shirataki Obsidian, Northern Hokkaido: Geological and Petrological Source Data for the Precise Obsidian Source Identification. *Kyusekki Kenkyu* 7, 57–73 (in Japanese with English Abstract).

YAHATA, M. 1997. Neogene Monbetsu-Kamishihoro Graben in the Northeast Hokkaido Metallogenic Province, Japan. *Chika Shigen Chosajo Hokoku* 68, 43–56 (in Japanese with English Abstract).

YAHATA, M., and H. NISHIDO. 1995. Neogene Volcanism and Tectonics in the Monbetsu–Engaru District in the Northeastern part of Central Hokkaido, Japan. *Journal of the Geological Society of Japan* 101, 685–704 (in Japanese with English Abstract).

YAHATA, M., J. TAJIKA, K. KUROSAWA, and T. MATSUNAMI. 1988. *"Maruseppu Hokubu (Abashiri-23)". Explanatory Text of the Geological Map of Japan (Scale 1:50,000).* Sapporo, Geological Survey of Hokkaido (in Japanese with English Abstract).

YAMADA, K., Y. TERAOKA, and M. ISHIDA. 1964. *"Ikutahara (Abashiri-35)" Explanatory Text of the Geological Map of Japan (Scale 1:50,000).* Sapporo, Geological Survey of Hokkaido (in Japanese with English Abstract).

YOKOYAMA, H., M. YAHATA, S. OKAMURA, and H. NISHIDO. 2003. Volcano-Stratigraphy and Geological Development of Akaigawa Caldera in Southwest Hokkaido, Japan. *Ganseki Kobutsu Kagaku* 32, 80–95 (in Japanese with English Abstract).

Chapter 6
THE NEUTRON ACTIVATION ANALYSIS OF VOLCANIC GLASSES IN THE RUSSIAN FAR EAST AND NEIGHBOURING NORTHEAST ASIA: A SUMMARY OF THE FIRST 20 YEARS OF RESEARCH

Yaroslav V. KUZMIN and Michael D. GLASCOCK

Abstract: *The results from Neutron Activation Analysis (NAA) of obsidian and other volcanic glasses from the Russian Far East conducted in the 1990s and 2000s are presented in an abridged form. Analytical work was performed at the Archaeometry Laboratory, University of Missouri Research Reactor Center, following the methodology of data acquisition and interpretation described in Glascock et al. (1998). As a result, secure information on the geochemical compositions of the major obsidian sources in the Russian Far East and adjacent Northeast Asia (Basaltic and Obluchie plateaus, Paektusan, Shirataki, and Oketo) was obtained. It is methodologically important that for the correlation of primary sources and archaeological assemblages the analysis of both kinds of obsidian ('geological' and 'archaeological') should be conducted by the same analytical technique, as it was done by our group. In this case, the geochemical data are fully comparable. The chemical compositions of obsidian from primary sources reported by NAA analysis may not necessarily be compatible with results obtained by other analytical methods. An example of possible misinterpretation is presented. The standardisation of analyses and cross-checking of the same samples in several laboratories by different analytical methods should be performed regularly. This is ongoing research for a selection of obsidian specimens obtained from the Hokkaido sources in 2011.*

Keywords: *Obsidian, Standardisation, Provenance Studies, Neutron Activation Analysis, the Russian Far East, Northeast Asia*

INTRODUCTION

In the course of the major themes of this volume, provenance of obsidian artefacts and the standardisation of data on obsidian sources in Northeast Asia, it is necessary to summarise the results of previous studies. The Russian Far East, comprising the Primorye [Maritime] Province, Amur River basin, Sakhalin Island, and the Kurile Islands (Figure 6.1), today is an important region for obsidian studies (see recent summaries: Kuzmin 2010, 2011; see also Kuzmin, this volume). In this chapter, a short review of the existing data on the geochemistry of volcanic glasses from this region is presented, with reference to original publications when applicable.

That obsidian exists in the southern Russian Far East has been known to geologists and archaeologists since the late nineteenth – early twentieth centuries (see Kuzmin, this volume). However, scientific provenance studies of obsidian artefacts did not begin until the 1980s. The first attempt to analyse obsidian from archaeological sites in the Primorye Province and the Amur River basin by instrumental geochemical methods was conducted in the mid-1980s by Dr Masao Suzuki, Rikkyo [St. Paul] University, in Tokyo (Japan). However, the limited number of elements measured (Sm, U, Th, Hf, S, Fe, and La; see Kuzmin and Popov 2000, 158–9) and the lack of proper interpretation and publication of these results (see brief information: Tabarev and Kuzmin 2000, 10–1; Vasilievsky and Gladyshev 1989, 101–2) have led to conclusions which today look awkward. For example, it was suggested that obsidian at the Upper Palaeolithic site of Ustinovka 1 in Primorye (Figure 6.1) is similar to the Paektusan source due to the high content of iron and low content of uranium (see Vasilievsky and Gladyshev 1989, 102). Our studies show that obsidian from the Zerkalnaya River basin in central Primorye, including the Ustinovka 1 site, belongs to the Basaltic Plateau source located much closer than Paektusan (e.g., Kuzmin *et al.* 2002a, 508).

The situation changed significantly when our study of the far eastern Russian obsidian was initiated in 1991–2. From the beginning, it was based on modern methodological principles (see Glascock *et al.* 1998). The first analytical data on the 'archaeological' obsidian (i.e., artefacts from prehistoric complexes) were obtained and initially interpreted in 1994, and delivered that year as an oral communication at the International Conference "Archaeology of the Northern Pacific" (Vladivostok, Russia). In 1996, another presentation was given as a poster at the International Symposium on Archaeometry (Urbana-Champaign, IL, USA). Publication of these reports soon followed (Glascock *et al.* 1996; Shackley *et al.* 1996). The first data on 'geological' obsidian (i.e., derived from the primary sources) were published during the following year (Popov and Shackley 1997).

The first comprehensive volume, containing information on both geological and archaeological obsidians from the Russian Far East, was released a few years later (Kuzmin and Popov 2000). However, it was written mainly in the Russian language, and the distribution of the book was quite limited. This is why data originally presented in this volume were published in the early 2000s in international journals (see Kuzmin *et al.* 2002a, 2002b), and detailed studies were released later (see Glascock *et al.* 2011; Kuzmin and Glascock 2007; Kuzmin *et al.* 2013). A second volume (see Kuzmin and Glascock 2010), based on the Symposium "Crossing the Straits: Prehistoric Obsidian Source Exploitation in the Pacific Rim" held in

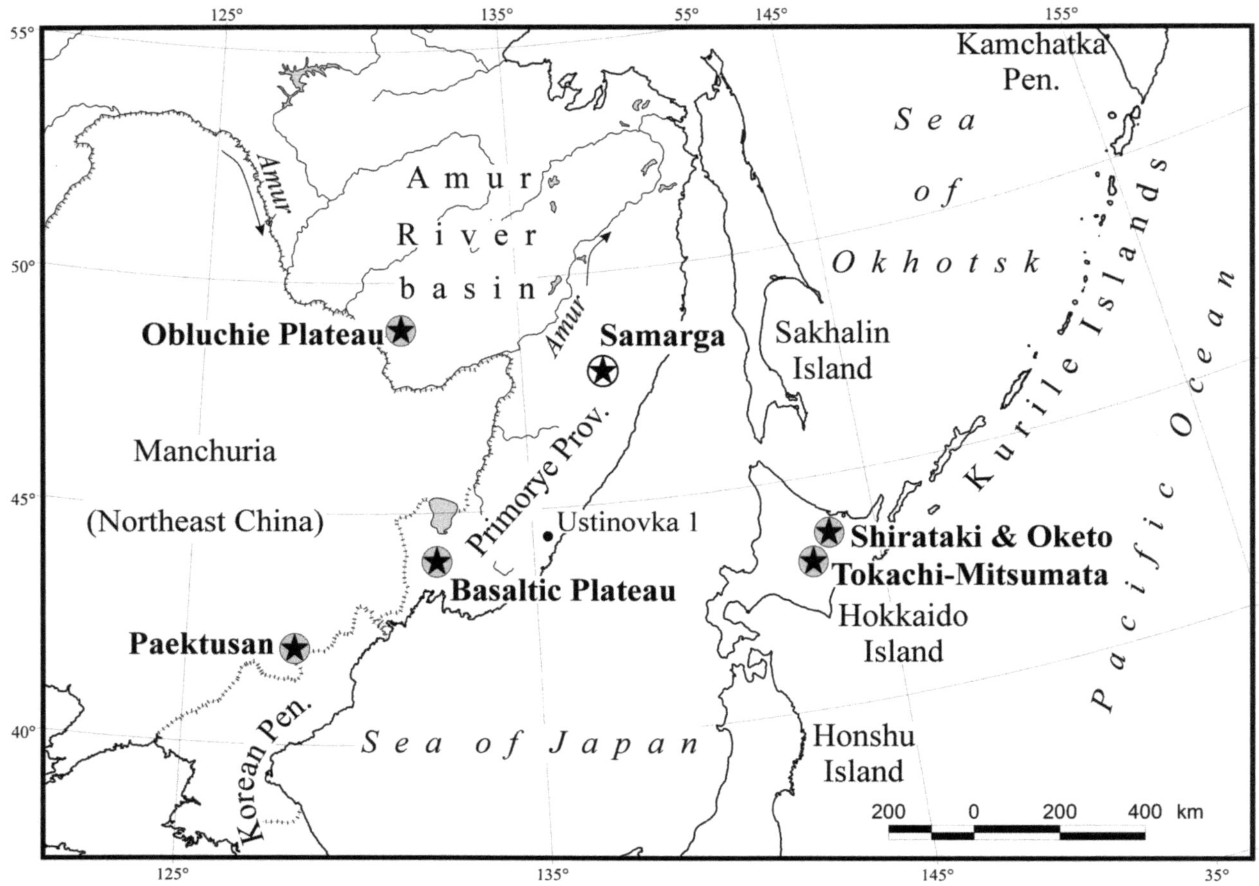

Figure 6.1. Main primary obsidian sources in mainland Northeast Asia and adjacent regions

2005 at the 70th Annual Meeting of the Society for American Archaeology (Salt Lake City, UT, USA), put our studies in a wider geographic context, touching upon not only the Russian Far East but also the Japanese Islands, Korean Peninsula, and Northeast China (Manchuria).

In the early 2000s, it became clear that some of the major obsidian sources for prehistoric complexes of the Russian Far East are situated in neighbouring Northeast Asia: at the Paektusan Volcano on the modern border between China and North Korea, and in Hokkaido Island of Japan (e.g., Kuzmin *et al.* 2002a, 2002b). Three research trips were undertaken to Chinese side of the Paektusan Volcano in 2002, 2004, and 2007; and two trips to Hokkaido obsidian sources in 1999 and 2003. The results of studies based on samples collected during these fieldwork campaigns have been published (see Kim, this volume; Kuzmin and Glascock 2007; Kuzmin *et al.* 2002b, 2013; Popov *et al.* 2005, 2008).

MATERIAL AND METHODS

The major sources of obsidian in mainland Northeast Asia, including the Russian Far East, Korean Peninsula, and Manchuria, are: 1) the Basaltic Plateau in southern Primorye; 2) the Obluchie Plateau in the middle course of the Amur River; and 3) the Paektusan Volcano on the border of China and North Korea. Another set of important obsidian sources, Shirataki and Oketo, are located in Hokkaido Island (Figure 6.1). Both the Basaltic Plateau and Obluchie Plateau contain high quality volcanic glass in hyaloclastites ("An aggregate of fine, glassy debris formed by the sudden contact of hot, coherent magma and either cold water or water-saturated sediment."; see Allaby and Allaby 2003, 268) and pillow lavas, both within the basaltic–andesitic lava flows and/or on the contact between lava and bedrock (see Glascock *et al.* 2011; Popov *et al.* 2009). This is the unique feature of the mainland Russian Far East where two important volcanic glass sources, Basaltic and Obluchie plateaus, originated in basic rocks, while most of the obsidian sources in the Pacific Rim region belong to acidic rocks in island arc settings. Obsidian at both the Shirataki and Oketo sources is related to acidic volcanic rocks, mainly rhyolites (see Izuho and Hirose 2010; Kuzmin *et al.* 2013). The Paektusan Volcano source has a complex structure, and its obsidian is related to alkali rocks, mainly trachytes and rhyolites (see Popov *et al.* 2005, 2008).

Obsidian is widely distributed in prehistoric cultural complexes of the Russian Far East. It was used as a valuable raw material in the Upper Palaeolithic (dated to ca. 19,000–12,000 BP) everywhere in the region, besides the Kurile Islands where reliable Palaeolithic sites are unknown. In the following Neolithic period (ca. 12,000–

Table 6.1. Obsidian samples from the Russian Far East and neighbouring Northeast Asia analysed by NAA in 1992–2011 (after Glascock *et al.* 2011; Kuzmin and Glascock 2007; Kuzmin and Popov 2000; Kuzmin *et al.* 2002a, 2002b, 2013; Popov *et al.* 2005; see also Kuzmin, this volume)

Region	Geological Samples		Archaeological Samples	
	Long-Irradiation	Short-Irradiation	Long-Irradiation	Short-Irradiation
Primorye Province	102	—	94	23
The Amur River basin	12	—	18	10
Sakhalin Island	—	—	113	62
Paektusan Volcano	14	—	36	13
Hokkaido Island	53	—	—	—
Subtotal	181	—	261	108
Total	181		369	
Grand Total	550			

3000 BP), obsidian was common throughout all of Northeast Asia (see Kuzmin, this volume). During the Palaeometal epoch, comprising Bronze and Early Iron ages in the Russian Far East and dated to ca. 3000–1500 BP, obsidian was occasionally used as a raw material but its importance significantly declined due to the introduction of metals. Only in the most remote part of the Russian Far East — the Kurile Islands — was obsidian widely used after the end of the Neolithic at ca. 2500–2000 BP (Phillips 2010; see chronology of the prehistoric Kuriles: Kuzmin *et al.* 2012 and references therein).

The Neutron Activation Analysis (hereafter – NAA) of volcanic glasses from the Russian Far East and neighbouring Northeast Asia, used in this volume as a commodity for studying prehistoric migrations and exchange, was performed at the Archaeometry Laboratory, University of Missouri Research Reactor Center (henceforth – MURR) in Columbia, Missouri (USA). The methodology used for the NAA obsidian has already been described in several of our papers (e.g., Glascock *et al.* 1998; Glascock *et al.* 2007, 346; Glascock *et al.* 2011, 1835–1836), and also in handbooks (e.g., Malainey 2011; Pollard *et al.* 2007), and we refer to them for details. The advantages of NAA are: 1) determination of a large number of chemical elements (up to 30), including the rare-earth elements which are important for separation of obsidian sources; 2) use of a single reference material for all analyses enables the generation of highly precise and accurate data; and 3) the possibility of analysing small sample sizes (e.g., 5–20 mg). The statistical analysis of results obtained was accomplished with the help of software written in GAUSS language; more details can be found in Glascock *et al.* (1998).

In the absence of reliable geochemical data for far eastern Russian obsidian prior to the early 1990s, it was first necessary to establish geochemical signatures for all of the important sources. NAA allows one to determine the content of up to 28 elements in most obsidians when using both short- and long-irradiation procedures. This was performed mainly on geological obsidians, and also on archaeological specimens in the regions where primary sources are either unknown or non-existent (e.g., Sakhalin Island). As a result, the geochemical compositions of primary obsidian sources were securely established. Afterwards, the determination of a reduced number of elements using short-irradiation NAA is usually successful, because the geochemical signatures of primary sources are already known. Thus, the content of seven elements was measured. The option of performing the long-irradiation is only necessary when the short-irradiation results are inadequate to differentiate between similar sources. Thus far, the total number of obsidian samples analysed by NAA (both short- and long-irradiations) from the region is 550, including 181 geological and 369 archaeological specimens. The distribution of samples analysed by region is shown in Table 6.1.

The standard reference material (SRM) used for all obsidian analyses performed at the Archaeometry Laboratory at MURR is SRM-278 Obsidian Rock issued by the National Institute of Standards and Technology (NIST; located in Gaithersburg, MD, USA). This SRM is ideal for NAA measurements on obsidian artefacts and geological materials because the concentrations for more than 30 elements have been certified by NIST (SRM-278 Obsidian Rock Certificate of Analysis 1992; Graham *et al.* 1982). Use of a single SRM for all obsidian analyses has the advantage of producing highly precise and accurate data, and this facilitates reliable comparisons between samples analysed at different times and between samples analysed in different NAA laboratories.

RESULTS AND DISCUSSION

The geochemical compositions for the major obsidian sources in the Russian Far East and neighbouring Northeast Asia are given in Table 6.2. These are the results of systematic studies conducted by our team over the past 20 years (see Glascock *et al.* 2011; Kuzmin and Glascock 2007; Kuzmin *et al.* 2002a, 2002b, 2013;

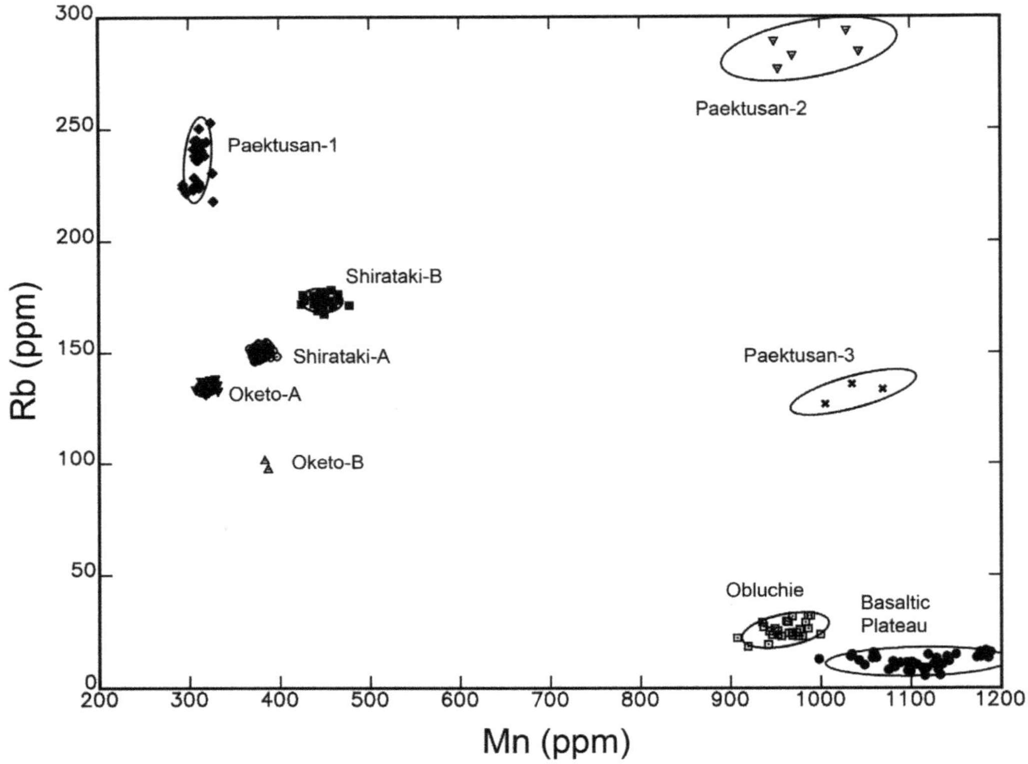

Figure 6.2. Bivariative plot (Mn vs. Rb) showing geochemical groups of northeast Asian obsidian (ellipses indicate 90% confidence level)

Popov *et al.* 2005). Measurement of the content of a large number of elements allowed us to demonstrate secure discriminations between obsidian sources with a high degree of confidence (Figure 6.2). These data can from now on be used as a basis for the identification of obsidian sources in archaeological collections from Northeast Asia. The geochemical compositions of obsidian from other sources in Northeast Asia not as widely used by prehistoric people are given elsewhere (see Glascock *et al.* 2011; Kuzmin and Glascock 2007; Kuzmin *et al.* 2002a, 2002b).

The issue of exact localisation of three volcanic glass sub-sources at the Paektusan Volcano (see Table 6.2, Figure 6.2) is still not completely solved (see also Ono *et al.*, this volume). While the position of Paektusan-2 [PNK-2 in Popov *et al.* (2005)] and Paektusan-3 [PNK-3 in Popov *et al.* (2005)] sources can now be securely pinpointed on the northern part of the volcano's caldera which administratively belongs to China (see Popov *et al.* 2005), none of the samples collected by our group on the Chinese side of Paektusan in 2002 and 2007 can be associated with the Paektusan-1 [PNK-1 in Popov *et al.* (2005)] group (see also Kim, this volume). It has been suggested that the exact location of the primary source of the Paektusan-1 obsidian is in the southern part of the caldera under North Korean control (Popov *et al.* 2005, 86), but we were not able to collect reliable geological samples for this geochemical group due to current political obstacles. Preliminary results of our 2007 trip to the Paektusan Volcano as well as additional information about the composition of obsidian boulders from North Korea found not far from Paektusan (see Kim, this volume) gives us hope that the exact location of the Paektusan-1 source will be determined in the near future.

In Hokkaido Island, the two major source clusters have been studied, Shirataki and Oketo (Figure 6.1). Both have two sub-sources that differ in their geochemistry (Table 6.2, Figure 6.2). This information was unknown before when the Hokkaido obsidian was analysed by NAA and X-ray Fluorescence (hereafter – XRF) techniques (Hall and Kimura 2002; Kuzmin *et al.* 2002b; see also Ferguson *et al.*, this volume; and Wada *et al.*, this volume). A more comprehensive report on these sources was recently published (Kuzmin *et al.* 2013). Samples from one of the Shirataki localities, previously analysed by XRF (Hall and Kimura 2002), were also tested by NAA, and the results are generally compatible (see Kuzmin *et al.* 2013).

The important methodological point of our research is that all samples used for geochemical studies (Table 6.2) were measured in the same laboratory (MURR), using the same analytical standards and equipment. This made possible direct comparison of the data obtained, in order to check the presence or absence of obsidian from a particular source in the given region. For example, the common belief about the transport of obsidian from the Hokkaido sources to the mainland Russian Far East existed in the 1980–90s was found to be unsubstantiated after comparison of the geochemical data for the Hokkaido sources and obsidian artefacts from the Amur

Table 6.2. Means and standard deviations in parts-per-million (ppm, unless otherwise indicated) of element concentrations for obsidian sources in the Russian Far East and neighbouring Northeast Asia measured by NAA; number of samples analysed is shown in brackets

Source	**Basaltic Plateau**	**Obluchie Plateau**	**Paektusan-1**	**Paektusan-2**	**Paektusan-3**
Element	(n = 40)	(n = 26)	(n = 38)	(n = 5)	(n = 3)
Na (%)	2.36 ± 0.10	3.00 ± 0.13	3.07 ± 0.09	3.69 ± 0.17	4.17 ± 0.11
Al (%)	7.90 ± 0.39	8.04 ± 0.29	6.76 ± 0.48	5.74 ± 0.63	8.19 ± 0.25
Cl	91 ± 38	81 ± 24	703 ± 87	1043 ± 48	429 ± 53
K (%)	0.41 ± 0.14	0.96 ± 0.16	4.19 ± 0.28	3.99 ± 0.29	4.51 ± 0.08
Sc	17.9 ± 0.9	12.3 ± 0.6	1.1 ± 0.1	1.6 ± 0.1	4.7 ± 0.7
Mn	1107 ± 47	960 ± 22	310 ± 7	989 ± 44	1036 ± 32
Fe (%)	7.22 ± 0.24	6.54 ± 0.17	1.08 ± 0.02	2.96 ± 0.10	3.70 ± 0.06
Co	37.6 ± 1.3	31.8 ± 1.1	0.3 ± 0.1	0.2 ± 0.2	1.2 ± 0.3
Zn	126 ± 21	132 ± 8	111 ± 138	245 ± 9	139 ± 15
Rb	11.6 ± 2.9	25.4 ± 3.6	236 ± 9	285 ± 6	132 ± 5
Sr	390 ± 93	658 ± 153	28 ± 6	n.d.	n.d.
Zr	96 ± 20	129 ± 17	251 ± 12	1247 ± 37	506 ± 19
Sb	n.d.*	0.03 ± 0.01	0.37 ± 0.03	0.27 ± 0.03	0.15 ± 0.01
Cs	0.24 ± 0.07	0.29 ± 0.08	3.90 ± 0.16	4.01 ± 0.10	1.40 ± 0.03
Ba	121 ± 29	375 ± 50	102 ± 36	56 ± 23	61 ± 26
La	6.35 ± 1.07	16.3 ± 1.6	67.7 ± 1.6	167 ± 5	80.4 ± 4.5
Ce	14.3 ± 2.0	33.6 ± 2.6	137 ± 4	327 ± 11	155 ± 7
Nd	9.0 ± 2.0	18.5 ± 0.8	49 ± 5	134 ± 14	65 ± 2
Sm	3.71 ± 0.29	5.11 ± 0.17	10.9 ± 0.5	22.7 ± 1.4	11.9 ± 0.3
Eu	1.47 ± 0.07	1.66 ± 0.08	0.27 ± 0.06	0.31 ± 0.01	0.51 ± 0.12
Tb	0.85 ± 0.27	0.66 ± 0.05	1.62 ± 0.12	2.91 ± 0.05	1.48 ± 0.01
Dy	3.86 ± 0.40	3.39 ± 0.45	10.3 ± 0.8	18.7 ± 1.2	8.07 ± 0.39
Yb	1.34 ± 0.1	1.15 ± 0.06	4.55 ± 0.34	8.63 ± 0.48	4.05 ± 0.48
Lu	0.26 ± 0.05	0.17 ± 0.02	0.72 ± 0.06	1.23 ± 0.03	0.56 ± 0.03
Hf	2.28 ± 0.17	3.44 ± 0.26	10.0 ± 0.3	35.1 ± 1.0	14.6 ± 0.3
Ta	0.29 ± 0.08	0.67 ± 0.11	6.77 ± 0.42	8.88 ± 0.26	4.28 ± 0.10
Th	0.77 ± 0.19	1.44 ± 0.17	27.5 ± 0.8	29.2 ± 0.8	12.7 ± 0.4
U	n.d.	0.46 ± 0.21	7.1 ± 3.0	9.8 ± 3.5	5.19 ± 0.46

Source	**Shirataki-A**	**Shirataki-B**	**Oketo-A**	**Oketo-B**
Element	(n = 59)	(n = 37)	(n = 26)	(n = 2)
Na (%)	2.87 ± 0.05	2.94 ± 0.06	2.78 ± 0.06	3.25 ± 0.04
Al (%)	6.86 ± 0.31	6.88 ± 0.3	6.72 ± 0.32	6.94 ± 0.18
Cl	561 ± 118	517 ± 88	464 ± 81	449 ± 130
K (%)	3.79 ± 0.24	3.81 ± 0.18	3.40 ± 0.19	3.05 ± 0.18
Sc	2.7 ± 0.1	2.9 ± 0.1	3.3 ± 0.1	3.3 ± 0.1
Mn	379 ± 6	447 ± 11	319 ± 6	385 ± 3
Fe (%)	0.79 ± 0.01	0.74 ± 0.01	0.73 ± 0.01	0.89 ± 0.03
Co	0.2 ± 0.2	0.1 ± 0.1	0.5 ± 0.1	0.5 ± 0.1
Zn	44 ± 40	39 ± 15	30 ± 4	37 ± 1
Rb	150 ± 2	174 ± 2	134 ± 2	99 ± 3
Sr	28 ± 8	n.d.	73 ± 23	79 ± 37
Zr	91 ± 9	87 ± 16	119 ± 9	128 ± 2
Sb	0.29 ± 0.04	0.35 ± 0.04	0.26 ± 0.06	0.40 ± 0.02

Source	**Shirataki-A**	**Shirataki-B**	**Oketo-A**	**Oketo-B**
Element	(n = 59)	(n = 37)	(n = 26)	(n = 2)
Cs	9.54 ± 0.11	11.8 ± 0.2	6.77 ± 0.09	5.34 ± 0.07
Ba	850 ± 17	192 ± 18	989 ± 16	722 ± 10
La	20.0 ± 0.3	13.3 ± 0.3	22.2 ± 0.4	21.0 ± 0.4
Ce	42.5 ± 0.7	31.34 ± 0.7	43.6 ± 0.6	41.8 ± 1.2
Nd	15.7 ± 1.8	12.2 ± 1.1	15.0 ± 1.6	15.5 ± 0.4
Sm	3.99 ± 0.06	3.83 ± 0.08	3.39 ± 0.07	3.24 ± 0.06
Eu	0.28 ± 0.01	0.13 ± 0.01	0.37 ± 0.01	0.54 ± 0.01
Tb	0.65 ± 0.03	0.73 ± 0.03	0.51 ± 0.02	0.50 ± 0.04
Dy	4.39 ± 0.32	5.08 ± 0.42	3.49 ± 0.40	3.35 ± 0.18
Yb	2.98 ± 0.16	3.55 ± 0.14	2.61 ± 0.16	2.64 ± 0.10
Lu	0.45 ± 0.01	0.53 ± 0.01	0.42 ± 0.01	0.39 ± 0.01
Hf	2.78 ± 0.09	2.69 ± 0.08	3.18 ± 0.05	3.66 ± 0.13
Ta	0.54 ± 0.01	0.66 ± 0.01	0.57 ± 0.01	0.52 ± 0.01
Th	11.1 ± 0.1	9.7 ± 0.2	11.9 ± 0.2	9.3 ± 0.2
U	3.19 ± 0.25	3.72 ± 0.31	3.40 ± 0.24	2.57 ± 0.28

* not detected.

River basin and Primorye (e.g., Kuzmin 2010, 148–9). Only very recently, obsidian from the Hokkaido source of Shirataki was identified at one Neolithic site in the lower course of the Amur River (see Glascock *et al.* 2011; Kuzmin *et al.* 2013). The use of geochemical signatures for obsidian sources in the Amur River basin and Primorye allowed us to identify volcanic glass from a still unknown source called 'Samarga' (Kuzmin *et al.* 2002a), which is located presumably in northern Primorye (see Figure 6.1), at some archaeological sites in the Lower Amur River basin (Glascock *et al.* 2011; Kuzmin 2010; see also Kuzmin, this volume). These two examples show the power of our Database which now should be used by the scientific community for enhancement of obsidian source studies in Northeast Asia.

It seems that in some cases scholars assume that their geochemical data generated by methods different from NAA can be used for direct comparison with the NAA results. This is the case with obsidian from archaeological collections recovered in the Kurile Islands. An initial study of 131 artefacts, based on the determination of ten elements (K, Mn, Fe, Ga, Th, Rb, Sr, Y, Zr, and Nb) by a portable variant of XRF, allowed the identification of several primary obsidian sources (see Phillips and Speakman 2009). It was assumed that '...bulk compositional data generated for obsidian by XRF analysis are comparable to data generated by NAA and LA-ICP-MS' (Phillips and Speakman 2009, 1258), with reference to several papers where this issue is discussed. In a subsequent study (Phillips 2010), more obsidian artefacts were selected (total of 774 samples), and the analytical technique used was the Laser Ablation – Inductive Couple Plasma Mass-Spectrometry (henceforth – LA–ICP–MS). Phillips (2010, 126) states 'At the beginning and end of each daily analysis session, a series of blanks, the NIST SRM 612 glass standard, and six well-characterised obsidian samples that previously had been analysed by Neutron Activation Analysis (NAA) were run to develop a set of calibration parameters and to monitor instrumental drift of the ICP-MS, which can affect count rates over several hours and/or days.' The results of this extended study were very close to the initial conclusions (see Phillips 2010).

The latest research of archaeological obsidian from the Kurile Islands is based on analysis of 1433 artefacts conducted by two methods, the LA–ICP–MS (most of samples) and the portable XRF (PXRF) (Phillips 2011). As far as we understand, this is essentially the same dataset as the one used in Phillips (2010). However, the main conclusions by Phillips (2011) are not the same as in Phillips (2010). The most striking difference between these two publications is the alleged wide use of the Tokachimisumata source [Tokachi-Mitsumata in Kuzmin and Glascock (2007); Kuzmin *et al.* (2002b)] in the Epi-Jomon and Okhotsk cultural complexes of the Kuriles (see Phillips 2011). The exploitation of obsidian from the Tokachi-Mitsumata source was not detected previously (see Phillips 2010; Phillips and Speakman 2009). It is unclear to us why the results of Phillips (2011) are so different from those of Phillips (2010). Unfortunately, the analytical details for data used by Phillips (2011) were not provided, besides the content of Rb, Sr, and Zr (see Phillips 2011, 127–33), and we can only guess at this stage about the cause of the discrepancy. Perhaps, it would be wise to run some of the Phillips' (2011) samples by NAA and see how their composition is related to data generated by both the PXRF and LA–ICP–MS methods. Until then, the results of obsidian sourcing on the Kuriles put forward by Phillips (2011) should not be accepted at face value (Kuzmin *et al.* 2013; see also Kuzmin, this volume).

The issue of data comparison is directly related to one of the main topics of the 2011 Symposium in Nagano

Prefecture and this volume, that is the standardisation of data on obsidian sources in Northeast Asia. This is especially important for sources which were repeatedly sampled by researchers from several countries, with different analytical methods used without inter-comparison of the results obtained. Obsidian sources from Hokkaido Island and the Paektusan Volcano are clear examples of this. Due to the different methods employed to study the geochemistry of primary sources and archaeological assemblages, it is practically impossible to use the results generated by each team for comparative analysis. This is why conclusions made by different scholars are often controversial: for example, see the results by Cho *et al.* (2010) and Kim *et al.* (2007) regarding the presence/absence of the Paektusan obsidian at some Korean sites. Nevertheless, some positive results also exist for the Hokkaido sources where the same samples were analysed using both the NAA and XRF methods (see Kuzmin *et al.* 2013). However, examples of carefully planned cross-checks are still rare. Cross-checking should be performed on a regular basis (i.e., Glascock 1999). The unification of data collection and analysis is therefore a vital aspect of obsidian provenance studies in Northeast Asia, and the joint Japanese–Russian–US–Korean fieldtrip in October 2011 to the major sources of obsidian in Hokkaido Island — Shirataki, Oketo, and Tokachi-Mitsumata (see Ono *et al.*, this volume) — with ongoing cross-checks of selected samples is a very positive step in this direction.

CONCLUSIONS

Based on data collected throughout the past 20 years in the Russian Far East and adjacent Northeast Asia in the field of obsidian geochemistry, geology, and archaeology, it is now possible to establish geochemical 'fingerprints' for the major sources of archaeological obsidian. The Database created by our informal team of scholars from Russia, USA, Japan, and South Korea, allows us to solve many of the current and future questions related to the provenance of obsidian artefacts in both mainland and insular parts of Northeast Asia. Obviously, more research is still needed for some primary sources, especially the Paektusan Volcano, and for regions with obsidian artefacts such as the Amur River basin and the Kurile Islands. Consistency checks of data generated by different analytical methods are also an urgent task.

Acknowledgements

This paper is a short résumé of research conducted over the last two decades, and numerous colleagues participated in it and also collaborated with us by supplying samples and giving information; we are grateful to all of them. This study was supported by grants from several foundations, including the US NSF (DBS-9205506, BNS-9102016, SBR-9503035 and 9802366) and CRDF (RG1-2538-VL-03 and RUG1-7097-NO-13); Russian RFFI (96-06-80688, 99-06-80348, 02-06-80282 and 06-06-80258), SB RAS (Programme IV.31.2, Project "GIS Modeling and Spatial Analysis of Environmental Systems and Their Components"), RFFI–FEB RAS (06-08-96012), and FEB RAS (06-III-A-08-319); Ministry of Education, Culture, Sport, Science, and Technology of Japan (MEXT) (2003); and Korea Foundation (2002). We are grateful to Prof. A. Ono and the Centre for Obsidian and Lithic Studies of Meiji University (Tokyo, Japan) for their invitation to visit Japan in October–November 2011, in order to participate in the fieldwork excursion in Hokkaido Island and in the Symposium "Methodological Issues of Obsidian Provenance Studies and the Standardisation of Geologic Obsidian" held at the Centre for Obsidian and Lithic Studies near Nagawa Town (Nagano Pref.) on 5–6 November 2011, and for financial support. We thank Dr V. K. Popov for clarifying some issues on the geology of obsidian sources.

References

ALLABY, A., and M. ALLABY (eds). 2003. *A Dictionary of Earth Sciences. Second edition.* Oxford, Oxford University Press.

CHO, N.-C., J.-C. KIM, and H.-T. KANG. 2010. Provenance Study of Obsidian Artefacts Excavated from Palaeolithic Sites on the Korean Peninsula. In *Crossing the Straits: Prehistoric Obsidian Source Exploitation in the North Pacific Rim* (B.A.R. International Series 2152), edited by Y. V. Kuzmin and M. D. Glascock, 73–87. Oxford, BAR Publishing.

GLASCOCK, M. D. 1999. An Inter-Laboratory Comparison of Element Compositions of Two Obsidian Sources. *International Association for Obsidian Studies Bulletin* 23, 13–15.

GLASCOCK, M. D., G. E. BRASWELL, and R. H. COBEAN. 1998. A Systematic Approach to Obsidian Source Characterization. In *Archaeological Obsidian Studies: Method and Theory*, edited by M. S. Shackley, 15–65. New York & London, Plenum Press.

GLASCOCK. M. D., A. A. KRUPIANKO, Y. V. KUZMIN, M. S. SHACKLEY, and A. V. TABAREV. 1996. Geochemical Characterization of Obsidian Artefacts from Prehistoric Sites in the Russian Far East: An Initial Study. In *Arkheologiya Severnoi Patsifiki*, edited by A. L. Ivliev, N. N. Kradin, and I. S. Zhushchikhovskaya, 406–410. Vladivostok, Dalnauka Press.

GLASCOCK, M. D., Y. V. KUZMIN, A. V. GREBENNIKOV, V. K. POPOV, V. E. MEDVEDEV, I. Y. SHEWKOMUD, and N. N. ZAITSEV. 2011. Obsidian Provenance for Prehistoric Complexes in the Amur River Basin (Russian Far East). *Journal of Archaeological Science* 38, 1832–1841.

GLASCOCK, M. D., R. J. SPEAKMAN, and H. NEFF. 2007. Archaeometry at the University of Missouri Research Reactor and the Provenance of Obsidian Artefacts in North America. *Archaeometry* 49, 353–367.

GRAHAM, C. C., M. D. GLASCOCK, J. J. CARNI, J. R. VOGT, and T. G. SPALDING. 1982. Determination of

Elements in the National Bureau of Standards' Geological Standard Reference Materials by Neutron Activation Analysis. *Analytical Chemistry* 54, 1623–1627.

HALL, M. E., and H. KIMURA. 2002. Quantitative EDXRF Studies of Obsidian Sources in Northern Hokkaido. *Journal of Archaeological Science* 29, 259–266.

IZUHO, M., and W. HIROSE. 2010. A Review of Archaeological Obsidian Studies on Hokkaido Island (Japan). In *Crossing the Straits: Prehistoric Obsidian Source Exploitation in the North Pacific Rim* (B.A.R. International Series 2152), edited by Y. V. Kuzmin and M. D. Glascock, 9–25. Oxford, BAR Publishing.

KIM, J.-C., D. K. KIM, M. YOON, C. C. YUN, G. PARK, H. J. WOO, M.-Y. HONG, and G. K. LEE. 2007. PIXE Provenancing of Obsidian Artefacts from Paleolithic Sites in Korea. *Bulletin of the Indo-Pacific Prehistory Association* 27, 122–128.

KUZMIN, Y. V. 2010. Crossing Mountains, Rivers, and Straits: A Review of the Current Evidence for Prehistoric Obsidian Exchange in Northeast Asia. In *Crossing the Straits: Prehistoric Obsidian Source Exploitation in the North Pacific Rim* (B.A.R. International Series 2152), edited by Y. V. Kuzmin and M. D. Glascock, 137–153. Oxford, BAR Publishing.

KUZMIN, Y. V. 2011. The Patterns of Obsidian Exploitation in the Late Upper Pleistocene of the Russian Far East and Neighbouring Northeast Asia. In *Natural Resource, Environment and Humans* (Proceedings of the Meiji University Centre for Obsidian and Lithic Studies No. 1), edited by A. Ono, 67–82. Tokyo, Centre for Obsidian and Lithic Studies, Meiji University.

KUZMIN, Y. V., and M. D. GLASCOCK. 2007. Two Islands in the Ocean: Prehistoric Obsidian Exchange between Sakhalin and Hokkaido, Northeast Asia. *Journal of Island & Coastal Archaeology* 2, 99–120.

KUZMIN, Y. V., and M. D. GLASCOCK (eds). 2010. *Crossing the Straits: Prehistoric Obsidian Source Exploitation in the North Pacific Rim* (B.A.R. International Series 2152). Oxford, BAR Publishing.

KUZMIN, Y. V., M. D. GLASCOCK, and M. IZUHO. 2013. The Geochemistry of the Major Sources of Archaeological Obsidian on Hokkaido Island (Japan): Shirataki and Oketo. *Archaeometry* 55, 355–369.

KUZMIN, Y. V., M. D. GLASCOCK, and H. SATO. 2002b. Sources of Archaeological Obsidian on Sakhalin Island (Russian Far East). *Journal of Archaeological Science* 29, 741–749.

KUZMIN, Y. V., and V. K. POPOV (eds). 2000. *Vulkanicheskie Stekla Dalnego Vostoka Rossii: Geologicheskie i Arkheologicheskie Aspekty* [Volcanic Glasses of the Russian Far East: Geological and Archaeological Aspects]. Vladivostok, Dalnevostochny Geologichesky Institut.

KUZMIN, Y. V., V. K. POPOV, M. D. GLASCOCK, and M. S. SHACKLEY. 2002a. Sources of Archaeological Volcanic Glass in the Primorye (Maritime) Province, Russian Far East. *Archaeometry* 44, 505–515.

KUZMIN, Y. V., O. V. YANSHINA, S. M. FITZPATRICK, and O. A. SHUBINA. 2012. The Neolithic of the Kurile Islands (Russian Far East): Current State and Future Prospects. *Journal of Island & Coastal Archaeology* 7, 234–254.

MALAINEY, M. E. 2011. *A Consumer's Guide to Archaeological Science: Analytical Techniques*. New York, Springer.

SRM-278 Obsidian Rock Certificate of Analysis. 1992. National Institute of Standards and Technology, Gaithersburg, MD, USA.

PHILLIPS, S. C. 2010. Bridging the Gap between Two Obsidian Source Areas in Northeast Asia: LA-ICP-MS Analysis of Obsidian Artefacts from the Kurile Islands of the Russian Far East. In *Crossing the Straits: Prehistoric Obsidian Source Exploitation in the North Pacific Rim* (B.A.R. International Series 2152), edited by Y. V. Kuzmin and M. D. Glascock, 121–136. Oxford, BAR Publishing.

PHILLIPS, S. C. 2011. Networked Glass: Lithic Raw Material Consumption and Social Networks in the Kuril Islands, Far Eastern Russia. Unpublished PhD Dissertation. Department of Anthropology, University of Washington, Seattle, WA, USA.

PHILLIPS, S. C., and R. J. SPEAKMAN. 2009. Initial Source Evaluation of Archaeological Obsidian from the Kurile Islands of the Russian Far East Using Portable XRF. *Journal of Archaeological Science* 36, 1256–1263.

POLLARD, M., C. BATT, B. STERN, and S. M. M. YOUNG. 2007. *Analytical Chemistry in Archaeology*. Cambridge, Cambridge University Press.

POPOV, V. K., V. G. SAKHNO, Y. V. KUZMIN, M. D. GLASCOCK, and B.-K. CHOI. 2005. Geochemistry of Volcanic Glasses of the Paektusan Volcano. *Doklady Earth Sciences* 403, 254–259.

POPOV, V. K., G. P. SANDIMIROVA, and T. A. VELIVETSKAYA. 2008. Strontium, Neodymium, and Oxygen Isotopic Variations in the Alkali Basalt–Trachyte–Pantellerite–Comendite Series of Paektusan Volcano. *Doklady Earth Sciences* 419, 329–334.

POPOV, V. K., and M. S. SHACKLEY. 1997. Obsidian of Primorye: First Results of Archaeological–Geological Correlation. *Vestnik Dalnevostochnogo Otdeleniya Rossiiskoi Akademii Nauk* 3(73), 77–85 (in Russian with English Abstract).

POPOV, V. K., V. A. SOLYANIK, and D. G. FEDOSEEV. 2009. Decorative Volcanic Glasses from Hyaloclastites of the Shkotovo Basaltic Plateau (Primorye, Russia). *The Journal of the Gemmological Association of Hong Kong* 30, 51–56.

SHACKLEY, M. S., M. D. GLASCOCK, Y. V. KUZMIN, and A. V. TABAREV. 1996. Geochemical Characterization of Archaeological Obsidian from the Russian Far East: a Pilot Study. *International Association for Obsidian Studies Bulletin* 17, 16–19.

TABAREV, A. V., and Y. V. KUZMIN. 2000. Obsidian in Archaeological Cultures of Primorye and the Sakhalin. In *Vulkanicheskie Stekla Dalnego Vostoka Rossii: Geologicheskie i Arkheologicheskie Aspekty*, edited by Y. V. Kuzmin and V. K. Popov, 7–13. Vladivostok, Dalnevostochny Geologichesky Institut (in Russian with English Abstract).

VASILIEVSKY, R. S., and S. A. GLADYSHEV. 1989. *Verkhnyi Paleolit Yuzhnogo Primorya* [The Upper Palaeolithic of the Southern Primorye]. Novosibirsk, Nauka Publishers.

Chapter 7
GEOCHEMISTRY OF VOLCANIC GLASSES AND THE SEARCH STRATEGY FOR UNKNOWN OBSIDIAN SOURCES ON KAMCHATKA PENINSULA (RUSSIAN FAR EAST)

Andrei V. GREBENNIKOV, Vladimir K. POPOV, and Yaroslav V. KUZMIN

Abstract: *The 2013 state-of-the-art in obsidian geology and geochemistry of Kamchatka Peninsula (northern Russian Far East) is presented. General information on volcanic geology of Kamchatka is given, and major obsidian sources studied by our team since 2003 are briefly described. A strategy for the search of the unknown primary sources of obsidian on Kamchatka (i.e., those for which the precise position is not established) is proposed. It is based on the following criteria: 1) general geological structure and the history of geological development; 2) geochemical typification of volcanic rocks and glasses; and 3) absolute age of both geological and archaeological obsidians. Using the data available and its analysis, possible primary sources for archaeological obsidian on Kamchatka are suggested. It now requires verification by the means of fieldwork.*

Keywords: *Obsidian, Geochemistry, Volcanic Geology, Source Identification, Neutron Activation Analysis, Potassium–Argon Dating, Kamchatka Peninsula, Russian Far East*

INTRODUCTION

Information on the geochemistry of volcanic rocks, specifically glasses which reflect the initial composition of "frozen" eruptive magmas, is very important for studying the conditions of the creation and evolution of magmatic melts. Also, geochemical data have a superior importance for studying the sources of archaeological obsidian. The determination of obsidian sources for prehistoric tool assemblages allows researchers to reconstruct the directions of ancient migrations and contacts. Recent progress in these fields in Northeast Asia can be found in the volume edited by Kuzmin and Glascock (2010).

In the northern Russian Far East, the Kamchatka Peninsula is one of the most promising areas for obsidian provenance studies due to an abundance of volcanic glass sources and their extensive use by prehistoric people. Since the early 1900s, obsidian artefacts have been discovered at more than 800 archaeological sites on Kamchatka (see Grebennikov *et al.* 2010, 91). Volcanic glasses of this region have been systematically studied by our group since 2003 (see Glascock *et al.* 2006; Kuzmin *et al.* 2008; Popov *et al.* 2007). As a result, an extensive Database containing information on the geochemical composition of about 500 samples was created for the first time. Based on the data obtained, volcanic glasses from Kamchatka were divided into 16 geochemical groups. Seven of them include both geological and archaeological samples, and correspond to the main obsidian sources used by prehistoric people (Figure 7.1). The other seven groups consist of 'archaeological' obsidian (i.e., artefacts) only, and the primary sources for these groups are unknown so far. Two groups are represented by solely geological source samples which were not used by ancient people (Figure 7.1) (see details in Grebennikov *et al.* 2010).

The Kamchatka Peninsula in the Northwestern Pacific stretches approximately 1200 km in SSW–NNE direction, and is flanked by the Bering Sea in the east, the Sea of Okhotsk in the west, and the open Pacific Ocean in the south (Figures 7.1–7.2). The main geomorphic features of the region are two mountain ranges, Central [Sredinniy] and Eastern [Vostochniy], with a sedimentary basin between them occupied by the Kamchatka River. From a tectonic point of view, the Kamchatka Peninsula sits on the boundary between the Pacific and Eurasian plates. This is one of the most active volcanic arcs in the world, and at least 28 modern volcanoes are known in the region (Fedotov and Masurenkov 1991). Most of the Kamchatkan terrain consists of Cenozoic volcanic rocks, with some sedimentary and volcanic-sedimentary formations (Khain 1994). Volcanic glasses (obsidians and perlites) are widely distributed on Kamchatka, and they are part of the dacitic–rhyolitic volcanic complexes of Neogene–Pleistocene age (Shevchuk 1981).

Nowadays, about 30 sources of high and medium quality volcanic glasses are known on Kamchatka (Figure 7.2). In the Central Kamchatkan Volcanic Belt corresponding to the Central Range obsidian-bearing volcanic formations are dated to Oligocene–Neogene times; and in the Eastern Kamchatkan Volcanic Belt occupied by the Eastern Range obsidian is known mainly among the Pleistocene rocks. In southern Kamchatka, obsidian-containing rocks are dated to the Pliocene–Pleistocene. Volcanic glasses on Kamchatka occur in extrusive domes, lava and pyroclastic flows, and in the pyroclastic products (as tephras and pumice tuffa fragments). According to their chemical composition, volcanic glasses correspond to dacites and rhyolites.

In this chapter, we summarise the results of research in obsidian geology and geochemistry of the Kamchatka Peninsula as of 2013, and present an attempt to create a search strategy for the still unknown primary sources by means of geochemical typification and geochronological examination of both archaeological and geological obsidians.

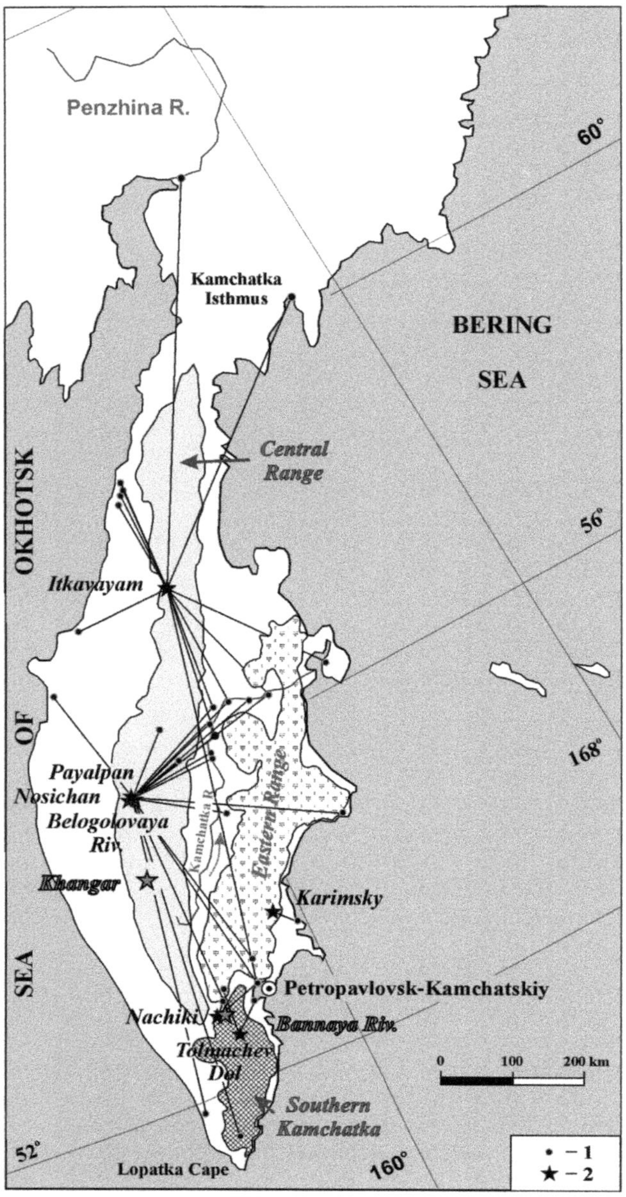

Figure 7.1. Selected obsidian sources on Kamchatka and archaeological sites with artefacts associated with them. Major tectonic zones of Kamchatka are indicated by grey shadings. 1 – archaeological sites; 2 – obsidian sources (black stars are primary sources of archaeological obsidian, and grey stars are geological sources not used in prehistory)

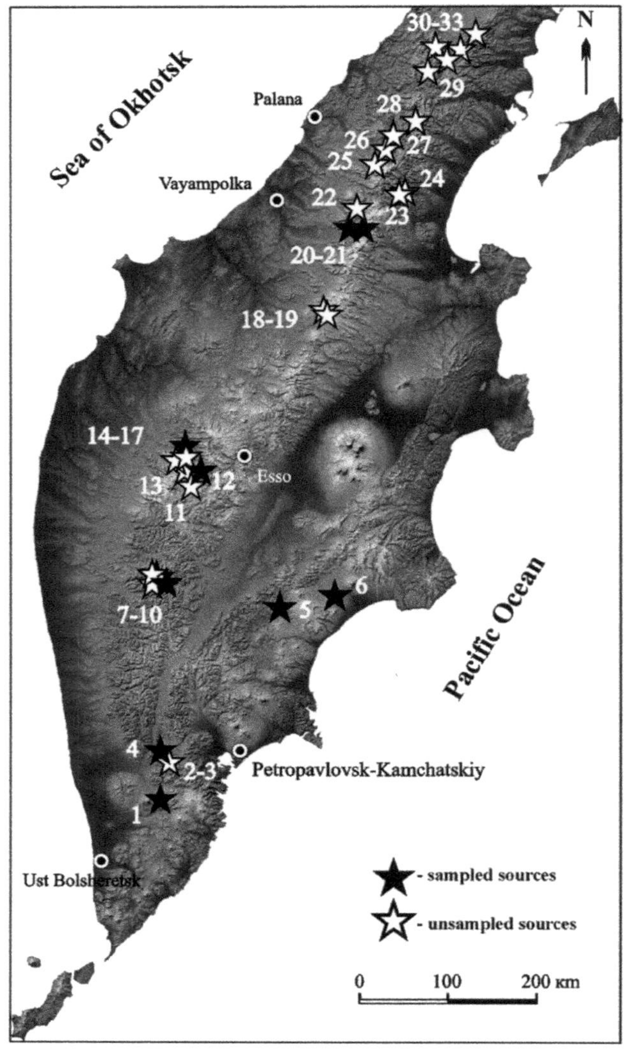

Figure 7.2. Major obsidian sources on the Kamchatka Peninsula (after Otchet ... 1992, modified). 1 – Tolmachev Dol (Chasha Maar); 2 – Nachiki Stream; 3 – Taburetka River; 4 – Nachiki (Shapochka Summit); 5 – Karimsky Volcano; 6 – Krasheninnikov Volcano; 7 – Khangar Volcano (Southern); 8 – Khangar Volcano (Central); 9 – Khangar Volcano (Eastern); 10 – Khangar Volcano (Northern); 11 – Gigigilen; 12 – Payalpan; 13 – Maly Payalpan; 14 – Nosichan; 15 – Polyarnaya Summit; 16 – Tynya Summit; 17 – Belogolovaya Vtoraya River; 18 – Kunkhilok; 19 – Sedanka; 20 – Itkavayam (Southern); 21 – Itkavayam (Northern); 22 – Kevenei (Northern); 23 – Kevenei (Western); 24 – Levye Nachiki; 25 – Levoe Khailuli Plateau; 26 – Maryavaam; 27 – Palana (Southern) and Korkavayam; 28 – Palana 1; 29 – Posledny Stream; 30 – Vanyavaam; 31 – Kichiga River, right side; 32 – Belaya River headwaters; and 33 – Kichiga River, left side

GEOLOGICAL HISTORY OF KAMCHATKA: A SHORT OUTLINE

Kamchatka Peninsula together with the Kurile Islands is usually considered as a single island arc system; however, the tectonic position of Kamchatka is more complicated. The Kurile–Kamchatkan Island Arc system consists of a typical island arc (the Kuriles) and an active continental margin (Kamchatka). In the north it meets the Aleutian Island Arc at an almost right angle. Three island arcs of different ages are distinguished on Kamchatka (Avdeiko *et al.* 2006). The most ancient one is the Western Kamchatkan Volcanic Arc of late Cretaceous age. In the Oligocene – early Miocene, the Central Kamchatkan and the Southern Kamchatkan island arc systems existed, with a wide Malki–Petropavlovsk Zone of tectonic dislocations between them. In the middle Miocene – early Pliocene, due to the displacement of the subduction zone to its present position, the Eastern Kamchatkan Volcanic Belt emerged. The late Pleistocene – Holocene volcanism progressed not only in the frontal and rear parts of the Eastern Kamchatkan Volcanic Belt, but also in the Central Range. In southern Kamchatka, the volcanic belt coincided with the Oligocene – early Miocene volcanic zone.

The Central Kamchatkan and Southern Kamchatkan Volcanic Belts

The geological evolution of the Kamchatkan Island Arc in the Oligocene–Miocene is connected to the formation of surface volcanism in the central and southern parts of the peninsula. The Central Kamchatkan Volcanic Zone (hereafter – the Central Range; see Figure 7.1) stretches from the Kamchatka Isthmus to the Malki–Petropavlovsk Zone. The Central Range appeared in the late Paleogene as the horst–anticline uplifts composed of late Mesozoic folded complexes.

The Southern Kamchatkan Volcanic Zone (see Figure 7.1, "Southern Kamchatka") extends for 230 km from the upper reaches of the Paratunka and Plotnikova rivers (the latitude of Petropalvovsk-Kamchatskiy) southward to the Lopatka Cape, and ranges in width from 10 to 45 km. In contrast to the Central Range, the formation of this zone extended to the middle Miocene and even up to the Pliocene in the extreme south, in the tectonic settings of a typical island arc. It existed as a chain of volcanic islands surrounded by shallow waters (Khanchuk 2006).

The geological history of both volcanic zones on Kamchatka is notable for two major stages of volcanism. The first stage is dated to the late Oligocene – early Miocene. Volcanic rock formations accumulated at that time in the Central Range are represented by two suites of the Anavgai Complex. According to the chemical composition of effusive rocks, the lower (Krapivina) suite is dominated by andesites and basaltic andesites, with some basalts and their tuffs, sand tuffs, and aleurolites. The upper (Berezovka) suite consists of two parts according to their composition: 1) a lower part, composed predominantly of andesite–dacite lavas and tuffs about 800 m thick; and 2) an upper part, represented by volcanic-sedimentary rocks about 500 m thick.

The second stage of rock formation is related to the most powerful flare-up of surface volcanism which occurred in the late Miocene – Pliocene, and resulted in the formation of the Alnei Series. Volcanic rocks are represented by basaltic, andesitic, and dacitic lava flows and subvolcanic bodies, ignimbrite sheets, and detached tuff horizons. According to Leonov *et al.* (2008), the age of the Central Range ignimbrites corresponds to the lower Pliocene, ca. 3.5–5.7 Ma.

It should be noted that an independent phase of volcanism within the Central Range in the Pleistocene–Holocene resulted in the formation of volcanic structures recognised as the Pleistocene Volcanic Belt. The most abundant primary occurrences of obsidian are associated with products of the Oligocene–Pliocene and Pleistocene–Holocene phases of acidic volcanism in the Central Range.

The Eastern Kamchatkan Volcanic Belt

The Eastern Kamchatkan Volcanic Belt stretches along the eastern coast of Kamchatka for about 850 km in NNE direction, and ranges in width from 50 to 100 km (Figure 7.1, "Eastern Range"). It was formed during the last 50,000–60,000 years, and is composed of large volcanic edifices such as shield volcanoes, stratovolcanoes, and calderas, as well as a multitude of smaller volcanic structures like maars, explosive funnels, and extrusive domes. The volume of volcanic products is ca. 14,000–15,000 km^3, and it covers a territory of about 15,000 km^2 (Fedotov and Masurenkov 1991). Shield volcanoes and stratovolcanoes consist mostly of basalts and andesites; dacites and rhyolites are also widespread. Volcanic glasses (obsidians and perlites) are related to some types of these rocks in the Uzon Caldera and the Karymsky Volcanic Centre.

MATERIALS AND METHODS

The geochemical study of volcanic glasses was conducted using Neutron Activation Analysis (henceforth – NAA). The NAA is one of the most advanced methods to study the chemical composition of volcanic glasses, with sensitivity limits for most elements in the parts-per-billion to parts-per-million ranges. The advantage of NAA is that one can use small samples (starting from 10–20 mg) to determine the content of more than 25 chemical elements, including rare-earth ones, in order to reveal a unique geochemical 'fingerprint' of a particular volcanic glass source. For our study, NAA analysis was performed at the Archaeometry Laboratory, University of Missouri Research Reactor Center (hereafter – MURR) in Columbia, Missouri (USA).

In total, about 500 samples of obsidian from Kamchatka were studied in 2003–11. Some of them were collected by our team during fieldwork campaigns in 2004–5. Several colleagues from the Institute of Volcanology and Seismology (Far Eastern Branch of the Russian Academy of Sciences) in Petropavlovsk-Kamchatskiy and from the Institute of Geochemistry (Siberian Branch of the Russian Academy of Sciences) in Irkutsk (both in Russia), contributed to our collection of obsidian from primary Kamchatkan sources. Archaeological samples (obsidian artefacts from the final Palaeolithic – Palaeometal sites) were collected by Dr A. V. Ptashinsky who participated in our studies (see Glascock *et al.* 2006; Grebennikov *et al.* 2010, 91; Kuzmin *et al.* 2008); other samples were obtained by us in 2005 from the collection of the late Prof. Nikolai N. Dikov stored at the Northeastern Multidisciplinary Research Institute (Far Eastern Branch of the Russian Academy of Sciences) in Magadan,

Russia, including the Ushki cluster (see Kuzmin *et al.* 2008).

The samples were cleaned to rid the surface of contaminants, and then divided into aliquots of 25–50 mg each. Specimens were exposed to thermal neutrons in the reactor; after that, the abundance of up to 28 chemical elements (Al, Cl, Dy, K, Mn, Na, Ba, La, Lu, Nd, Sm, U, Yb, Ce, Co, Cs, Eu, Fe, Hf, Rb, Sb, Sc, Sr, Ta, Tb, Th, Zn, and Zr) was determined. Data about the content of all these elements can be obtained using long-irradiation of samples. For some samples, only short-irradiation was applied, and the content of seven elements was measured: Al, Ba, Cl, Dy, K, Mn, and Na. Details on the NAA method can be found elsewhere (e.g., Glascock *et al.* 2007; Malainey 2011; Pollard *et al.* 2007; see also Grebennikov *et al.* 2010, 91).

Statistical grouping, based on bivariate and triangle plots, and cluster and discriminant classification analyses, was performed with the help of GAUSS software (available from the MURR; Internet access: http://www.missouri.edu/~glascock/profile.htm), to indicate within a 95% degree of probability the major geochemical groups reflecting obsidian sources (see details in Glascock *et al.* 1998). As a result, the individual sources were securely identified.

The potassium–argon (K–Ar) dating of obsidian samples was carried out by the Laboratory of Stable Isotopes of the Far Eastern Geological Institute, Far Eastern Branch of the Russian Academy of Sciences (hereafter – FEB RAS) in Vladivostok, Russia. In order to measure the content of radiogenic Ar, an original Continuous-Flow Gas Chromatography/Isotope-Ratio Mass Spectrometry (CF–GC–IRMS) with laser ablation method, developed by the staff of this Laboratory, was applied (Ignatiev *et al.* 2010; Budnitsky 2013). The isotope composition of Ar was investigated using a MAT-253 Mass Spectrometer (Thermo Scientific Co.); the content of K was measured with the Atomic Absorption Spectrophotometry method at the Laboratory of Analytical Chemistry, Far Eastern Geological Institute FEB RAS.

PRIMARY SOURCES OF OBSIDIAN ON KAMCHATKA: RECENT STATUS

Seven geochemical groups containing both archaeological and geological obsidians were established by our team on Kamchatka (Figure 7.1). Four groups are directly attributed to primary sources in the Central Range, with precisely known locations: 1) Itkavayam Volcano (group KAM-03); 2) Payalpan Volcano (KAM-05); 3) Belogolovaya Vtoraya River source (KAM-07); and 4) Nosichan Volcano (KAM-16). The Karymsky Volcano source (group KAM-09) was identified in the Akademii Nauk [Academy of Sciences] Caldera which is a part of the Karymsky Volcanic Centre in the Eastern Range. Two primary sources were identified in southern Kamchatka, the Nachiki Volcano (group KAM-06) and the Tolmachev Dol Volcano (KAM-11).

The Central Range

The majority of obsidian sources on Kamchatka are located in the Central Range (Figures 7.1–7.2). They are situated at higher elevations within the mountain plateaus covered with tundra, as open-air scatters of colluvial origin. Dacitic and rhyolitic obsidians are presented by blocks and big chunks. The colour and structure of obsidian varies, and in some places the iridescent kinds are found.

The Itkavayam Volcano Source Group

In the northern part of the Central Range, more precisely the headwaters of the Vayampolka and Kutina rivers, the Itkavayam obsidian source group is located (Figure 7.2, Nos. 20–21). It consists of at least four sub-sources on the slopes of Ritman Volcano (Otchet ... 1992; Shevchuk 1981). The first sub-source, at the headwaters of the Itkavayam River, is a 200 m thick deposits which originated from the Obsidianovy Volcano due to a circular fracture on the slopes of the Ritman Volcano. The obsidian-bearing rocks form a plateau-like upland. The thickness of several alternating obsidian layers ranges from 0.3–0.4 to 15 m. Obsidians are black-coloured, transparent in thin sections, and sometimes banded due to light and dark streaks. Obsidians have a strong glassy lustre and shell-like fracture with sharp edges. Red and mahogany coloured obsidians, with the colours due to the presence of dusty and flaky hematite, were also observed.

The second sub-source is located 2.5 km SW of the first one. It is composed of sub-horizontally bedded lava flows, with its thickness ranging from 3–5 to 5–15 m; the aggregated beds are up to 200 m thick. The lower part of the obsidian-bearing stratum (70 m in thickness) consists of mainly black obsidian, with thin red mahogany coloured bands. The upper part of the stratum is composed of pure black volcanic glass with small plagioclase impregnations. Black obsidians are characterised by brittleness, a dull lustre, and an almost absolute lack of the distinct shell-like fracture. It is worth noting that the obsidians of the second sub-source differ from obsidians of the first one by their lower silica contents (Otchet ... 1992).

The third sub-source of the Itkavayam obsidian group was found on the southeastern slope of the Ritman Volcano, in a valley of one of the Keveneivayam River tributaries. The obsidian flows occupy a territory of 4 km^2; several of them are up to 30–60 m thick. The obsidians have a distinct black colour, a glassy lustre, and a shell-like fracture, with small plagioclase inclusions occupying 5–10% of the rock volume.

The fourth sub-source is located on the southeastern slope of the Ritman Volcano in the valley of a small stream, a tributary of the Keveneivayam River, and is represented by blocks of eluvium consisting of black glassy obsidian. This locality covers an area of up to 4 km^2. N. T. Demidov and his colleagues (see Otchet ... 1992, 421) suggested that these obsidians initially formed the top of a series of lava flows which later underwent intensive glacial weathering. The thickness of the flows remains unknown.

Obsidian Sources of the Ichinsky Volcanic Centre

The Ichinsky cluster of obsidian sources is situated in the middle sector of the Central Range (Figure 7.2, Nos. 11–17). This is the largest single area with volcanic glasses on Kamchatka, and it covers a territory of ca. 700 km^2. In the vicinity of the Ichinsky Volcano with an elevation of 3621 m above sea level (Figure 7.3, see lower left corner), situated on the western slope of the Central Range, there are at least 11 distinct volcanic glass sources, namely Belogolovaya Vtoraya River, Tynya, Polyarnaya, Nosichan, Payalpan, Maly Payalpan, Galdavit, Studeny, Zemnoy Creek, Zemnoy Summit, and Gigigilen (Figure 7.3). This information is based on the inventory of volcanic glass sources generated by the Geological Survey of Russia (Otchet ... 1992).

Obsidians from three of these sources — Payalpan (group KAM-05), Belogolovaya Vtoraya River (KAM-07), and Nosichan (KAM-16) — have been identified in archaeological collections (Figure 7.1; see also Grebennikov *et al.* 2010). The volcanic glass corresponds mainly to dacite–rhyolites of the upper part of the Alnei Series of late Miocene age (Sheimovich and Patoka 2000). Some sources, such as the Belogolovaya Vtoraya River, may have been created approximately 2.5 Ma ago.

The Payalpan is the major volcanic glass source in Kamchatka, and it is famous because of a rare decorative variety of blue obsidian. The Payalpan source of Pliocene age is located 25 km NE of the Ichinsky Volcano, on the western slope of the Maly Payalpan Volcano (Figure 7.3, No. 5). Our microscopic study shows that the blue shade of this obsidian is defined by the colour of fluidal lines of crystallites rather than by the ore dust impurity, as it was previously suggested by Eremina (2009). Obsidian deposits at the Payalpan source are associated mainly with the upper Pliocene sub-volcanic rhyolitic domes and effusive formations like sheets and covers of lava, tuff, tuff breccia, and ignimbrite (Figure 7.3). The obsidian layers here are generally not thick, at 1–2 m on average. Obsidian occurs mainly as individual layers, and also as sets of extended, squeezed, and contiguous lenses in rhyolites; small obsidian flows are also known. The richest in blue obsidians are colluvial deposits on the banks of the Obsidianovy Creek, with a diameter of pieces from 3–5 to 50 cm. These deposits are located so close to the primary rhyolitic extrusion that some researchers suggested an extrusive genesis of obsidian (see Otchet ... 1992). However, this issue is still uncertain due to insufficient data on the depositional conditions and characteristics of obsidian bodies, and the abundance of lenses of brecciated obsidians in pyroclastic deposits. The obsidians vary in colour and texture: there are dark-grey, black, and greenish massive obsidians; and finely fragmented reddish-brown and bluish-banded ones.

The Belogolovaya Vtoraya River source (Figure 7.3, No. 1) is a set of lenses and layers of obsidians embedded in effusive and pyroclastic rocks of moderately acidic composition. Obsidian lenses are found in the upper part of the dacitic lava flow outcrop of middle Pleistocene age, as scatters of colluvium on steep slopes of the Tynya Summit (Figure 7.3, No. 2). Volcanic glass is mostly black and with cracks, and contains plagioclase phenocrysts. Other varieties are represented by brownish and mahogany coloured obsidians with black inclusions.

On the right bank of the Nosichan River and in the vicinity of the Polyarnaya Summit (Figure 7.3, Nos. 3–4), there are small obsidian scatters and rare primary outcrops in the form of lenticular beds and dikes of Pliocene age. They are from 4–5 to 8 m thick (at bulges), and up to 270 m long. All locales are genetically related to rhyolitic extrusions, and are characterised by multi-coloured volcanic glasses: mostly black glass (both transparent and smoky), and, more rarely, amber, brown, and lilac ones. The size of non-rounded obsidian fragments is predominantly 0.5–2.5 cm; pieces of up to 10 cm in diameter are rare.

The Khangar Volcano Source

South of the Ichinsky cluster, there is another obsidian source in the caldera of the Khangar Volcano (Figure 7.2, Nos. 8–10). The recent crater was formed as a result of an eruption at about 6900 BP (Melekestsev *et al.* 1996). According to data collected by the Geological Survey of Russia (see Otchet ... 1992), there are several obsidian outcrops inside the caldera. They are related to extrusive dacites and represented by lenses, dikes, and beds of Pleistocene volcanic rocks, created after apical explosion of the caldera but before the formation of the recent crater in the mid-Holocene. Fissured and massive volcanic glasses are mainly black and fallow in colour; brown and blue kinds are rarer.

The Eastern Range

This part of Kamchatka has several active volcanoes, such as Krasheninnikov, Karymskaya Sopka, Avachinskaya Sopka, and others (see Fedotov and Masurenkov 1991). Several volcanic glass sources are known in the central part of the range, and they are correlated to the Pleistocene phase of the acidic ignimbrite volcanism. Glasses are mainly perlites, and one of the best-known sources is situated in the Uzon Caldera.

The Krasheninnikov Volcano obsidian source (Figure 7.2, No. 6) is a series of lava flows 100–200 m long and 5–20 m thick of early-to-middle Pleistocene age. Obsidian occurs as thin layers inside these volcanic rocks, and it is mainly dark-grey and rarely brick-red in colour.

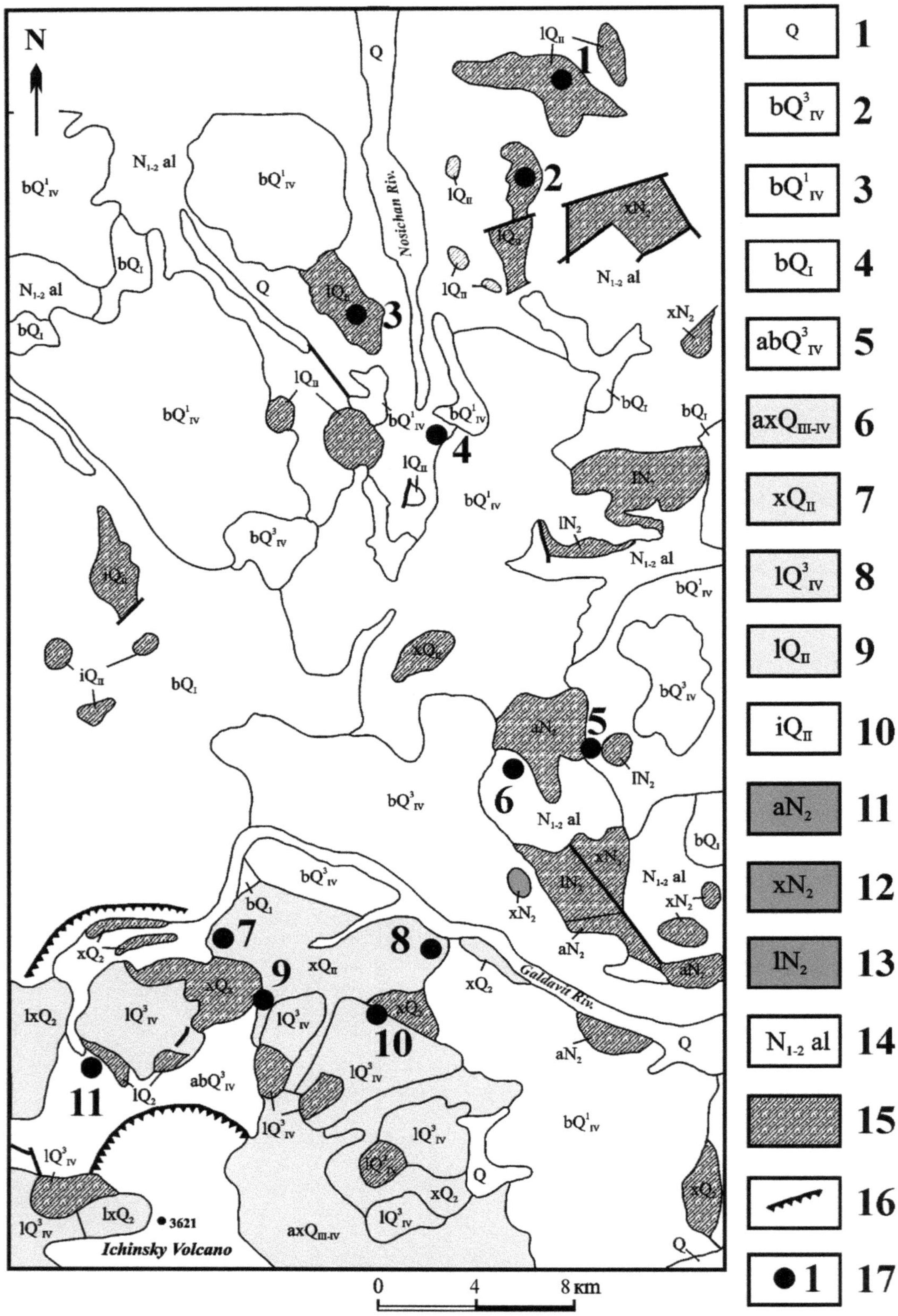

Figure 7.3. Schematic map of volcanic glass sources of the Ichinsky Volcanic Centre (after Otchet ... 1992, modified). 1–3 – Holocene deposits: 1 – alluvial and glacial deposits; 2 – upper basaltic strata; 3 – lower basaltic strata; 4 – lower Pleistocene basalts; 5 – upper Pleistocene basaltic andesites and scoriae; 6 – upper Pleistocene – Holocene andesites and dacites; 7 – middle Pleistocene dacites and ignimbrites; 8 – Holocene rhyolites and dacites; 9 – middle Pleistocene rhyolites and dacites; 10 – trachytes and trachyandesites; 11 – andesites; 12 – dacites; 13 – rhyolites; 14 – Miocene – Pliocene basalts, basaltic andesites, andesites, dacites, tuffs, tuff breccias, and ignimbrites of the Alnei Series; 15 – subvolcanic bodies; 16 – borders of volcano-tectonic calderas; 17 – volcanic glass sources: 1 – Belogolovaya Vtoraya River; 2 – Tynya Summit; 3 – Polyarnaya Summit; 4 – Nosichan; 5 – Payalpan; 6 – Maly Payalpan; 7 – Galdavit; 8 – Studeny; 9 – Zemnoy Creek; 10 – Gigigilen; and 11 – Galdavit Creek

Seven primary outcrops of obsidian are found, 0.8–2 m thick and up to 5 m long, and they are black, massive, and dotted with transparent insets (Otchet ... 1992).

Another obsidian source is known in the Karymsky Volcanic Centre (Figure 7.2, No. 5). Here pure obsidians are embedded in the pumice "noble" tuffa of the Odnoboky Volcano, dated to ca. 100,000 years ago (Masurenkov 1980). The younger pyroclastic pumices of the Akademii Nauk Caldera, dated to ca. 28,000–40,000 years ago, also contain obsidian fragments. The texture is massive, with a transparent thin edge, and black in colour. Obsidian is plentiful on the modern shore of the caldera lake.

Southern Kamchatka

South of the city of Petropavlovsk-Kamchatskiy, several volcanic glass sources of Neogene and Pleistocene age are known. One of the largest is the Nachiki source of perlites and pure obsidians as a part of extrusive dome of the Shapochka Volcano (Figure 7.1; Figure 7.2, No. 4), dated to the Pliocene. A distinct feature of the Nachiki source is the presence of plagioclase and pyroxene phenocrysts in the black glass matrix. Natural outcrops of this source are visible at the top of the Nachiki Summit as a sheet of volcanic glasses representing the selvage of chilled extrusive rhyolites. The obsidian and perlite deposits have a complex structure. The obsidian is black, whereas the perlites have a lighter colour (greyish-black) and they are weakly transparent and gently banded. Perlite from the Nachiki source is a subject of commercial mining due to its superior quality for the production of thermal insulation materials.

The second obsidian source is situated on the watershed of the Bannaya and Plotnikova rivers, and is named by us the Bannaya River (Figure 7.1). Numerous nodules of high quality transparent obsidian (up to 3–7 cm in diameter) are visible in the rhyolites and perlites of the extrusive dome of the Yagodnaya Summit.

The third source, Chasha Maar, is located to the west of the Gorelaya Sopka Volcano in the Tolmachev Dol River valley (Figure 7.2, No. 1). It originated from the slaggy lava cone of the Chasha Maar dated to ca. 4600 BP (Dirksen *et al.* 2002). Among the fragments of volcanic tephra, large (up to 30 cm long) pieces of obsidian of unusual light grey colour were found.

SEARCH STRATEGY FOR UNKNOWN OBSIDIAN SOURCES ON KAMCHATKA PENINSULA

Prior to the description of the search strategy for still unknown obsidian sources, it should be highlighted that the majority of obsidian sources on Kamchatka consist of several sub-sources separated from each other by often considerable distances, and they occur within volcanoes in a variety of forms: as extrusive bodies, lava sheets, and pyroclastic deposits. They reflect, as a rule, different formation stages of the volcano and may also differ in chemical composition. Unfortunately, we had no opportunity for detailed sampling of obsidians (including assignment of samples to particular strata) from many sub-sources due to their inaccessibility without helicopter transport. Also, obsidian samples collected by other researchers without precise pinpointing to a particular layer in the volcanic deposits are frequently used in our study. As a result, these circumstances limit to some extent the possibility to identify the exact primary sources of obsidian artefacts.

As was discussed earlier (see Glascock *et al.* 2006; Grebennikov *et al.* 2010; Kuzmin *et al.* 2008; Popov *et al.* 2007), primary ('geological') sources of obsidian on Kamchatka have been studied insufficiently in comparison to obsidian artefacts. There are at least seven geochemical groups of obsidian on Kamchatka identified only in prehistoric assemblages (Grebennikov *et al.* 2010; Kuzmin *et al.* 2008), and the location of their sources is still unknown. A similar situation can be observed in neighbouring Alaska where geochemical analysis of obsidian artefacts shows that they are derived from at least 25–30 primary sources, but less than ten have been precisely located so far (see Reuther *et al.* 2011).

Due to the large number of obsidian sources on Kamchatka, more research is still needed to correlate the archaeological volcanic glasses with the sources. On the one hand, geochemical data and the spatial distribution of archaeological sites allowed us to suggest possible localisation for the unknown sources; the first attempt was done by Grebennikov *et al.* (2010, 104). On the other hand, additional information is still necessary to determine the position of primary sources with a satisfactory degree of precision, in order to be able to visit them in the future and to conduct detailed studies. This situation, along with the fact that many of the obsidian sources on Kamchatka are still not sampled (see Figure 7.2), is the reason why we put forward the task of identification of the currently unknown primary obsidian locales with a minimal amount of labour and finance expenditure necessary for fieldwork on the large and logistically extremely difficult Kamchatkan terrain.

For this purpose, we have developed a search strategy based upon the existing geochemical data of geological and archaeological obsidians from Kamchatka (Grebennikov *et al.* 2010). In the quest for unknown archaeological obsidian sources, we use two important parameters: 1) the geochemical typification of volcanic rocks and glasses; and 2) the absolute age of archaeological and geological obsidians. For the first task, all possible geological data were analysed. As for the second task, age determination for archaeological obsidians and their assignment to volcanic glass outcrops of appropriate age was undertaken. Chronological study is also necessary to confirm that obsidian artefacts belong to a particular volcanic structure. To solve the problem of age correlation between volcanic glasses from geologic sources and archaeological sites, K–Ar dating of obsidians was conducted (Budnitsky 2013).

GEOCHEMICAL CRITERIA FOR IDENTIFICATION OF UNKNOWN OBSIDIAN SOURCES

The KAM-01, KAM-04, KAM-10, and KAM-14 Groups

The KAM-01, KAM-04, KAM-10, and KAM-14 groups have a Nb/Zr ratio of around 0.04, which is typical for volcanic rocks of the Eastern Kamchatkan Volcanic Belt. It should be noted that archaeological sites with artefacts belonging to these groups (see Grebennikov *et al.* 2010, 107–10) are in the vicinity of the Karymsky source (group KAM-09), which has a similar Nb/Zr ratio. Additionally, the chemical composition of these groups is very similar to the KAM-09 group (Figures 7.4–7.6). Based on current information, we assume that the primary obsidian sources for all of these groups are situated somewhere in eastern Kamchatka.

The KAM-02 Group

The most enigmatic geochemical group of Kamchatkan archaeological obsidian is KAM-02. Artefacts belonging to it are found at the majority of sites in southern and southeastern Kamchatka (Figure 7.7). For this group, the high concentrations of K, high-field-strength elements (Zr, Hf, Ta, and Th), and rare-earth elements, are considered typical for within-plate granites. Volcanic glass of the KAM-02 group plotted on the Hf–Rb–Ta diagram corresponds to the Eastern Kamchatkan Volcanic Belt (Figure 7.6). Besides this, volcanic glasses from eastern Kamchatka (Uzon Caldera and Odnoboky Volcano in the Karymsky Volcanic Centre) are characterised by minimal Nb/Zr ratios, and this is similar to the KAM-02 group. However, based on the majority of geochemical plots the obsidian of this group is closer geochemically to volcanic glasses from southern Kamchatka or falls into outliers. In Figures 7.4–7.5, it is clear that volcanic glasses from eastern and southern Kamchatka, and the Central Range create separate clusters (divided by empiric lines, see Figure 7.5), depending on the content of chemical elements. Probably, this feature reflects the specifics of primary magma sources that determine the formation of obsidians in these regions.

One of the specific features of the Kamchatka Peninsula is that besides the proper island arc volcanism in the boundary region between two convergent tectonic plates, a different type of volcanic activity is also known, and it is petrologically and geochemically close to the within-plate geodynamic settings (Khanchuk 2006). The Bakening Volcano, located between the Central Kamchatkan Basin and the Eastern Range, is one of such structures (Figure 7.7). Here, volcanic rocks are represented by the Miocene–Pliocene calc-alkali basalts of the Paratunka and Alnei series, and by the late Pliocene – early Pleistocene sub-alkali rocks enriched with high-charged elements (Dorendorf *et al.* 2000). The

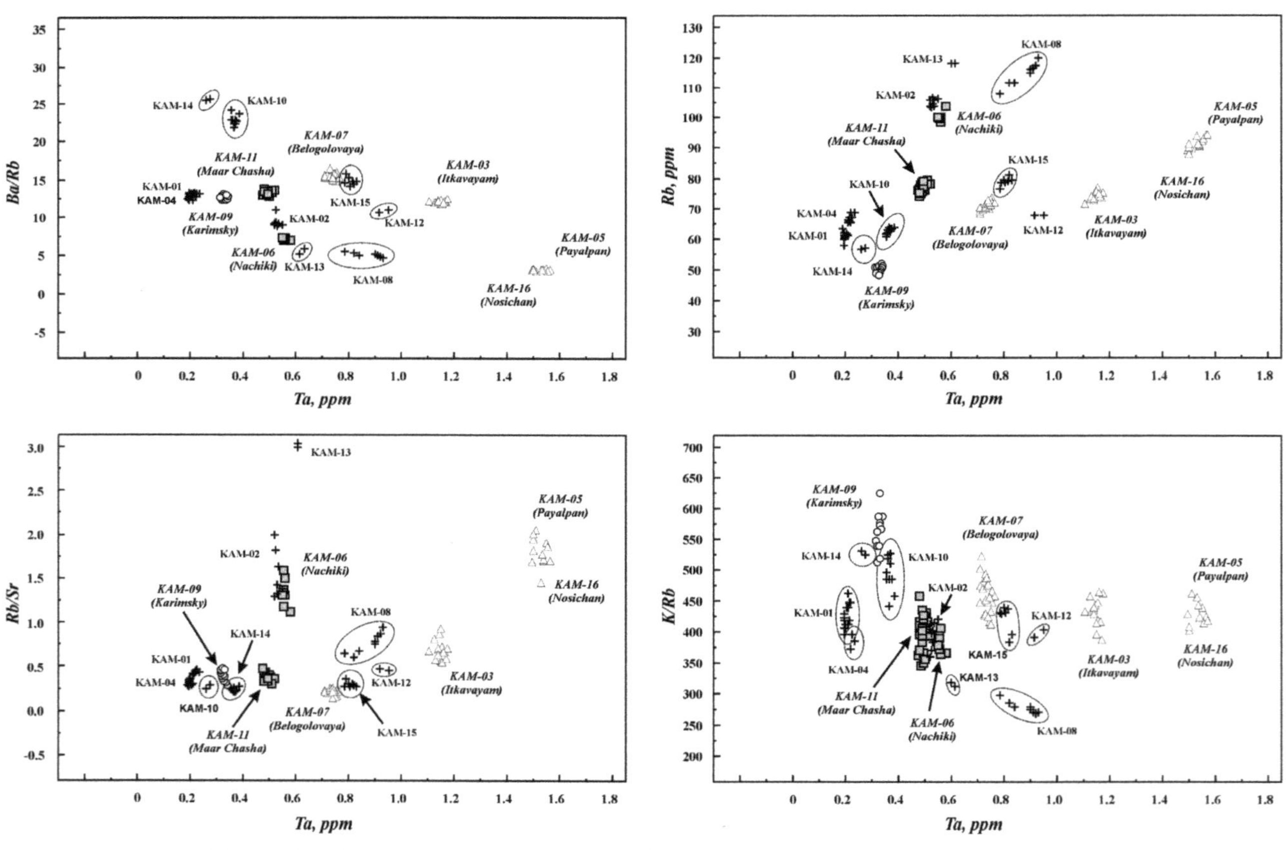

Figure 7.4. Bivariate plots showing geochemical groups of Kamchatkan obsidian (ellipses indicate 95% confidence level)

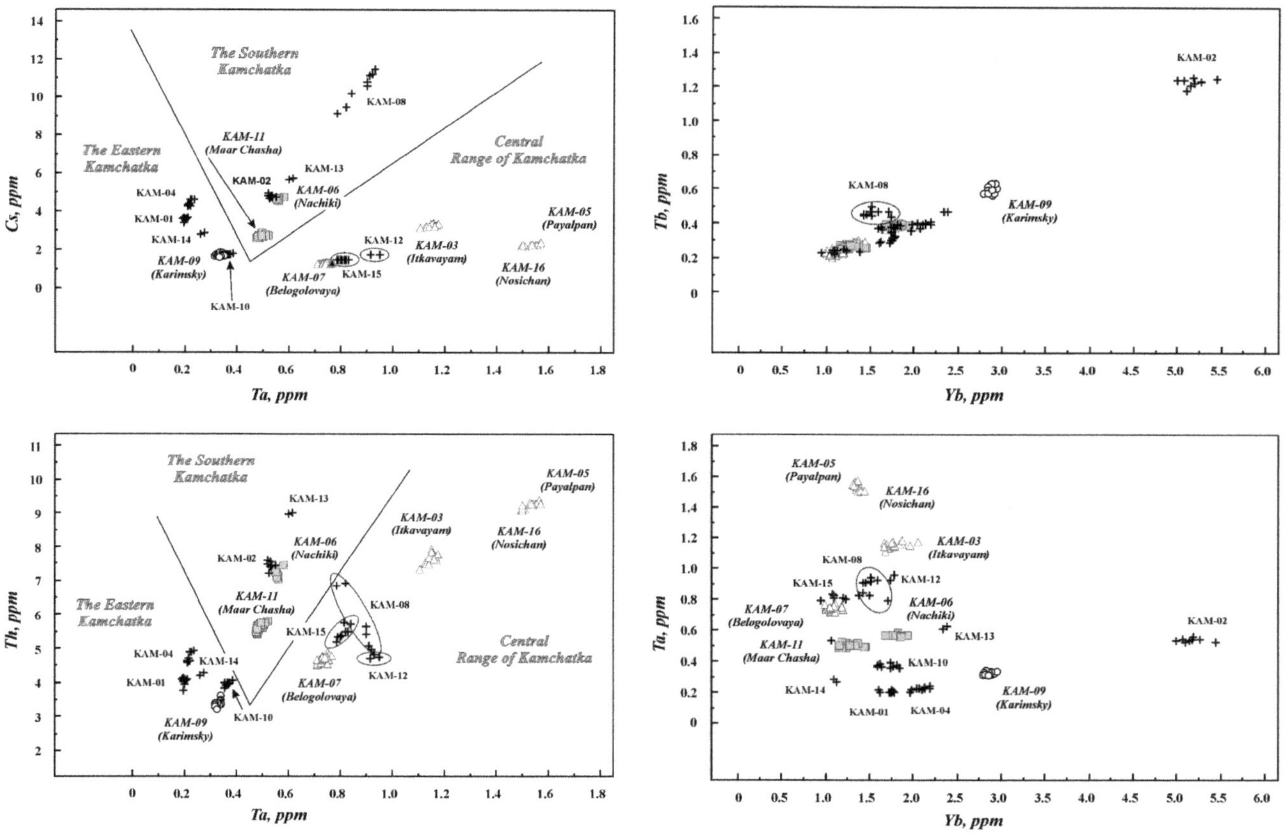

Figure 7.5. Bivariate plots showing geochemical groups of Kamchatkan obsidian (ellipses indicate 95% confidence level)

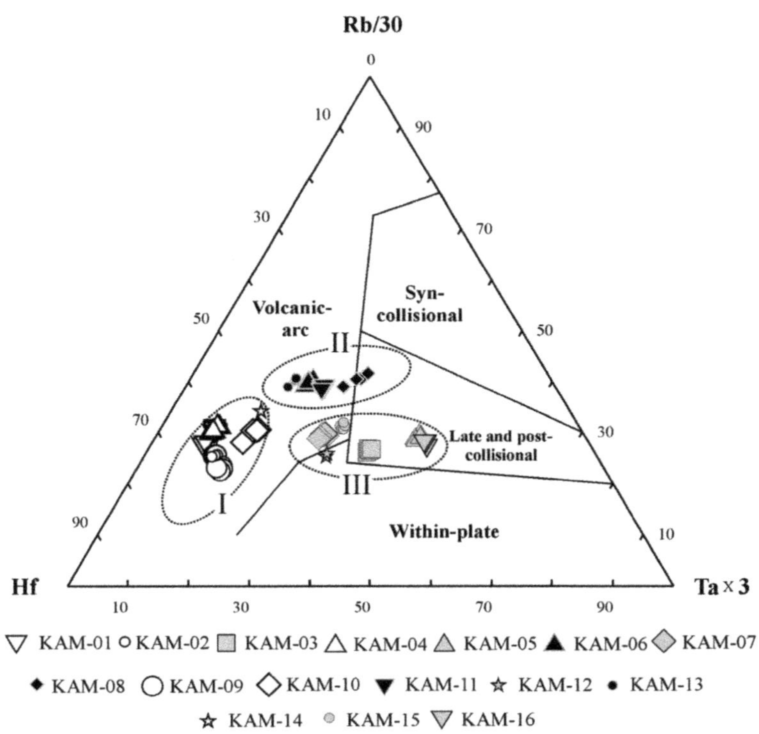

Figure 7.6. The Hf – Rb / 30 – Ta x 3 discriminant diagram for obsidian source samples and artefacts (groups KAM-01 through KAM-16) from Kamchatka and their tectonic position (after Grebennikov et al. *2010, modified). I – volcanic glasses from the Eastern Range; II – volcanic glasses from southern Kamchatka; III – volcanic glasses from the Central Range*

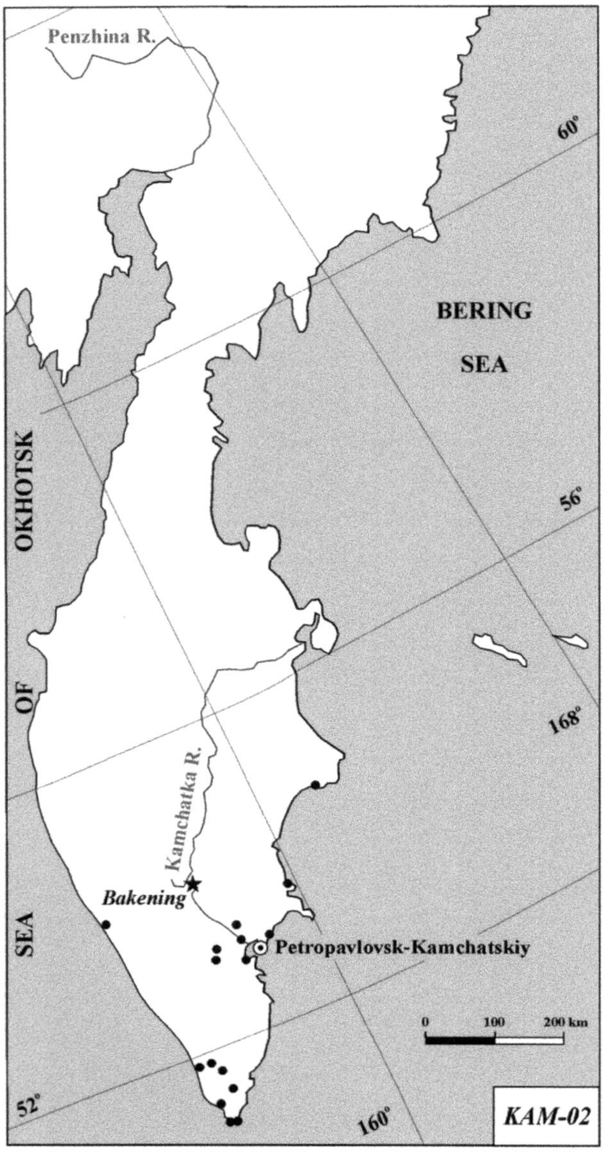

Figure 7.7. The distribution of archaeological sites with obsidian artefacts of the KAM-02 group (black circles) on Kamchatka, and location of the Bakening Volcano (black star) as their possible source (after Grebennikov et al. *2010, modified)*

volcano is situated in the headwaters of the Srednaya Avacha River. Several archaeological sites with obsidian of the KAM-02 group are located at the mouth of the Avacha River which originates from the confluence of the Srednaya Avacha and Levaya Avacha rivers (Figure 7.7). Along with the stratovolcano of Bakening, the watershed of these rivers has a multitude of monogenic rhyodacite volcanic domes, so-called *tuyas*. Dorendorf *et al.* (2000, 135) recorded the black phenocryst-free obsidian among the tuya rocks. It is therefore quite possible that the source of the KAM-02 group is located in this area.

The KAM-15 Group

According to the geochemical data, the KAM-15 group has some resemblance to the Belogolovaya Vtoraya River source (group KAM-07) and corresponds to obsidians of the Central Range (Figure 7.4). This fact can be used to tentatively place the source of the KAM-15 group in the Ichinsky Volcano region.

AGE OF OBSIDIAN AS A CRITERION FOR THE SEARCH OF UNKNOWN SOURCES

Geological Samples

The obsidians for age determination were selected from the Ichinsky Volcanic Centre, Lake Palana, Nachiki, and the Yagodnaya Summit, which are among the most important obsidian outcrops of Kamchatka. The late Pleistocene – Holocene volcanic glasses from the Chasha Maar, Karymsky, Khangar, and other volcanoes were excluded from this study because of their relatively young geological age, which cannot be determined by the K–Ar method with the necessary precision.

The obsidian outcrops mentioned above (except for the Yagodnaya Summit) have not been dated by isotopic methods before, and their ages remained the subject of dispute because of a lack of geochronological data for young volcanic rocks on Kamchatka. In recent years, however, the situation has changed considerably. New argon-argon (Ar–Ar) age determinations for ignimbrites were obtained for the eastern and southern parts of Kamchatka as well as for the Central Range (Bindeman *et al.* 2010; Leonov and Rogozin 2010; Leonov *et al.* 2008). According to these studies, the oldest ignimbrites are now known in the Central Range (dated to ca. 3.5–5.7 Ma ago), and ignimbrites from eastern and southern Kamchatka are younger than 3 Ma. This is an important fact because ignimbrite volcanism was generally accompanied by and ended with the intrusion of extrusive rhyolites; and the outcrops of volcanic glasses (obsidians and perlites) are often found in direct association with ignimbrites.

The Lake Palana obsidian outcrop is located in the Palana River headwaters (northern part of the Central Range), and is represented by obsidian dikes and beds in volcanic rocks of the Alnei Series (Shevchuk 1981). The age of the obsidian obtained, ca. 5.57 Ma (Table 7.1), confirms the earlier conclusion by Leonov *et al.* (2008) about the early Pliocene acidic volcanism in the Central Range.

The Ichinsky Volcanic Centre in the middle section of the Central Range has more than ten primary obsidian outcrops (see above). Although the geology of igneous rocks in this region is well-studied, it was still necessary to obtain an age determination for the obsidian. We analysed five obsidian samples from the main outcrops. Two of them, from the Payalpan and the Belogolovaya Vtoraya River sources, revealed excess air argon. Therefore, these samples are unsuitable for age determination by the K–Ar method. Three other samples (from the Dimshikan, Bystraya River, and Nosichan outcrops) were successfully dated.

Table 7.1. Ages determined by K–Ar method for obsidians from Kamchatka (after Budnitsky 2013)

Sample ID	Age, Ma	Source Name and/or Geochemical Group No.	Location of Source
Geological samples			
G-7	5.57 ± 0.16	Lake Palana	Central Range (north)
G-1	2.78 ± 0.14	Bystraya River	Central Range (middle)
KTE-4	2.77 ± 0.085	Dimshikan	Central Range (middle)
KLD-2	1.31 ± 0.087	Nosichan (KAM-16)	Central Range (middle)
PK-3-1	3.06 ± 0.19	Bannaya River	Southern Kamchatka
Archaeological samples			
KP-3-1	5.90 ± 0.18	KAM-08	Central Range (north) or Kamchatka Isthmus?
KP-3-1	6.32 ± 0.18	KAM-08	Central Range (north) or Kamchatka Isthmus?
KP-20-6	3.27 ± 0.09	KAM-10	Eastern Kamchatka? [Stol Summit, age of 3.71 ± 0.08 Ma; Bindeman *et al.* (2010); Leonov *et al.* (2008)]
KP-1-1	1.94 ± 0.057	KAM-01	Southern Kamchatka? [Karymshina Caldera, age of ca. 1.78–1.81 Ma; Leonov *et al.* (2008)]

The Dimshikan outcrop is located in the headwaters of the Dimshikan River, a tributary of the Bystraya River in the Central Range. It is represented by lens-like obsidian layers within the rhyolite lava flows of the Alnei Series. The result of K–Ar dating, ca. 2.77 Ma (Table 7.1), testifies that rhyolites and the related volcanic glasses were formed in the late Pliocene, and this corresponds to the Alnei Phase of Kamchatkan volcanism (Bindeman *et al.* 2010). The Bystraya River obsidian outcrop is situated in the upper reaches of the river basin. Volcanic glasses can be found as a sheet in the sequence of acidic rocks of the Alnei Series. These rocks of unusual light-brown colour contain up to 10% of plagioclase phenocrysts. The absolute age of ca. 2.78 Ma (Table 7.1) corresponds to the Alnei Phase. The Nosichan obsidian outcrop belongs to the middle Pleistocene extrusive rhyolites according to the available data (Otchet ... 1992). However, dating of this obsidian revealed an early Pleistocene age of the rhyolite extrusion, ca. 1.31 Ma (Table 7.1).

Leonov *et al.* (2008) determined the age of ignimbrites in the Nosichan Volcano area as 4.02 ± 0.12 Ma (an obsidian fragment in welded tuff was dated). The results obtained by our group testify that extrusive rhyolites are not connected to the Pliocene phase of the ignimbrite volcanism that occurred in the Central Range. The time of origin for obsidians at the Nosichan Volcano corresponds to ignimbrites of the Kronotsk–Gamchen volcanic structure in the Eastern Kamchatkan Volcanic Belt and of the Bannaya River basin in southern Kamchatka (Leonov *et al.* 2008). We suggest that obsidians of the Nosichan Volcano were formed during the early Pleistocene phase of acidic volcanism in the Central Range.

Archaeological Samples

Out of seven geochemical groups consisting entirely of archaeolcgical obsidian, four samples were analysed (Table 7.1). The data obtained reflect ages of three volcanic glass sources from unknown locales. The KAM-01 group is the largest in terms of the number of samples: 113 artefacts from 20 archaeological sites, with most of them from southern Kamchatka (see Grebennikov *et al.* 2010, 107). The KAM-01 obsidians correspond to volcanic glasses from the Eastern Kamchatkan Volcanic Belt (see above). For dating, an obsidian artefact from the Avacha site was selected (KP-1-1, Table 7.1). The age of this specimen, ca. 1.94 Ma, indicates that the primary source of volcanic glass is of early Pleistocene age. This epoch is known for powerful ignimbrite eruptions which affected both eastern and southern Kamchatka. In particular, ignimbrites of the Karymshyna Caldera in southern Kamchatka were formed during the period of 1.78–1.81 Ma ago (Table 7.1). According to information provided by V. L. Leonov (personal communication, 2012), older rhyolite extrusions dated to ca. 2–3 Ma ago are known in the vicinity of this caldera. All these data help our group to plan a fieldwork campaign within the boundaries of the Karymshyna Caldera in the near future.

The KAM-10 group of archaeological obsidians is represented by 52 samples from 10 sites, compactly distributed along the Kamchatka River; the only exception is the Lisy site at the southern end of the Kronotsk Peninsula (Grebennikov *et al.* 2010). An obsidian artefact from the Anavgai 2 site was chosen for dating as typical for this group. The age obtained, ca. 3.27 Ma (KP-2-6; Table 7.1), testifies that the primary obsidian source is dated to the late Pliocene. Close correlation exists between the geochemical composition of obsidians of this group and volcanic glasses of the Eastern Kamchatkan Volcanic Belt (see above). The data about the age of the KAM-10 obsidian allows us to suggest a possible connection between the primary source and the Pliocene ignimbrite volcanism of the Karymsky Volcanic Centre. For example, the ignimbrite strata of the Stol Summit situated on the watershed of the Levaya

Zhupanova and Pravaya Zhupanova rivers contain welded tuffs and obsidian fragments dated to ca. 3.71 Ma ago (Table 7.1). Besides this, near the Anavgai 2 site on the western slopes of the Eastern Range there are outcrops of obsidian of the same age (V. L. Leonov, personal communication 2012).

The KAM-08 group consists of nine volcanic glass samples from the Pakhachi, Vaimitangin, and Penzhina archaeological sites in northern Kamchatka (also called Koryak Region) (Grebennikov *et al.* 2010, 113). We analysed two kinds of obsidian artefacts from the Pakhachi site: 1) translucent volcanic glass; and 2) poorly translucent (with abundant ore mineral spots) volcanic glass, and determined two ages corresponding to the late Miocene: ca. 5.90–6.32 Ma (KP-3-1; Table 7.1). According to the data obtained, the primary source of obsidian for this group may occur at one of the volcanic structures of the Central Range. However, based on the overall distribu-tion of artefacts assigned to this source (see Grebennikov *et al.* 2010, 113), it seems very likely that it is located north of the Kamchatka Isthmus, or within this area.

CONCLUSIONS

Knowledge of geology and volcanology in regions with abundant obsidian artefacts is one of the major require-ments for understanding the genesis and exact location of primary volcanic glass sources as well as the patterns of obsidian exploitation by prehistoric people. Using the information on the geological structure of the territory under investigation, it is possible to establish the criteria for analysis of geochemical data for both geological and archaeological obsidians, in order to pinpoint the original outcrops which were used by ancient populations to acquire valuable raw material (see also Wada *et al.*, this volume). This is especially necessary for regions with a complex geological structure and a multitude of obsidian sources, and the Kamchatka Peninsula of the northern Russian Far East is one of these territories.

The age determinations for both geological and archaeological obsidians used in our study are in good accordance with the results of geochemical typification of volcanic glasses from Kamchatka. The combination of these approaches along with features of spatial distribution of obsidian with different geochemical signatures at archaeological sites allows us to propose a strategy for the search of the unknown sources of archaeological obsidian. As a result of its application, we suggest the localisations of primary sources for different geochemical groups of archaeological obsidian: 1) the Karymshina Caldera and its vicinity (southern Kamchatka) for the KAM-01 group; 2) the vicinity of the Stol Summit (Karymsky Volcanic Centre, the Eastern Range) for the KAM-04, KAM-10, and KAM-14 groups; 3) rhyodacitic domes (tuyas) of the Bakening Volcano (between the Eastern and Central ranges in the upper reaches of the Kamchatka River) for the KAM-02 group; 4) the Kamchatka Isthmus (northern part of the Central Range) for the KAM-08 group; and 5) the vicinities of the Ichinsky Volcano (middle part of the Central Range) for the KAM-15 group.

We are fully aware that only extensive fieldwork, including collection and analyses of additional samples of volcanic glass from the above mentioned localities, can prove the reliability of our suggestions. We are strongly aiming to complete this task in the near future with the help of colleagues from Russia, USA, and Japan.

Acknowledegements

This research, initiated by the authors in 2003, is the logical continuation of earlier studies of obsidian sources in the Russian Far East which began in 1992. This study was supported throughout the last 20 years by grants from several foundations, including the US NSF (DBS-9205506, BNS-9102016, SBR-9503035 and 9802366) and CRDF (RG1-2538-VL-03); Russian RFFI (96-06-80688, 99-06-80348, 02-06-80282 and 06-06-80258), RFFI–FEB RAS (06-08-96012), and FEB RAS (06-III-A-08-319). We are especially thankful to Dr M. D. Glascock for continuous help and support in studying the obsidian sources of the Russian Far East, and to the MURR staff members for providing assistance with sampling and processing of obsidian samples from Kamchatka. We are also grateful to Drs A. B. Perepelov, V. M. Okrugin, E. Y. Baluev, A. A. Gorbach, V. L. Leonov, E. N. Grib, V. I. Andreev, S. M. Moskaleva, and M. M. Pevzner for supplying us with samples from obsidian sources of Kamchatka and for valuable discussions. We thank Prof. M. A. Kiryak (Dikova) and Dr A. A. Lebedintsev for providing access to obsidian collections from Kamchatka stored at Magadan. Dr S. Y. Budnitsky generously provided the results of K–Ar dating, and Dr A. V. Ptashinsky submitted obsidian artefacts for analysis; and we appreciate it. We are grateful to Dr V. L. Leonov and Prof. A. E. Izokh for useful comments and remarks on the earlier draft. Finally, our thanks go to Prof. A. Ono for his invitation to visit Japan in October–November 2011 and financial support. The core of this paper is the presentation given at the Workshop "Methodological Issues of Obsidian Provenance Studies and the Standardisation of Geologic Obsidian", which took place at the Centre for Obsidian and Lithic Studies, Nagawa Town (Nagano Pref., Japan), on 5–6 November 2012. Dr S. G. Keates kindly helped with improving the grammar of this paper.

References

AVDEIKO, G. P., A. A. PALUEVA, and O. A. KHLEBORODOVA. 2006. Geodynamic Conditions of Volcanism and Magma Formation in the Kurile–Kamchatka Island Arc System. *Petrology* 14, 230–246.

BINDEMAN, I. N., V. L. LEONOV, P. E. IZBEKOV, V. V. PONOMAREVA, K. E. WATTS, N. K. SHIPLEY, A. B. PEREPELOV, L. I. BAZANOVA, B. R. JICHA, B. S.

SINGER, A. K. SCHMITT, M. V. PORTNYAGIN, and C. H. CHEN. 2010. Large-Volume Silicic Volcanism in Kamchatka: Ar–Ar and U–Pb Ages, Isotopic, and Geochemical Characteristics of Major Pre-Holocene Caldera-Forming Eruptions. *Journal of Volcanology and Geothermal Research* 189, 57–80.

BUDNITSKY, S. Y. 2013. *Novaya Kalii–Argonovaya Geokhronologiya Redkikh i Unikalnykh Geologicheskikh i Arkheologicheskikh Obyektov* [The New Potassium–Argon Geochronology of Rare and Unique Geological and Archaeological Objects]. Synopsis of the Candidate of Geological–Mineralogical Sciences (PhD-equivalent) Dissertation. Vladivostok, Dalnevostochny Geologichesky Institut.

DIRKSEN, O. V., V. V. PONOMAREVA, and L. D. SULERZHITSKY. 2002. The Chasha Crater (South Kamchatka) – Unique Example of Mass Explosion of Acidic Pyroclastics in the Field of Areas Basalt Volcanism. *Volcanology and Seismology* 24(5), 3–10.

DORENDORF, F., T. CHURIKOVA, A. KOLOSKOV, and G. WÖRNER. 2000. Late Pleistocene to Holocene Activity at Bakening Volcano and Surrounding Monogenetic Centers (Kamchatka): Volcanic Geology and Geochemical Evolution. *Journal of Volcanology and Geothermal Research* 104, 131–151.

EREMINA, T. S. 2009. Geneticheskaya Interpretatsiya Geokhimicheskikh i Petrographicheskikh Osobennostey Obsidianov Proyavleniya Payalpan (Kamchatka) [Genetic Interpretation of Geochemical and Petrographic Features of Obsidians from Payalpan Outcrop (Kamchatka)]. In *Vulkanism i Geodinamika, Tom 1*, edited by E. I. Gordeev, 166–169. Petropavlovsk-Kamchatskiy, Institut Vulkanologii i Seismologii.

FEDOTOV, S. A., and Y. P. MASURENKOV (eds). 1991. *Active Volcanoes of Kamchatka*. Moscow, Nauka Publishers.

GLASCOCK, M. D., G. E. BRASWELL, and R. H. COBEAN. 1998. A Systematic Approach to Obsidian Source Characterization. In *Archaeological Obsidian Studies: Method and Theory*, edited by M. S. Shackley, 15–65. New York & London, Plenum Press.

GLASCOCK, M. D., V. K. POPOV, Y. V. KUZMIN, R. J. SPEAKMAN, and A. V. PTASHINSKY. 2006. Obsidian Sources and Prehistoric Obsidian Use on the Kamchatka Peninsula: Initial Results of Research. In *Archaeology in Northeast Asia: On the Pathway to Bering Bridge*, edited by D. E. Dumond and R. L. Bland, 73–88. Eugene, University of Oregon Museum of Natural and Cultural History.

GLASCOCK, M. D., R. J. SPEAKMAN, and H. NEFF. 2007. Archaeometry at the University of Missouri Research Reactor and the Provenance of Obsidian Artefacts in North America. *Archaeometry* 49, 353–367.

GREBENNIKOV, A. V., V. K. POPOV, M. D. GLASCOCK, R. J. SPEAKMAN, Y. V. KUZMIN, and A. V. PTASHINSKY. 2010. Obsidian Provenance Studies on Kamchatka Peninsula (Far Eastern Russia): 2003–9 Results. In *Crossing the Straits: Prehistoric Obsidian Source Exploitation in the North Pacific Rim* (B.A.R. International Series 2152), edited by Y. V. Kuzmin and M. D. Glascock, 89–120. Oxford, BAR Publishing.

IGNATIEV, A. V., T. A. VELIVETSKAYA, and S. Y. BUDNITSKII. 2010. A Method for Determining Argon Isotopes in a Continuous Helium Flow for K/Ar Geochronology. *Journal of Analytical Chemistry* 65, 1347–1355.

KHAIN, V. E. 1994. *Geology of Northern Eurasia (Ex-USSR). Part 2. Phanerozoic Fold Belts and Young Platforms*. Berlin & Stuttgart, Gebrüder Borntraeger.

KHANCHUK, A. I. (ed.). 2006. *Geodinamika, Magmatizm i Metallogeniya Vostoka Rossii. Kniga 1* [Geodynamics, Magmatism, and Metallogeny of the Eastern Russia. Book 1]. Vladivostok, Dalnauka Publishers.

KUZMIN, Y. V., and M. D. GLASCOCK (eds). 2010. *Crossing the Straits: Prehistoric Obsidian Source Exploitation in the North Pacific Rim* (B.A.R. International Series 2152). Oxford, BAR Publishing.

KUZMIN, Y. V., R. J. SPEAKMAN, M. D. GLASCOCK, V. K. POPOV, A. V. GREBENNIKOV, M. A. DIKOVA, and A. V. PTASHINSKY. 2008. Obsidian Use at the Ushki Lake Complex, Kamchatka Peninsula (Northeastern Siberia): Implications for Terminal Pleistocene and Early Holocene Human Migrations in Beringia. *Journal of Archaeological Science* 35, 2179–2187.

LEONOV, V. L., I. N. BINDEMAN, and A. N. ROGOZIN. 2008. New Ar–Ar Age Dating of Kamchatkan Ignimbrites. In *Proceedings of the Conference on Volcanologist's Day 2007*, edited by E. I. Gordeev, 187–196. Petropavlovsk-Kamchatskiy, Institut Vulkanologii i Seismologii (in Russian with English Summary).

LEONOV, V. L., and A. N. ROGOZIN. 2010. Occasional Character of Acidic Volcanism in Southern Kamchatka (Based on Results of K–Ar, Ar–Ar, and U–Pb Datings) In *Proceedings of the Conference on Volcanologist's Day 2009*, edited by E. I. Gordeev, 62–74. Petropavlovsk-Kamchatskiy, Institut Vulkanologii i Seismologii (in Russian with English Summary).

MALAINEY, M. E. 2011. *A Consumer's Guide to Archaeological Science: Analytical Techniques*. New York, Springer.

MASURENKOV, Y. P. (ed.). 1980. *Vulkanichesky Tsentr: Stroenie, Dinamika, Veshchestvo (Karymskaya Struktura)* [Volcanic Centre: Feature, Dynamics, and Substance (Karymskaya Structure)]. Moscow, Nauka Publishers.

MELEKESTSEV, I. V., O. A. BRAITSEVA, L. I. BAZANOVA, V. V. PONOMAREVA, and L. D. SULERZHITSKY. 1996. A Particular Type of Catastrophic Explosive Eruptions with Reference to the Holocene Subcaldera Eruptions at Khangar, Khodutka Maar, and Baraniy Amfiteatr Volcanoes in Kamchatka. *Volcanology and Seismology* 18(2), 3–24.

OTCHET ... 1992. Otchet po Revizionno-Poiskovym Rabotam na Proyavleniakh Obsidiana Gory Skalistoi, Ichinskogo Vulkana i Predvaritelnaya Geologo–Ekonomicheskaya Otsenka Proyavleny Obsidiana Kamchatki za 1991–1992 Gody [Report on Revision and Survey Study of the Obsidian Localities of the Skalistaya Summit, Ichinsky Volcano, and Preliminary Geological–Economical Estimate of the Obsidian Localities at the Kamchatka in 1991–1992]. Manuscript on File. Petropavlovsk-Kamchatskiy, Kamchatgeologiya.

POLLARD, M., C. BATT, B. STERN, and S. M. M. YOUNG. 2007. *Analytical Chemistry in Archaeology*. Cambridge, Cambridge University Press.

POPOV, V. K., A. V. GREBENNIKOV, A. B. PEREPELOV, Y. V. KUZMIN, M. D. GLASCOCK, and R. J. SPEAKMAN. 2007. Geokhimicheskaya Tipizatsiya Kislykh Vulkanicheskikh Stekol Kamchatki [Geochemical Typification of the Acidic Volcanic Glasses from Kamchatka Peninsula]. In *Problemy Geokhimii Endogennykh Protsessov i Okruzhayushchei Sredy. Tom 1. Geokhimiya Magmaticheskikh, Metamorficheskikh i Metasomaticheskikh Protsessov*, edited by A. Y. Medvedev, 201–206. Irkutsk, Institut Geokhimii.

REUTHER, J. D., N. S. SLOBODINA, J. T. RASIC, J. P. COOK, and R. J. SPEAKMAN. 2011. Gaining Momentum: Late Pleistocene and Early Holocene Archaeological Obsidian Source Studies in Interior and Northeastern Beringia. In *From the Yenisei to the Yukon: Interpreting Lithic Assemblage Variability in Late Pleistocene/Early Holocene Beringia*, edited by T. Goebel and I. Buvit, 270–286. College Station, Texas A&M University Press.

SHEIMOVICH, V. S., and M. G. PATOKA. 2000. *Geologicheskoe Stroenie Zon Aktivnogo Kainozoisko-go Vulkanizma* [Geological Structure of Zones of the Active Cenozoic Volcanism]. Moscow, GEOS Press.

SHEVCHUK, V. D. 1981. Geneticheskie Tipy Mestorozhdeniy Kislykh Vulkanicheskikh Stekol na Kamchatke [Genetic Types of the Acidic Volcanic Glass Deposits in Kamchatka]. In *Perlity*, edited by V. V. Nasedkin and V. P. Petrov, 43–51. Moscow, Nauka Publishers.

PROVENANCE STUDIES OF ARCHAEOLOGICAL OBSIDIAN IN NORTHEAST ASIA: CURRENT PROGRESS

Chapter 8
IDENTIFICATION OF ARCHEOLOGICAL OBSIDIAN SOURCES IN KANTO AND CHUBU REGIONS (CENTRAL JAPAN) BY ENERGY DISPERSIVE X-RAY FLUORESCENCE ANALYSIS

Nobuyuki IKEYA

Abstract: *This paper describes the methods used for the identification of obsidian sources based on non-destructive Energy Dispersive X-ray Fluorescence analysis. The case study targets obsidian artefacts from Upper Palaeolithic sites at the piedmont of Mount Ashitaka in the Chubu Region of Japan. One of the most important outcomes is the confirmation of the presence of obsidian from the Kozushima insular source on the main landmass of Japan during the Upper Palaeolithic. Kozushima is a small volcanic island located in the Pacific Ocean approximately 50 km SE of the Izu Peninsula. It was never connected to the main islands of Japan. This fact raises extremely important questions in world prehistory regarding the sea voyages and maritime transport in the Upper Palaeolithic. Also, because many of the Kozushima obsidian artefacts belong to the early Upper Palaeolithic, they provide a new perspective regarding the means and routes used by modern humans to reach the Japanese Islands. In addition, one of the most important results, namely the presence of the Kozushima obsidian in central Japan, is now confirmed by the Neutron Activation Analysis.*

Keywords: *Energy Dispersive X-ray Fluorescence Analysis, Neutron Activation Analysis, Upper Palaeolithic, Discrimination Diagram, Kanto Region, Chubu Region, Kozushima Island, Japan, Sea Voyage*

INTRODUCTION

In Japan, X-ray Fluorescence (XRF) analysis which targets archaeological materials was put into practical use in the 1970s. Based on the results from XRF, the chemical characteristics of sanukite (glassy andesite) in Japan were determined. This information was applied to establish the primary sources of prehistoric sanukite implements (Higashimura and Warashina 1975). Subsequently, the chemical composition of 60 obsidian sources in Japan was studied by XRF, and the identification of sources for obsidian artefacts was also attempted (Kamaki *et al.* 1984).

At almost the same time, an effort to determine the origin of the obsidian by the Neutron Activation Analysis (NAA) was initiated (Suzuki and Tomura 1983). But this method had the drawback of requiring the sample to be destroyed and also being expensive. Consequently, Energy Dispersive XRF (EDXRF) analysis, which is completely non-destructive, has become the main way of identifying obsidian sources in Japan up until the present.

In the Hokkaido, Kanto, and Chubu regions of Japan, obsidian was a major raw material in prehistory from the Palaeolithic through to the end of the Jomon period. A significant number of obsidian artefacts were analysed by EDXRF from the late 1980s to the early 1990s. However, at each archaeological site the number of obsidian artefacts analysed by EDXRF ranged from just a few items to several dozens. The results obtained did not provide sufficient data to contribute to the discussion about the trade in raw materials or migrations of modern humans.

In 1994, a large-scale project was launched in order to study by XRF the composition of obsidian artefacts, sometimes ranging from several hundreds to thousands of pieces per site (Mochizuki *et al.* 1994). Subsequently, obsidian sourcing studies rapidly increased in Japan (e.g., Ikeya and Mochizuki 1998; Izuho and Sato 2007; Tsutsumi 2007, 2010). Such results manifest the beginning of advanced research related to the exchange and trade of stone materials, human migrations, and other issues.

One of the most important discoveries was the presence of Kozushima [Kozu-shima] insular obsidian on the main landmass of Japan during the Upper Palaeolithic. Kozushima is a small volcanic island measuring 18.48 km^2, located in the Pacific Ocean approximately 50 km SE of the Izu Peninsula, Honshu Island (Figure 8.1). It appeared as a result of volcanic activity, that occurred several tens of thousands of years ago, and was never connected to the Izu Peninsula. Therefore, the existence of Kozushima obsidian on one of the main islands of Japan raises extremely important questions in world prehistory regarding the possibility of seafaring and maritime transport in the Upper Palaeolithic. In addition, it provides a new perspective regarding the means and routes used by modern humans to reach the Japanese Islands during the early Upper Palaeolithic.

This paper describes the methods used for identification of obsidian sources based on EDXRF, and presents relevant information concerning Upper Palaeolithic obsidian artefacts from the sites located at the piedmont of Mount Ashitaka. In addition, it validates the identification of Kozushima obsidian based on the use of NAA.

OBSIDIAN SOURCES IN THE CHUBU AND KANTO REGIONS

In the inland part of the Kanto Region, only Mount Takahara [Takaharayama] located in the northern area

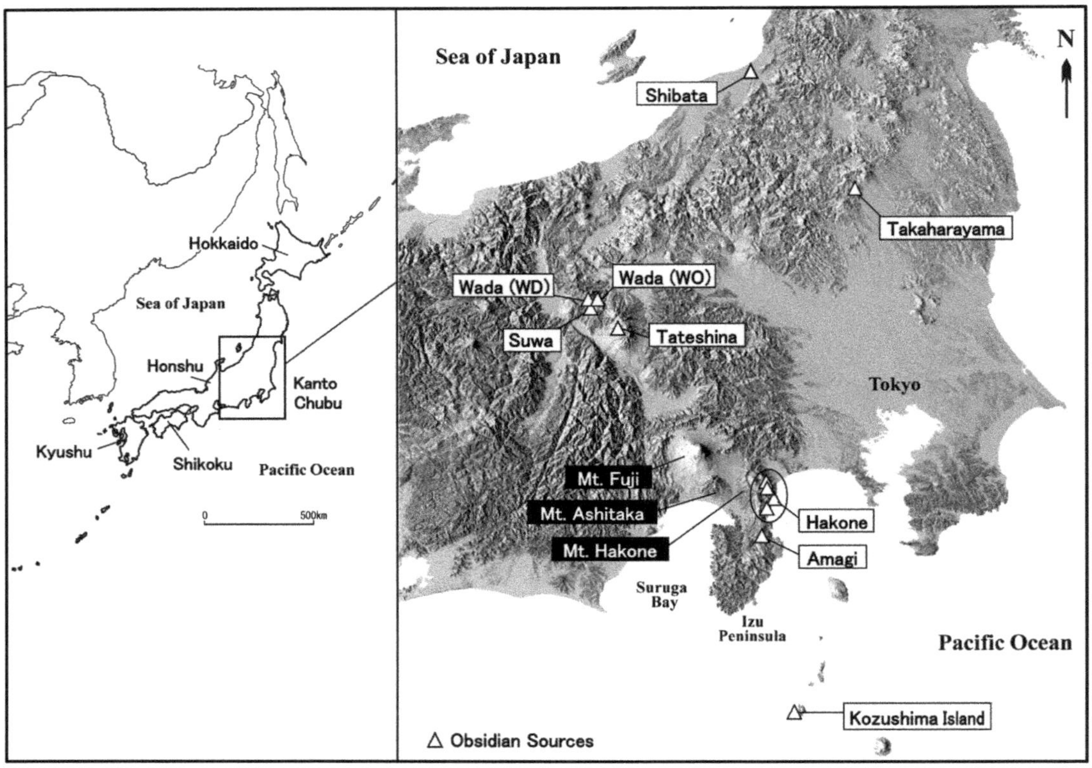

Figure 8.1. Major obsidian sources and areas in the Chubu and Kanto regions of Japan

produces obsidian (Figure 8.1). Orogenic activities that began 20 Ma ago created a series of mountains known as the Central Highlands in the middle part of the Chubu Region. In this area, the obsidian occurs in close association with the rhyolitic volcanism. Notable sources of obsidian in the Central Highlands (a.k.a. the Shunshu Region) include Suwa, Wada (WD), Wada (WO), and Tateshina (Figure 8.1). All of these are large in scale, and contain high quality (i.e., glassy and phenocryst-free) obsidian.

The obsidian-producing areas on the Pacific Ocean side of the Chubu and Kanto regions include Hakone, Amagi, and Kozushima. The Kozushima source area is composed almost entirely of rhyolitic ejecta, and consists of four well-known individual sources: Onbasejima, Sanukazaki, Sawajiri, and Nagahama. The Onbasejima source is a reef situated approximately 5 km W–SW of the main island of Kozushima. The seabed located east of the reef is several tens of metres deep and contains a large amount of relatively high quality crude obsidian blocks, with some samples larger than a human head. The reef is composed of an eroded rhyolitic lava dome. Big obsidian outcrops were likely visible above the water during the Upper Palaeolithic when the sea level was much lower than today.

MATERIALS AND METHODS

Energy Dispersive X-ray Fluorescence Analysis

An atom's nucleus is surrounded by orbiting electron shells known as the K, L, M, N, O, and P shells, in the order of their proximity to the nucleus. Outer orbits have higher energy levels than those of inner ones. When an incident X-ray encounters an atom, it knocks out an electron from the inner shell of the nucleus, and an electron from the outer shell falls down into the vacant inner shell by releasing energy which is in the form of X-ray fluorescence. Because the energy differences between orbits are specific to each atom, the resulting fluorescence X-rays have a specific energy for each element. Furthermore, because more X-ray fluorescence is produced when the concentration of the element in the sample is high, calculation of the amount of the element present in the sample is possible. The analytical method that utilises this principle is known as the XRF.

Two types of spectrometers can be used for this technique: Wavelength Dispersive XRF (WDXRF) and Energy Dispersive XRF (EDXRF). Although the resolution of EDXRF is lower than of WDXRF, the former is capable of performing non-destructive analysis at lower cost and by a simpler measurement method. Two EDXRF spectrometers, SII Nano Technology's SEA–2001 and SEA–2110L, are available at the Mochizuki Laboratory; and a SEA–2110 EDXRF spectrometer is located at the Ikeya Laboratory. Both types of instruments were used in this study.

Analytical Conditions

The intensities (number of counts of the electronic signal) of the Kα lines for 11 elements (Al, Si, K, Ca, Ti, Mn, Fe, Rb, Sr, Y, and Zr), were analysed by the EDXRF instruments. All the EDXRF devices used in this study are equipped with a Si (Li) semiconductor detector and the Rh-anode X-ray tube. The load voltage is 50 kV, and the current is automatically set (5–7 mA). The conditions of atmosphere, beam diameter, and measurement time are vacuum, 10 mm, and 300 seconds, respectively.

The Fundamental Parameter (FP) method is one of the ways to perform the quantitative analysis by EDXRF. However, the results obtained are quite variable due to the conditions of each sample's surface and shape. Most of the artefacts made of obsidian have irregular surfaces and shapes. Thus, only the intensities of the measurement lines were used for the evaluation and geochemical characterisation of the obsidian artefacts. Furthermore, the analyses were conducted using the Rb-Kα, Sr-Kα, Y-Kα, and Zr-Kα lines, which are adequate for geochemical purposes. The Y-Kα and the Zr-Kα lines are overlapp with the Rb-Kβ and Sr-Kβ lines, respectively. The calibration related to the effect of the overlapped spectral interferences was not performed.

Discrimination Diagrams

The discrimination diagrams proposed in this study are based on the following four indicators: Indicator 1, Rb x 100 / (Rb + Sr + Y + Zr); Indicator 2, Mn x 100 / Fe; Indicator 3, Sr x 100 / (Rb + Sr + Y + Zr); and Indicator 4, log (Fe/K) (see also Kannari *et al.*, this volume). Two discrimination diagrams, the first one with Indicator 1 vs. Indicator 2 (Figure 8.2), and the second one with Indicator 3 vs. Indicator 4 (Figure 8.3), were created by plotting data from obsidian artefacts against 'geological' obsidian.

Indicators 1–3 are the ratios or sums of the neighbouring elements on the periodic table. Indicator 4 uses elements that are far apart on the periodic table; Fe decreases during the process in which magma becomes basaltic, andesitic, dacitic, and then rhyolitic, while K conversely increases in the same process. Because obsidian is formed from dacitic-to-rhyolitic magmas, one can assume that this indicator shows the stage in crystallisation differentiation of magma from which the obsidian is derived.

The sources of obsidian in the Chubu and Kanto regions are divided into nine areas and 26 sub-areas on the basis of geographic distribution and chemical characteristics (Table 8.1). Four sources in the Hakone area (HNAY, HNHJ, HNKI, and HNKT) are shown as overlapping in Figure 8.2; however, they are clearly separated in Figure 8.3. Furthermore, while the Kashiwatoge source (AGKT) in the Amagi area and the Amayuzawa source (THAY) in the Mt. Takahara area are shown as partly overlapping in Figure 8.3, they can be separated by using the information in Figure 8.2.

Discrimination Analysis

The identification achieved with the help of scatterplots is further examined by discrimination analysis, which is a kind of multivariate analysis. The final determination of

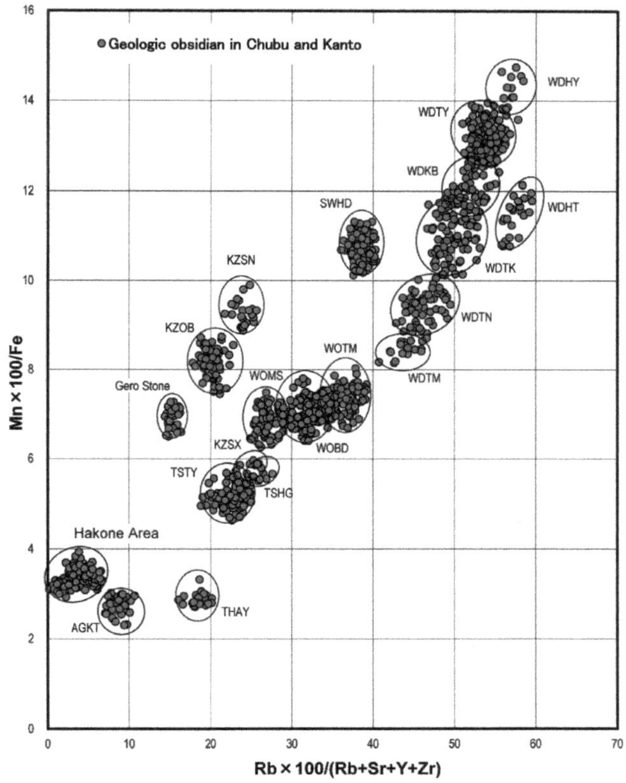

Figure 8.2. Discrimination diagram of Rb x 100 / (Rb + Sr + Y + Zr) vs. Mn x 100 / Fe, using the Kα intensity data from EDXRF for geological obsidian

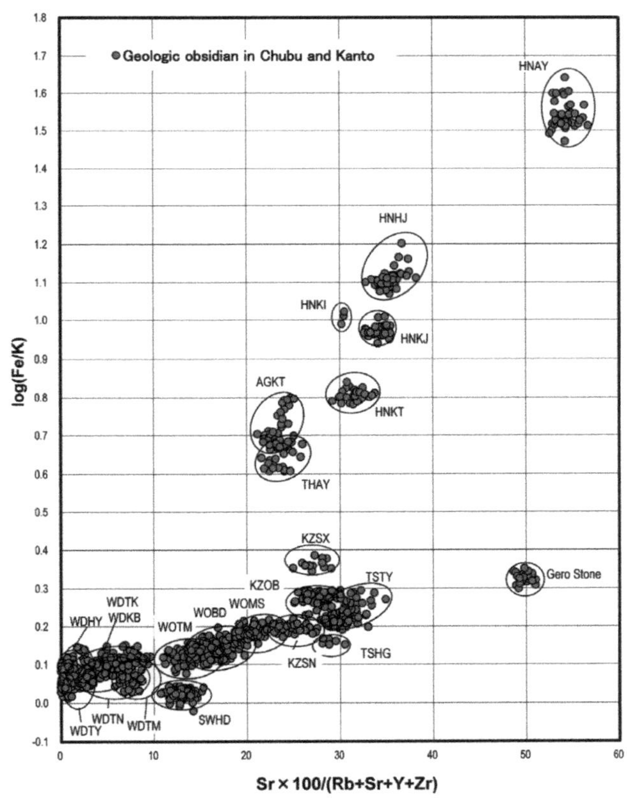

Figure 8.3. Discrimination diagram of Sr x 100 / (Rb + Sr + Y + Zr) vs. log (Fe/K), using the Kα intensity data from EDXRF for geological obsidian

Table 8.1. Areas and sub-areas of obsidian sources in the Chubu and Kanto regions

Area		Sub-Area	Abbreviations
Shinshu Region	Wada (WD)	Fuyo-raito	WDHY
		Takayama	WDTY
		Kobukazawa	WDKB
		Tsuchiyabashi-kita	WDTK
		Tsuchiyabashi-nishi	WDTN
		Tsuchiyabashi-minami	WDTM
		Furutoge	WDHT
	Wada (WO)	Takamatsuzawa	WOTM
		Budozawa	WOBD
		Makigasawa	WOMS
	Suwa	Hoshigadai	SWHD
	Tateshina	Tsumetayama	TSTY
		Futagoyama	TSHG
		Suribachiyama	TSSB
Amagi		Kashiwatoge	AGKT
Hakone		Hatajuku	HNHJ
		Kajiya	HNKJ
		Kuroiwabashi	HNKI
		Kamitaga	HNKT
		Ashinoyu	HNAY
		Hakone-X	HN-X
Kozushima		Onbasejima	KZOB
		Sanukazaki	KZSN
		Sanukazaki-X	KZSN-X
Takaharayama		Amayuzawa	THAY
Shibata		Itayama	SBIY

the source is made when both results agree. The consistency between the scatterplot and discrimination analysis is usually quite high. However, the results may disagree on rare occasions for sources with a similar chemical composition. In such cases, the sample is cleaned again, and the measurement is repeated until both results match. In some cases, part of the sample is broken off to allow measurement of a section of the sample which is not weathered.

Samples of Geological Obsidian

As reference samples used for identification of primary obsidian sources, 1241 specimens from the following source areas were collected and measured: 1) the Mt. Takahara Area: Sakurazawa source, 24 samples; 2) the Wada (WD) Area: Fuyo-raito, Choji-goryo, Takayama, Kobukazawa, Higashimochiya, Tsuchiyabashi-kita, Tsuchiyabashi-higashi, Tsuchiyabashi-nishi, Tsuchiyabashi-minami, Washigamine, Utsugisawa, Furutoge, and Wadatoge-nishi sources, 485 items; 3) the Wada (WO) Area: Budozawa, Makigasawa-shita, Makigasawa-ue, Takamatsuzawa, and Honzawa-shita sources, 183 samples; 4) the Suwa Area: Hoshigadai, Hoshigato, Suigetsu-Reien, Higashimata, and Yashima sources, 117 items; 5) the Tateshina Area: Mugikusatoge, Mugikusatoge-higashi, Shibunoyu, Tsumetayama, and Futagoike sources, 113 samples; 6) the Hakone Area: Ashinoyu, Hatajuku, Kuroiwabashi, Amazakebashi, Kajiya, and Kamitaga sources, 161 items; 7) the Amagi Area: Kashiwatoge source, 52 samples; and 8) the Kozushima Area: Onbasejima, Nagahama, Sawajiri, and Sanukazaki sources, 106 items.

RESULTS

Geological Background and Stratigraphic Sequence of Mount Ashitaka

The Ashitaka Volcano (or Mt. Ashitaka) emerged about 400,000 years ago. After the cessation of volcanic activity about 100,000 years ago, ash from the Palaeo-Fuji Volcano accumulated on Mt. Ashitaka. This thick cover of ash formed gentle ridges on the southeast piedmont of Mt. Ashitaka. This zone is called 'Ashitaka–Onoue Hill', and it was continuously occupied by humans in the Upper Palaeolithic.

The ash from the Palaeo-Fuji Volcano formed the Ashitaka Loam deposits which are sub-divided into three parts: Lower, Middle, and Upper members. The Upper Member of the Ashitaka Loam is overlain by intercalated aeolian deposits of wind-blown scoria (SCI, SCII, SCIII-s1, s2, s3, s4, s5, and SCIV), and palaeosol layers called "Black Bands" (BB0, BBI, BBII, BBIII, SCIII-b1, b2, BBIV, BBV, BBVI, and BBVII). Most of the archaeological assemblages are found in the Black Band layers (Figure 8.4). The age of the lowest members of the Black Bands, BBVI and BBVII, is estimated as ca. 36,000–38,500 cal BP (Miyoshi 2011; Nakamura 2011).

The middle part of the Upper Member of the Ashitaka Loam contains Layer NL which consists of weathered scorias and volcanic glasses. The latter is the Aira–Tn (AT) marker tephra dated to ca. 30,000 cal BP and derived from the Aira Volcano in the southern part of Kyushu Island. 'The Stone Tool Chronology of the Ashitaka–Hakone Area' based on such favourable stratigraphy is recognised as one of the most detailed and highly-developed Palaeolithic chronologies in Japan.

Identification of Sources for Obsidian Artefacts from the Mount Ashitaka Area

Most of the obsidian artefacts which are suitable for the EDXRF (minimum size is approximately 8 mm in diameter and 1.5 mm in thickness) were sampled from the main cultural assemblages of the Mt. Ashitaka area, and subsequently analysed. The bar graph (Figure 8.5) shows the EDXRF results for phases I–V of the Upper Palaeolithic and for the Incipient Jomon (Ikeya *et al.* 2010). It also indicates that the main obsidian sources of each phase roughly corresponded to chronological

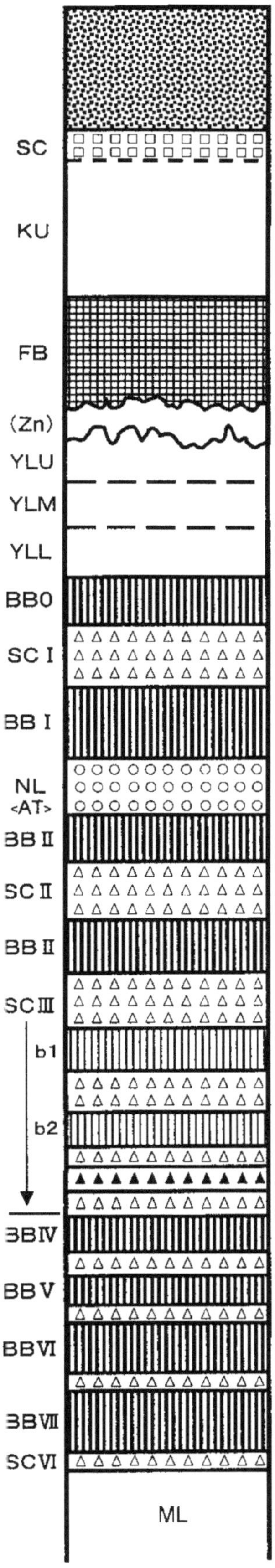

Figure 8.4. General stratigraphic sequence of the upper members of Ashitaka Loam at the piedmont of Mt. Ashitaka (see text for layers' labels)

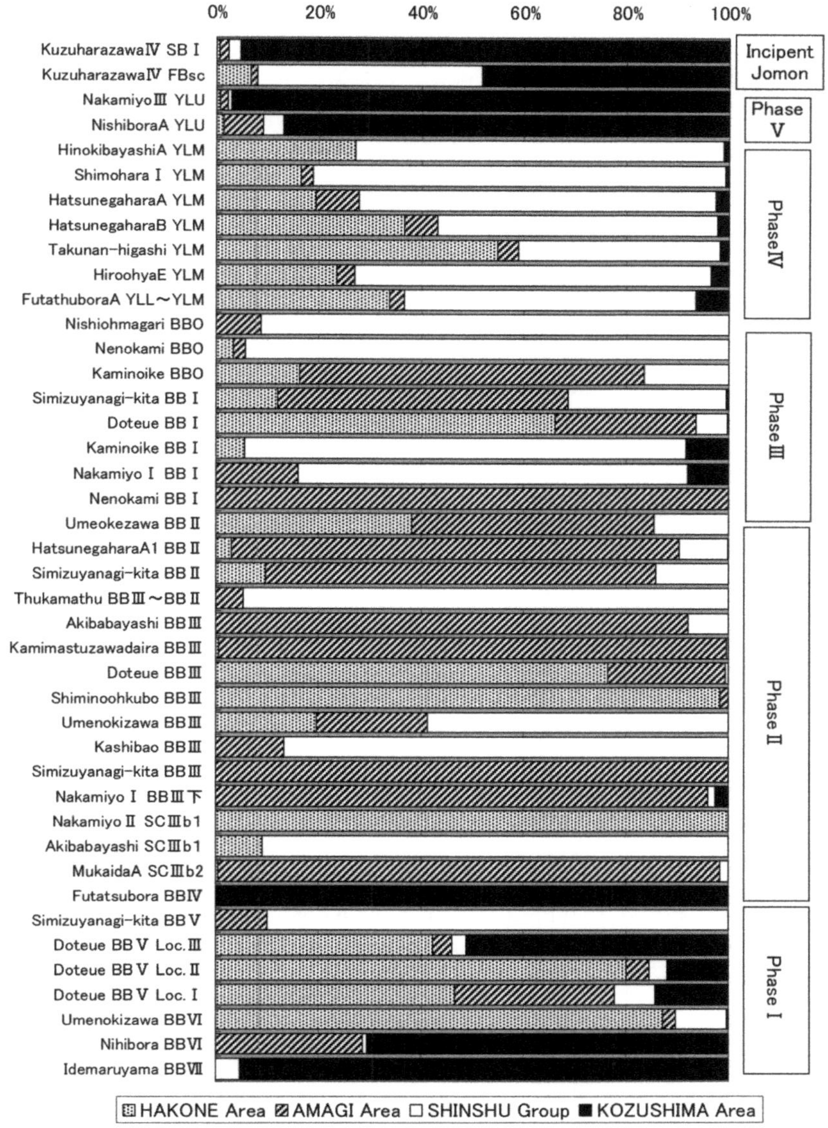

Figure 8.5. Amount of obsidian from different sources in the cultural layers at the piedmont of Mt. Ashitaka

changes (Phase I to Phase V of the Upper Palaeolithic, and the Incipient Jomon) of the 'The Stone Tool Chronology of Ashitaka–Hakone Area'. Even more important is the fact that a certain quantity of obsidian was transported from the Kozushima insular source across the Pacific Ocean in phases I and V of the Upper Palaeolithic, and in the Incipient Jomon.

DISCUSSION

History of Kozushima Obsidian Research

Kozushima is a small volcanic island located approximately 50 km SE of the tip of the Izu Peninsula (Figure 8.6). However, a land bridge from the main landmass of the Japanese Islands to Kozushima Island was never formed, even during the Last Glacial Maximum (ca. 26,000–19,000 cal BP; see Clark *et al.* 2009). On the

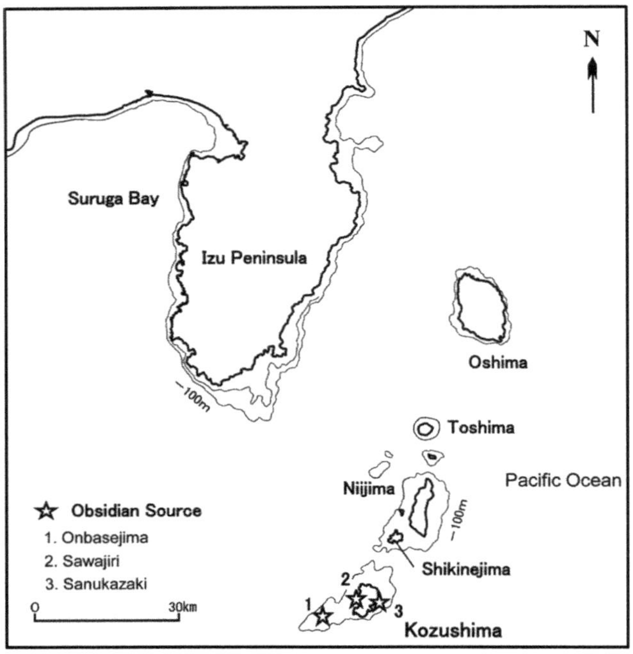

Figure 8.6. Location of obsidian sources in the northern part of the Izu Islands and the Izu Peninsula (the 100 m isobaths are indicated)

basis of fission-track dating, the results of provenance analysis initially showed that a number of Kozushima obsidian pieces are present in the collection of stone artefacts excavated from the Tsukimino and Nogawa sites in the Tokyo area of the Kanto Region (Suzuki 1973a, 1973b). Subsequently, Kozushima obsidian was identified at the Musashidai site (Kamaki *et al.* 1984, 342) and the Hashimoto site (Suzuki *et al.* 1985) in the Tokyo area.

The possibility of seafaring during the Upper Palaeolithic was recognised by Oda (1981) who pursued obsidian research along with Masao Suzuki. However, many other Palaeolithic archaeologists were inclined to disagree or they avoided making positive comments regarding this issue. This is because the reliability of the fission-track method itself remained questionable since it does not characterise the location of the obsidian, and only one sample from the Musashidai site was identified as originating from the Kozushima source. Considering the importance of this problem, it was necessary to validate the EDXRF results by using another analytical method with higher resolution.

Neutron Activation Analysis

Of the obsidian artefacts excavated from the Doteue site, 18 samples whose sources were identified by EDXRF, including Suwa Area 3 (SWHD 3), Tateshina Area 1 (TSTY 1), Amagi Area 4 (AGKT 4), Hakone Area 7 (HNHJ 2, HNKI 3, and HN-X 2), and Kozushima Area 3 (KZOB 3) (digits after abbreviations mean the number of samples), were submitted for NAA without prior knowledge of the EDXRF results.

The samples were irradiated by thermal neutrons at a flux of 5×10^{11} n cm^{-2} s^{-1} for six hours twice a day on the

Table 8.2. NAA measurements for obsidian from Doteue site, Numazu City; in ppm, unless otherwise indicated (after K. Tomura and S. Koshimizu)

No.	Fe (%)	Rb	La	Ce	Sm	Th	Hf	Sc
1	2.3	19.5	7.5	17.4	5.82	1.07	3.75	13.2
2	1.91	39.7	12	26.7	7.3	3.38	6.39	10.3
3	1.9	3430	10.8	24.5	6.9	2.76	5.95	9.68
4	0.63	141	14.2	38.2	5.7	9.4	3.29	3.06
5	0.6	136	13.1	27.2	6.43	8.96	3.13	2.92
6	0.7	55.5	17.5	28.9	4.22	3.88	1.94	3.2
7	246	19.9	8.33	20.2	6.4	1.12	4.3	15.5
8	0.87	97.3	20.7	36.5	5.06	7.43	3.46	2.18
9	0.59	147	14.6	28.6	6.57	9.67	2.93	3.07
10	1.68	45.1	10.7	23.5	6.11	2.96	4.83	8.41
11	1.8	53.6	10.6	25.3	7.51	2.73	5.67	9.21
12	0.75	62.9	18.4	32.1	4.73	4.69	2.34	3.63
13	0.9	76.8	20.4	36.9	4.88	4.74	2.8	3.95
14	2.11	20.8	7.66	19.7	7.77	1.24	4.13	14.5
15	2.12	23.4	8.37	20.1	6.85	1.42	4.68	15.2
18	2	21.4	7.43	18.5	6	1.25	4.3	14.4
19	2.06	3.61	6.84	20.6	6.66	1.32	4.4	14.2
20	2.23	10.3	8.13	19.4	6.5	1.48	4.53	14

rotary specimen rack of a 100 kw TRIGA Mark II Reactor operated by the Atomic Energy Institute of Rikkyo University (Tokyo). By using a gamma-ray spectrometer, gamma-rays were measured for 1800 seconds per sample over a period of ten days after the irradiation was completed. The geological standard, GSP-1 from the USGS, was used as a standard reference material. The amounts of Sc, Fe, Rb, La, Ce, Sm, Hf, Th, and U were quantified through comparison with the measured values for GSP-1 (Tomura *et al.* 1994). Table 8.2 shows the result of the quantitative analysis for the 18 samples from Doteue site.

The geochemical signatures of the main obsidian sources in Japan measured by NAA have been published (Osawa 1991). Table 8.3 shows the measured values of geological obsidian in the Chubu and Kanto regions. Identification of the sources was conducted by comparing the measured values of the Doteue artefacts against the quantitative values using Fe/Th and Sm/Th ratios as indicators. The results of the NAA, including the Kozushima obsidian, are in accordance with the EDXRF data (Table 8.4).

Therefore, the reliability of EDXRF in terms of the Kozushima obsidian on mainland Japan is confirmed. Figure 8.7 illustrates a newly proposed scatterplot from this study by using the data from Tables 8.2–8.3. Measured values for the obsidian artefacts from the Doteue site were overlaid on the geological obsidian. The results of the cross-check in this scatterplot (Figure 8.7) are also identical to the provenance of obsidian as determined by EDXRF.

Table 8.3. NAA measurements for obsidian from the Chubu and Kanto regions; in ppm, unless otherwise indicated (after M. Osawa)

Source	Sample No.	Fe (%)	Rb	La	Ce	Sm	Th	Hf	Sc
Wada (WD)	Kobukazawa01	0.54	260	20	46	6.8	29	5	5.7
	Kobukazawa02	0.66	270	19	44	6.8	27	4.7	6.7
	Kobukazawa03	0.75	260	19	45	6.7	28	4.6	6.3
	Kobukazawa04	0.56	260	20	45	6.9	29	4.8	5.8
	Kobukazawa05	0.47	260	19	44	6.7	28	4.5	5.6
	Kobukazawa06	0.54	260	19	44	6.7	28	4.7	5.6
	Kobukazawa07	0.58	270	20	47	6.7	28	4.6	5.9
	Kobukazawa08	0.53	260	19	44	6.6	28	4.6	5.5
	Kobukazawa09	0.5	270	20	46	6.8	29	4.7	5.6
	Kobukazawa10	0.89	250	19	44	6.4	26	4.9	6.3
	Kobukazawa11	0.7	260	19	43	6.6	28	4.6	6.3
	Kobukazawa12	0.7	250	19	44	6.5	28	4.8	5.8
	Kobukazawa13	0.54	250	19	43	6.5	27	4.4	5.4
	Kobukazawa14	0.5	270	19	45	6.8	29	4.6	5.5
	Kobukazawa15	0.48	250	18	42	6.6	28	4.2	5.5
	Kobukazawa16	0.51	260	18	45	6.5	28	4.6	5.5
	Kobukazawa17	0.48	250	18	44	6.5	28	4.7	5.5
	Kobukazawa18	0.47	260	18	43	6.5	28	4.6	5.4
	Kobukazawa19	0.49	240	18	43	6.6	27	4.7	5.4
	Kobukazawa20	0.45	250	18	41	6.3	27	4.5	5.2
	Kobukazawa21	0.43	260	19	42	6.1	25	4.8	4.9
	Kobukazawa22	0.48	270	19	42	6.2	26	4.9	5.1
	Kobukazawa23	0.48	290	20	46	6.6	28	4.8	5.4
	Kobukazawa24	0.46	280	21	45	6.9	28	4.6	5.4
	Kobukazawa25	0.48	280	20	46	6.7	28	4.7	5.3
	Kobukazawa26	0.51	290	21	47	7	29	5.1	5.5
	Kobukazawa27	0.47	290	20	45	6.9	28	4.7	5.4
	Kobukazawa28	0.48	280	21	46	6.9	29	4.9	5.5
	Kobukazawa29	0.45	290	19	44	6.5	26	4.5	5.1
	Kobukazawa30	0.65	280	20	45	6.9	27	4.8	5.8
	Kobukazawa31	0.47	280	21	45	6.9	28	4.9	5.3

Source	Sample No.	Fe (%)	Rb	La	Ce	Sm	Th	Hf	Sc
Wada (WD)	Kobukazawa32	0.45	290	20	45	6.7	28	4.7	5.2
	Kobukazawa33	0.49	300	20	48	6.9	29	4.8	5.4
	Kobukazawa34	0.5	280	20	48	6.7	27	4.8	5.3
	Kobukazawa35	0.64	280	20	47	7	28	4.8	6.3
	Kobukazawa36	0.5	290	21	48	6.9	28	4.9	5.5
	Kobukazawa37	0.49	290	19	45	6.9	28	5.1	5.5
	Kobukazawa38	0.46	300	20	46	6.9	28	4.9	5.4
	Kobukazawa39	0.47	300	19	44	6.6	27	4.9	5.3
	Kobukazawa40	0.45	270	19	43	6.3	25	4.8	5.0
	Kobukazawa41	0.55	280	21	43	7	28	4.7	5.6
	Kobukazawa42	0.51	300	22	43	7	28	5.1	5.4
	Kobukazawa43	0.53	280	21	41	6.9	28	4.6	5.4
	Kobukazawa44	0.5	280	20	44	6.8	28	4.8	5.5
	Kobukazawa45	0.52	300	21	43	7	28	4.6	5.3
	Kobukazawa46	0.68	260	21	45	7	26	4.8	5.9
	Kobukazawa47	0.63	250	21	43	6.8	25	4.6	5.6
	Kobukazawa48	0.80	270	21	44	6.7	25	4.5	6.3
	Kobukazawa49	0.70	250	21	44	6.7	25	4.8	5.7
	Kobukazawa50	0.52	290	22	44	7.1	28	4.7	5.5
	Kobukazawa51	0.5	290	21	45	7	28	4.6	5.4
	Kobukazawa52	0.54	260	21	45	7	29	4.8	5.6
	Kobukazawa53	0.51	280	21	45	7.2	30	4.9	5.6
	Kobukazawa54	0.58	290	22	45	6.9	28	4.7	5.5
	Kobukazawa55	0.53	290	21	45	6.9	28	4.5	5.5
	Kobukazawa56	0.51	300	21	45	6.9	28	4.9	5.5
	Kobukazawa57	0.48	270	23	46	7	28	5	5.3
	Kobukazawa58	0.48	270	22	44	6.9	27	5.1	5.2
	Kobukazawa59	0.46	280	21	43	7	28	4.9	5.3
	Kobukazawa60	0.51	280	22	46	7	28	5.4	5.4
Wada (WD)	Chojigoryo01	0.65	300	19	35	7.1	31	5.5	6.1
	Chojigoryo02	0.55	310	23	33	7.4	32	5.7	6.2
	Chojigoryo03	0.52	320	18	39	6.3	30	5.0	5.7
	Chojigoryo04	0.56	340	14	35	6.4	32	5.1	6.3
	Chojigoryo05	0.54	320	16	36	6.3	30	4.8	5.8
	Chojigoryo06	0.54	340	16	34	6.4	31	4.9	5.9
	Chojigoryo07	0.55	340	15	34	6.6	32	5.6	6.2
Wada (WD)	Azeria01	0.51	260	25	44	7.5	27	5	5.4
	Azeria02	0.47	250	26	45	7.4	30	5.1	5.4
	Azeria03	0.52	270	22	46	7	28	4.7	5.6
	Azeria04	0.48	260	21	47	6.8	27	5	5.5
	Azeria05	0.48	290	21	44	6.6	27	4.6	5.4
	Azeria06	0.52	320	20	46	6.7	28	4.9	5.4
	Azeria07	0.50	290	20	45	6.6	28	4.8	5.4
	Azeria08	0.48	290	20	44	6.7	28	4.7	5.5
Wada (WD)	Higashimochiya01	0.65	270	24	41	6.8	26	4.6	5.8
	Higashimochiya02	0.86	220	23	44	6.9	24	5.2	6.3
	Higashimochiya03	0.59	240	23	44	6.7	25	5	5.7

Source	Sample No.	Fe (%)	Rb	La	Ce	Sm	Th	Hf	Sc
Wada (WD)	Higashimochiya04	0.68	230	22	43	7.1	26	4.8	5.8
	Higashimochiya05	1.15	210	24	44	6.8	23	5	7.4
	Higashimochiya06	0.80	220	24	45	6.8	23	4.9	6.3
	Higashimochiya07	0.69	250	21	46	6.8	26	5	6
	Higashimochiya08	1.03	240	22	46	6.6	24	4.8	7.5
	Higashimochiya09	1.03	240	22	47	6.7	25	5	7.3
	Higashimochiya10	0.83	260	22	47	6.7	25	5.2	6.4
	Higashimochiya11	0.79	260	22	47	6.8	25	5.3	6.3
	Higashimochiya12	0.97	240	21	47	6.8	24	5	7.1
	Higashimochiya13	1.01	250	22	47	6.6	24	5.2	6.8
Wada (WD)	Toho-Pearite01	0.66	240	25	45	7.1	26	5.3	6.1
	Toho-Pearite02	0.58	250	23	43	7	27	5.2	5.7
	Toho-Pearite03	0.74	220	23	45	7.1	25	5.8	6.3
	Toho-Pearite04	0.56	250	24	48	7.2	26	5.2	5.7
	Toho-Pearite05	0.62	230	24	43	7.1	26	5.2	5.9
	Toho-Pearite06	0.70	260	25	48	7.4	27	5.3	6.3
Wada (WO)	Omegura01	0.64	170	27	49	5	16	4.3	3.1
	Omegura02	0.73	160	26	48	4.8	14	4.7	3.1
	Omegura03	0.67	190	28	47	5.1	17	4.2	3.2
	Omegura04	0.73	160	25	49	4.9	15	4.6	3.1
	Omegura05	0.73	160	27	49	4.9	15	4.6	3.1
Suwa	Hoshigato01	0.46	150	16	31	4.7	10	3.5	3
	Hoshigato02	0.49	160	16	31	5	10	3.3	2.9
	Hoshigato03	0.47	160	16	32	4.8	10	3.5	3
	Hoshigato04	0.44	170	16	34	4.7	10	3.3	2.9
	Hoshigato05	0.48	160	17	33	4.8	11	3.4	2.9
	Hoshigato06	0.44	150	15	30	4.8	10	3.3	2.9
Tateshina	Mugikusatoge01	0.65	100	25	42	3.9	8.9	3.7	2.2
	Mugikusatoge02	0.65	100	24	45	3.8	8.9	3.8	2.2
	Mugikusatoge03	0.67	100	26	42	3.8	8.7	3.9	2.2
	Mugikusatoge04	0.63	110	24	43	3.8	8.9	3.8	2.3
	Mugikusatoge05	0.73	100	25	47	4.5	8.5	4.6	2.5
	Mugikusatoge06	0.7	100	24	44	4	8.7	4	2.3
	Mugikusatoge07	0.69	100	25	46	4.3	8.6	4.4	2.3
	Mugikusatoge08	0.66	100	24	44	3.8	8.8	3.7	2.3
	Mugikusatoge09	0.7	110	25	46	4	8.9	3.8	2.4
	Mugikusatoge10	0.64	120	25	44	3.8	9.2	3.8	2.3
	Mugikusatoge11	0.66	110	24	45	3.9	9.1	3.8	2.3
	Mugikusatoge12	0.71	100	25	45	4.2	9	4	2.4
	Mugikusatoge13	0.67	110	25	44	4	9	3.8	2.3
	Shirakomaike01	0.71	110	21	42	3.5	9	3.6	2.1
	Shirakomaike02	0.68	110	21	42	3.4	8.7	3.8	2.2
Hakone	Kajiya01	1.75	20	11	24	4.9	2	5.1	13
	Kajiya02	1.63	30	11	25	5	2	5	13
	Kajiya03	1.57	30	11	26	4.8	2.1	5.2	13
	Kajiya04	1.74	30	11	27	4.7	2.1	5.2	13
	Kajiya05	1.79	30	12	27	4.9	2.1	5.2	13
	Kajiya06	1.78	30	11	25	4.9	1.9	5	13

Source	Sample No.	Fe (%)	Rb	La	Ce	Sm	Th	Hf	Sc
Hakone	Hatajuku1	1.77	20	8	20	4.2	1.3	4.2	14
	Hatajuku2	1.77	30	8	19	4.2	1.3	4.1	14
	Hatajuku3	1.76	30	9	17	4.9	1.2	4.5	15
	Hatajuku4	1.9	30	8	21	4.6	1.3	4.4	14
	Hatajuku5	1.81	30	9	18	4.8	1.5	4.4	15
	Hatajuku6	1.92	20	9	21	5.2	0.9	4.8	15
	Hatajuku7	1.88	10	9	20	5.1	1	4.6	15
	Hatajuku8	1.78	20	9	20	5	1.1	4.9	14
	Hatajuku9	1.90	20	8	22	5.2	1.2	4.8	15
	Hatajuku10	1.88	20	9	21	5.1	1.1	4.7	15
	Hatajuku11	1.95	20	9	21	5.2	1.1	4.6	15
	Hatajuku12	1.84	10	8	21	5	0.9	4.6	15
	Hatajuku13	1.85	20	9	22	5.1	1.1	4.5	15
	Hatajuku14	1.9	30	8	20	5.3	1	4.8	15
	Hatajuku15	1.83	20	8	20	5	1	4.6	15
	Fuezuka01	3.57	10	6	15	3.9	0.7	3.1	24
	Fuezuka02	3.58	20	6	16	4	0.6	3.0	23
	Ashinoyu01	3.19	30	4	16	4.1	0.4	3	22
	Ashinoyu02	3.53	20	6	16	3.9	0.5	3	24
	Ashinoyu03	3.55	20	7	16	4.2	0.6	3.1	24
	Ashinoyu04	3.72	10	6	16	4.1	0.5	3.2	24
	Ashinoyu05	3.5	10	7	18	4.1	0.7	2.9	23
	Kamitaga01	1.25	40	11	23	3.9	2.1	5	8.1
	Kamitaga02	1.17	40	11	24	3.9	2.2	4.6	8.2
Hakone	Kamitaga03	1.23	40	11	25	4	2.2	4.8	8.3
	Kamitaga04	1.29	30	11	23	4	2.2	5	8.4
	Kamitaga05	1.35	20	11	22	4	2	5.1	8.3
	Kamitaga06	1.39	30	11	23	4	2.3	5.2	9.3
	Kamitaga07	1.45	30	11	23	4.1	2	5	8.6
	Kamitaga08	1.4	30	10	23	4.1	2	5.1	8.7
Amagi	Kashiwatoge01	1.24	50	12	26	5.1	3	5.8	7.9
	Kashiwatoge02	1.96	50	11	24	5	2.6	5.7	12
	Kashiwatoge03	1.26	50	12	26	5.2	2.9	5.5	7.9
	Kashiwatoge04	1.91	40	11	24	4.8	2.4	5.8	11
	Kashiwatoge05	1.51	40	11	25	4.7	2.9	5.7	9.5
	Kashiwatoge06	1.31	40	13	27	5.1	3.1	5.8	8.3
	Kashiwatoge07	1.59	40	11	24	5	2.9	5.9	10
	Kashiwatoge08	1.33	60	10	28	5	3.1	6.6	8.8
	Kashiwatoge09	1.47	30	11	24	4.9	2.9	5.7	8.8
Kozushima	Sanukazaki01	0.52	70	20	35	3.5	5.2	2.3	3.6
	Sanukazaki02	0.54	70	21	34	3.5	5.5	2.3	3.6
	Sanukazaki03	0.54	70	22	35	3.5	5.3	2.5	3.6
	Sanukazaki04	0.52	80	21	34	3.2	5.4	2.3	3.5
	Sanukazaki05	0.54	70	22	38	3.6	5.2	2.6	3.7
	Sawajiriwan01	0.53	80	22	35	3.7	5.6	2.5	3.7
	Sawajiriwan02	0.59	70	21	34	3.6	4.9	2.6	3.6
	Sawajiriwan03	0.55	70	21	34	3.5	4.9	2.8	3.4

Source	Sample No.	Fe (%)	Rb	La	Ce	Sm	Th	Hf	Sc
Kozushima	Sawajiriwan04	0.63	70	22	34	3.6	4.9	2.8	3.4
	Sawajiriwan05	0.6	60	20	34	3.4	4.5	2.5	3.4
	Sawajiriwan06	0.55	60	20	34	3.4	4.7	2.4	3.3
	Sawajiriwan07	0.58	60	21	38	3.6	5	2.7	3.5
	Sawajiriwan08	0.6	80	22	38	3.5	5.2	2.6	3.5
	Sawajiriwan09	0.53	60	21	36	3.5	4.9	2.8	3.5
	Sawajiriwan10	0.57	70	21	35	3.1	4.8	2.7	3.6
	Sawajiriwan11	0.63	70	21	37	3.5	4.9	2.8	3.7
	Sawajiriwan12	0.57	60	21	36	3.3	4.8	2.5	3.5
	Sawajiriwan13	0.57	60	21	37	3.5	4.7	2.7	3.5
	Sawajiriwan14	0.62	70	22	37	3.3	4.8	2.5	3.5
	Sawajiriwan15	0.57	60	20	35	3.9	4.8	2.7	3.4
	Sawajiriwan16	0.58	60	20	37	3.3	4.8	2.6	3.4
	Sawajiriwan17	0.58	60	21	38	3.4	5	2.6	3.5
	Nagahama01	0.65	70	17	32	3.5	4.6	2.3	3.9
	Nagahama02	0.57	70	16	29	3.3	4.4	2.6	3.8
	Nagahama03	0.57	80	20	37	3.5	5	2.5	3.8
	Nagahama04	0.64	60	20	37	3.7	5	2.7	3.6
	Nagahama05	0.58	60	16	26	2.7	3.6	2.3	3.6
	Nagahama06	0.59	60	21	36	3.1	4.7	2.5	3.5
	Nagahama07	0.6	60	22	36	3.3	5.3	2.2	3.7
	Maehama01	0.55	70	20	37	3.5	5.1	2.1	3.7
	Maehama02	0.52	50	20	36	3.2	4.8	2.4	3.5
	Maehama03	0.52	60	20	34	3.3	4.5	2.3	3.3
	Maehama04	0.57	60	19	33	3.4	4.4	2.5	3.2
	Maehama05	0.59	60	21	34	3.4	4.8	2.6	3.3
	Maehama06	0.58	60	20	34	3.5	4.6	2.4	3.3
	Maehama07	0.62	70	20	35	3.5	5.1	2.8	3.4
	Maehama08	0.61	60	20	35	3.5	5	2.6	3.4
	Maehama09	0.58	60	20	35	3.5	4.6	2.3	3.2
	Onbasejima01	0.61	60	20	36	3.5	4.8	2.8	3.5
	Onbasejima02	0.61	70	20	36	3.5	4.7	2.3	3.5
	Onbasejima03	0.63	70	21	36	3.6	4.9	2.6	3.5
Takaharayama	Takaharayama01	1.24	110	25	48	5.8	3.9	5.1	7.4
	Takaharayama02	1.33	120	25	47	5.7	3.9	4.9	7.6
	Takaharayama03	1.3	110	26	47	5.9	4	5.1	7.6
	Takaharayama04	1.37	110	25	47	5.7	3.9	5.4	8
Takaharayama	Takaharayama05	1.37	110	26	48	6	4.1	5.3	8.0
	Takaharayama06	1.3	100	26	48	5.9	3.8	5	7.5
	Takaharayama07	1.28	110	26	48	5.9	4.1	5	7.8
	Takaharayama08	1.45	110	25	48	5.8	3.8	5.1	8.8
	Takaharayama09	1.51	100	25	46	5.8	3.7	5.3	9.1
	Takaharayama10	1.44	100	26	47	5.9	4	5.4	7.9
	Takaharayama11	1.35	110	26	50	6.1	4	5.4	7.7
	Takaharayama12	1.52	100	26	47	5.9	4	5.3	8.3
Shibata	Itayama01	0.53	160	18	33	3.6	2.8	2.3	3.2
	Itayama02	0.51	150	18	35	3.5	2.9	2.3	3.3
	Itayama03	0.52	150	18	34	3.6	3	2.2	3.2

Source	Sample No.	Fe (%)	Rb	La	Ce	Sm	Th	Hf	Sc
Shibata	Itayama04	0.49	150	18	36	3.6	3	2.2	3.2
	Itayama05	0.52	140	18	35	3.6	3.2	2.3	3.3
	Itayama06	0.53	150	19	35	3.7	2.5	2.5	3.2

Table 8.4. Comparison of identification the obsidian sources by NAA and EDXRF methods (after Tomura *et al.* 1994; Mochizuki *et al.* 1994)

	NAA (Tomura *et al.* 1994)		EDXRF (Mochizuki *et al.* 1994)
No.	***Area / Sub-Area***	***No.***	***Area / Sub-Area***
1	Hakone / Hatajuku	011	Hakone / Hatajuku
2	Amagi / Kashiwatoge	019	Amagi / Kashiwatoge
3	Amagi / Kashiwatoge	025	Amagi / Kashiwatoge
4	Suwa / Hoshigadai	027	Suwa / Hoshigadai
5	Suwa / Hoshigadai	029	Suwa / Hoshigadai
6	Kozushima	039	Kozushima / Onbase
7	Hakone / Hatajuku	040	Hakone / Hatajuku
8	Tateshina	050	Tateshina / Tsumetayama
9	Suwa / Hoshigadai	056	Suwa / Hoshigadai
10	Amagi / Kashiwatoge	062	Amagi / Kashiwatoge
11	Amagi / Kashiwatoge	077	Amagi / Kashiwatoge
12	Kozushima	093	Kozushima / Onbase
13	Kozushima	138	Kozushima / Onbase
14	Hakone / Hatajuku	147	Hakone / Kuroiwabashi
15	Hakone / Hatajuku	144	Hakone / Kuroiwabashi
18	Hakone / Hatajuku	159	Hakone / Kuroiwabashi
19	Hakone / Hatajuku	727	Hakone / X
20	Hakone / Hatajuku	728	Hakone / X

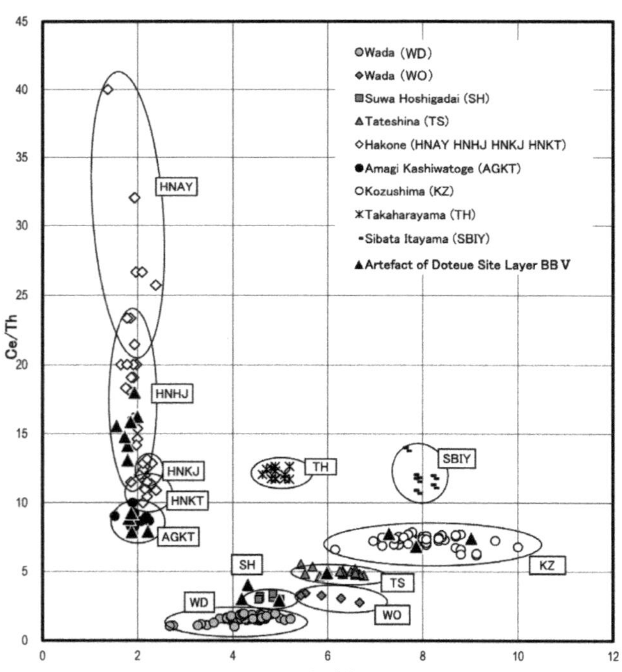

Figure 8.7. Discrimination diagram of La/Hf vs. Ce/Th based on NAA for geological samples and artefacts of Doteue site (Layer BBV)

CONCLUSIONS

There are 11 elements in obsidian that can be measured by EDXRF (Al, Si, K, Ca, Ti, Mn, Fe, Rb, Sr, Y, and Zr) compared to approximately 30 elements that NAA can detect. Among the EDXRF-analysed elements, there were only nine that are efficient and reliable for use in the identification of obsidian sources: Si, K, Ti, Mn, Fe, Rb, Sr, Y, and Zr. However, by using four new indicators [Rb x 100 / (Rb + Sr + Y + Zr); Mn x 100 / Fe; Sr x 100 / (Rb + Sr + Y + Zr); and log (Fe/K)] and a scatterplot based on the EDXRF trace element analysis, provenance studies of obsidian artefacts can be carried out with sufficient accuracy. In addition, the results of the Kozushima obsidian analysis based on the EDXRF method were confirmed by NAA, and this can be used to increase the likelihood that obsidian maritime transport occurred during the Upper Palaeolithic.

Acknowledgements

I wish to express my appreciation to Prof. A. Ono for his support and encouragement. Special thanks go to Dr Y. Suda for valuable suggestions. I am grateful to Drs Kenji Tomura and Satoshi Koshimizu for analysis of artefacts from the Doteue site. My thanks also go to Drs Y. V. Kuzmin and M. D. Glascock, and to Ms C. C. Lindsey, for help with correcting the grammar of the original manuscript.

References

CLARK, P. U., A. S. DYKE, J. D. SHAKUN, A. E. CARLSON, J. CLARK, B. WOHFARTH, J. X. MITROVICA, S. W. HOSTETLER, and A. M. MCCABE. 2009. The Last Glacial Maximum. *Science* 325, 710–714.

HIGASHIMURA, T., and T. WARASHINA. 1975. Sourcing of Sanukite Stone Implements by X-ray Fluorescence Analysis. *Journal of Archaeological Science* 2, 169–178.

IKEYA, N., and A. MOCHIZUKI. 1998. Ashitaka Sanroku Niokeru Sekizaisosei no Hensen [Composition and Transition of Raw Material at Mount Ashitaka during the Upper Palaeolithic]. *Shizuoka-ken Kokogaku Kenkyu* 36, 21–44.

IKEYA, N., T. TOGASHI, and K. MAGARA. 2010. Dai4sho Kyusekkibunka no Hennen to Chiikisei Tokai Hokuriku Chiho [Chapter 4. Locality and Chronology of Palaeolithic Culture, Tokai and Hokuriku Regions]. In *Koza Nihon no Kokogaku 2 Kyusekki Gidai Ge,* edited by K. Inada and H. Sato, 473–504. Tokyo, Aokishoten Publishers.

IZUHO, M., and H. SATO. 2007. Archaeological Obsidian Studies in Hokkaido, Japan: Retrospect and Prospects. *Bulletin of the Indo-Pacific Prehistory Association* 27, 114–121.

KAMAKI, Y., T. HIGASHIMURA, T. WARASHINA, and H. MIYAKE. 1984. Kokuyouseki, Sanukaitoseisekki no Sanchisuitei Niyoru Kobunkakoryu no Kenkyu [Study of Prehistoric Cultural Exchanges Based on Identification of Obsidian and Sanukite]. In *Kobunkazai no Shizenkagakuteki Kenkyu*, edited by N. Watanabe, 333–359. Kyoto, Dohoushiya Publishers.

MIYOSHI, M. 2011. Ashitaka Hakone Sanroku Niokeru Kyusekkigidai no Jinruikatsudou to Kankyo [Upper Palaeolithic Human Activities and Its Environments, at the Piedmont of Mount Ashitaka and Mount Hakone]. In *Nihon Kokogaku Kyokai Dai77kai Sokai Kenkyuhappyo Yoshi*, 18–19. Tokyo, Kokugakuin Daigaku.

MOCHIZUKI, A., N. IKEYA, K. KOBAYASHI, and Y. MUTO. 1994 Isekinai niokeru Kokuyosekiseisekki no Gensanchibestubunpu Nitsuite: Numazushi Doteueiseki BBVsou no Gensanchisuitei kara [Intra-Site Distribution Patterns of Obsidian Tools by Different Sources: A Case of Provenance Analysis of the BBV Layer, Doteue Site, Numazu City]. *Shizuoka-ken Kokogaku Kenkyu* 26, 1–24.

NAKAMURA, Y. 2011. Ashitakasanroku Saiko no Sekkigun no Shomondai [Various Problems about Stone Implements in the Incipient Upper Palaeolithic at the Piedmont of Mount Ashitaka]. *Sekki Bunka Kenkyu* 17, 76–94.

ODA, S. 1981. Kozushimasan no Kokuyoseki: Sono Senshijidai Niokeru Denpa [Obsidian from Kozushima Island: Its Diffusion in Prehistory]. *Rekishi Techo* 9(6), 11–17.

OSAWA, M. 1991. Kokuyoseki no Kagakusosei: Isekishuthudo Kokuyoseki Sekki no Gensanchisuitei no Kiso Toshite [Chemical Composition of Obsidian: The Basis for Identification of Obsidian Implements]. In *Heisei2nendo Kagakukenkyuhi Hojokin (Ippankenkyu B) Kenkyuseika Hokokusho*, 1–69. Tokyo (Self-Publishing).

SUZUKI, M. 1973a. Chronology of Prehistoric Human Activity in Kanto, Japan. Part 1: Framework for Reconstructing Prehistoric Human Activity in Obsidian. *Journal of the Faculty of Science, University of Tokyo. Section 5 (Anthropology)* 4(3), 241–318.

SUZUKI, M. 1973b. Chronology of Prehistoric Human Activity in Kanto, Japan. Part 2: Time-Space Analysis of Obsidian Transportation. *Journal of the Faculty of Science, University of Tokyo. Section 5 (Anthropology)* 4(4), 396–469.

SUZUKI, M., Y. KANAYAMA, Y. AOKI, and K. TOMURA. 1985. Kanagawaken Sagamiharashi Hashimotoiseki no Kokuyosekibunseki [Chemical Analysis for Obsidian Implements at the Hashimoto Site in Sagamihara City, Kanagawa Prefecture]. *Kyusekki Kokogaku* 30, 118–117.

SUZUKI, M., and K. TOMURA. 1983. Basic Data for Identification the Source of Archaeological Obsidian by Activation Analysis and Discriminant Analysis. *St. Paul's Review of Science* 4(4), 99–110.

TOMURA, K., T. KOSHIMIZU, N. IKEYA, and A. MOCHIZUKI. 1994. Hoshakabunseki Niyoru Numazushi Doteue Iseki Shustudo Kokuyoseki no Sanchisuitei [Identification of Obsidian Implements from Doteue Site, Numazu City, Using NAA]. In *Dai38kai Hoshakagaku Toronkai Koenyokoshu*, 91–92. Shizuoka, Shizuoka Daigaku.

TSUTSUMI, T. 2007. The Dynamics of Obsidian Use by the Microblade Industries of the Terminal Late Pleistocene. *Daiyonki Kenkyu* 46, 179–186.

TSUTSUMI, T. 2010. Prehistoric Procurement of Obsidian from Sources on Honshu Island (Japan). In *Crossing the Straits: Prehistoric Obsidian Source Exploitation in the North Pacific Rim* (B.A.R. International Series 2152), edited by Y. V. Kuzmin and M. D. Glascock, 27–55. Oxford, BAR Publishing.

Chapter 9
INTEGRATION OF OBSIDIAN COMPOSITIONAL STUDIES AND LITHIC REDUCTION SEQUENCE ANALYSIS AT THE UPPER PALAEOLITHIC SITE OF OGACHI-KATO 2, HOKKAIDO, JAPAN

Masami IZUHO, Jeffrey R. FERGUSON, Michael D. GLASCOCK, Noriyoshi ODA, Fumito AKAI, Yuichi NAKAZAWA, and Hiroyuki SATO

Abstract: *Here we discuss the obsidian raw material procurement of hunter–gatherers at the Upper Palaeolithic site of Ogachi-Kato 2 in Hokkaido, Japan, based on integration of obsidian compositional studies using X-ray Fluorescence and Neutron Activation Analysis, and lithic reduction sequence analysis. Refit analysis reveals three reduction sequences in the Ogachi-Kato 2 lithic assemblage: 1) a microblade reduction sequence related to Rankoshi-type microblade cores; 2) a flake reduction sequence identified as a Shimaki assemblage, and 3) a small flake reduction sequence. In the microblade assemblage, the obsidians from the Oketo (Oketoyama), Oketo (Tokoroyama), Ikutahara, Rubeshibe, Tokachi-Mitsumata, and Tokachi (Shikaribetsu) sources are present indicating procurement from a variety of local and non-local sources. A few microblades were removed at the site from cores made of Oketo (Oketoyama) and Ikutahara obsidians, as well as some secondary microblade reduction. Microblades made from Oketo (Tokoroyama), Rubeshibe, Tokachi-Mitsumata, and Tokachi (Shikaribetsu) obsidians have been recovered from the site, but there is no evidence of microblade production from these sources at the site. In the flake industry, primary and secondary reductions were undertaken, and at the site, several cores and tools were discarded after being worked. Nearby gravel from the local source of Oketo (Tokoroyama) was used for cores. Only tools made of obsidian from the local Rubeshibe and non-local Tokachi-Mitsumata sources are present, although several cores were likely transported away from the site. In the small flake industry, three high quality local sources —Oketo (Oketoyama), Oketo (Tokoroyama), and Rubeshibe —were used for manufacture of cores and tools. Primary and secondary small flake reductions were carried out at Ogachi-Kato 2, and several tools and cores were discarded. There is evidence of the transport of several cores from the site. Although all three industries used high quality obsidian, the procurement and reduction strategies varied by industry with the diversity of sources used increasing through time.*

Keywords: *Obsidian, Ogachi-Kato 2 Site, Reduction Sequence Analysis, Handheld X-ray Fluorescence Analysis, Neutron Activation Analysis, Upper Palaeolithic, Hokkaido Island, Japan*

INTRODUCTION

This chapter focuses on the first of two key issues in obsidian research in Northeast Asia: 1) the dispersal and adaptation processes of anatomically modern humans and their regional variability; and 2) the significance of obsidian exchange in the extent and formation of distribution networks (Izuho and Hirose 2010). We examine the obsidian raw material procurement of hunter–gatherers at the site of Ogachi-Kato 2 located in Hokkaido, Japan (Figure 9.1). Within a human ecosystems approach (Butzer 1982; Clarke 1968; Waters 1992), human behaviour is related to climate, landscape, flora, fauna, and culture, and here we focus on human behaviour as seen through lithic raw material procurement. Although other raw materials are present in the lithic assemblage, this study focuses only on obsidian due to the ability to accurately determine the specific geological source.

This chapter includes: 1) a description of Ogachi-Kato 2; 2) a summary of the individual refits; 3) the separation of artefacts by industry through the analysis of the reduction sequence; and 4) assignment of artefacts to specific geologic sources of obsidian. In conclusion, we address how the timing of the hunter–gatherer lithic procurement was related to production and use.

OGACHI-KATO 2 SITE

Ogachi-Kato 2 (geographic coordinates 43°44′18″ N, 143°39′10″ E; elevation of 210–225 m above sea level, hereafter – a.s.l.) is an open-air site on the south bank of the Kunneppu River (a tributary of the Tokoro River) in its middle course (Figures 9.1–9.2). The Tokoro River flows into the Okhotsk Sea. Ogachi-Kato 2 is situated approximately 35 m above the current Kunneppu River bed on an erosional terrace with a base of chert, a slightly metamorphized formation called "green rocks", overlaid by the Kuccharo 4 pyroclastic flow deposit (Kc-4, dated to 115,000–120,000 years ago) (Machida and Arai 2003; Okumura 1991; Soda 2012) (Figure 9.3).

A total of 43.8 m^2 was excavated between 1999 and 2002, revealing an Upper Palaeolithic cultural deposit with 494 stone artefacts and one natural charcoal concentration within the unconsolidated Quaternary sediments. The excavation report (see summary: Izuho and Tsurumaru 2012) includes a multifaceted reconstruction of natural site formation processes based on geoarchaeological studies (Izuho 2012a, 2012b, 2012c; Izuho and Hayashi 2012; Soda 2012); the statistical analysis of artefact spatial distribution (Izuho *et al.* 2012a); the analysis of the reduction sequence based on refitting analysis (Oda 2012); obsidian source

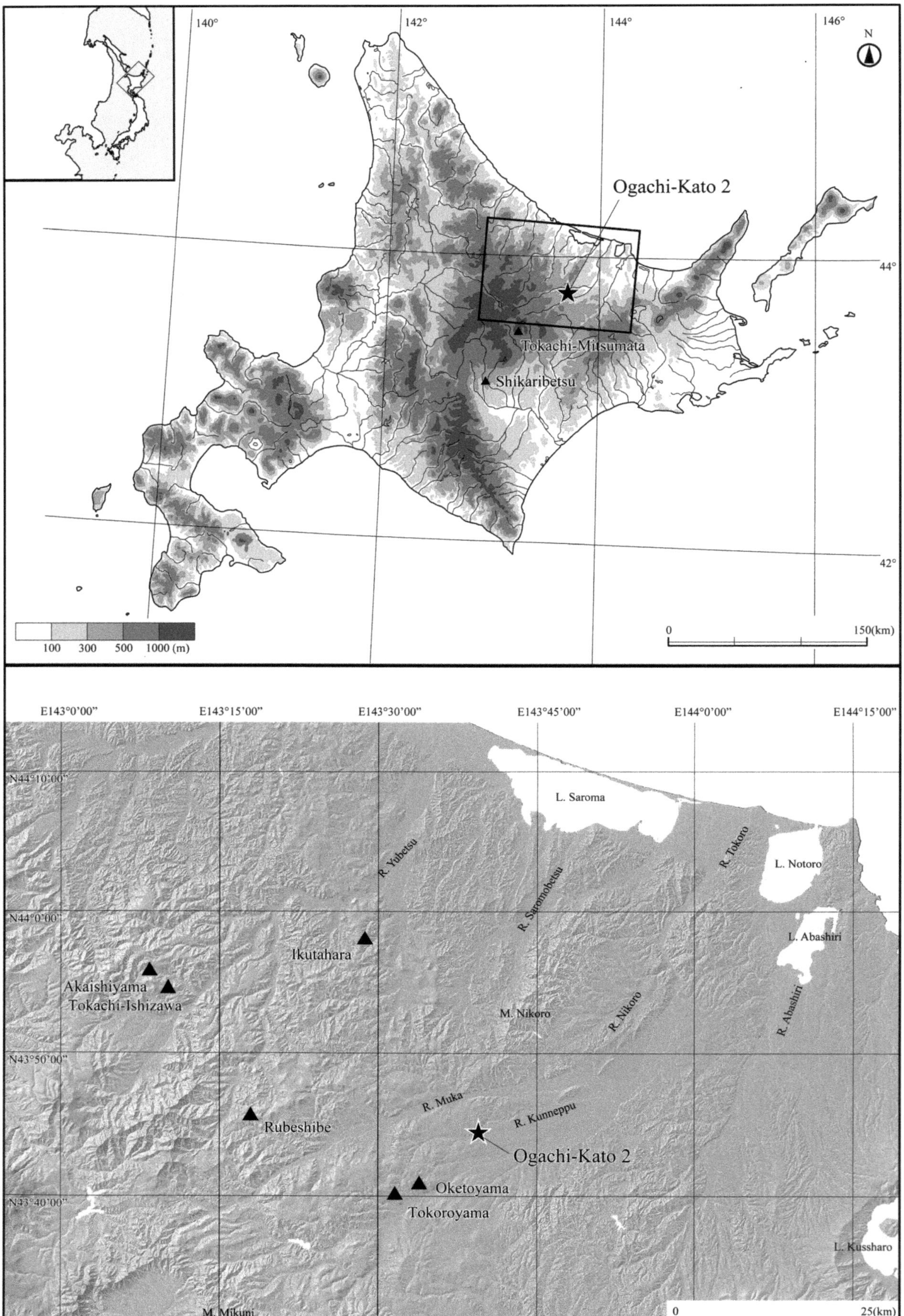

Figure 9.1. Topography of Hokkaido Island and location of the Ogachi-Kato 2 site (triangles are obsidian primary sources)

Figure 9.2. Distant view of the Ogachi-Kato 2 site

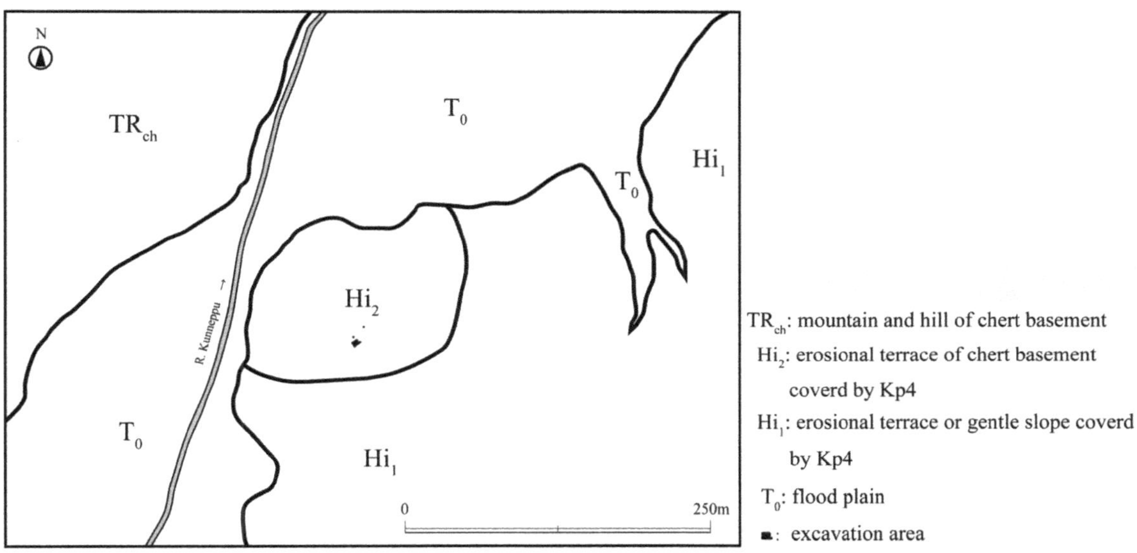

Figure 9.3. Geomorphological map of the Ogachi-Kato 2 site

assignment (Ferguson *et al.* 2012); flaking technique determination based on the gull-wing fracture analysis (Takakura 2012); and micro use-wear analysis (Iwase and Takase 2012).

The stratigraphic profile of Ogachi-Kato 2 is shown in Figure 9.4. The unconsolidated Quaternary deposits are approximately 132 cm in depth at the southeastern edge of the excavation. The eight stratigraphic divisions of the profile are labelled VIII to I from bottom to top. Unit VIII is "red weathered soil" covering the generally flat bedrock, although it has small depressions (less than 50 cm in depth) with unconformity. The soil is very compact with reticulated mottling, and was formed during Oxygen Isotope Stage (OIS) 5e (Okumura 1991; Soda 2012). Unit VII is the Kc-4 deposit that erupted from the Kuccharo Caldera (Soda 2012). Unit VI clearly covers the lower Kc-4 non-uniformly in many parts of the excavation area indicating it was deposited after some erosion of the primary Kc-4 deposit. An ice-wedge pseudomorph formed by freezing or drying developed at the boundary between units VII and VI. Unit Vb is a palaeosol formed below Unit Va. The border with the underlying unit VI is clear and wavy. Unit Vb is only found in a small area in the southern portion of the excavation area suggesting widespread erosion before the palaeosol of Unit Va was formed across the whole excavation area with a polygon structure caused by cryogenic processes. Unit IV is a soft loam that levelled out the irregular upper surface of Unit Va. Units III and II are Holocene Kuroboku soils ("black humus"), each sub-divided into two layers. Unit I is heavily disturbed by modern cultivation with some intrusions extending into units II and III.

One lithic concentration consisting of 488 lithic artefacts in total was identified during the excavation and shown in

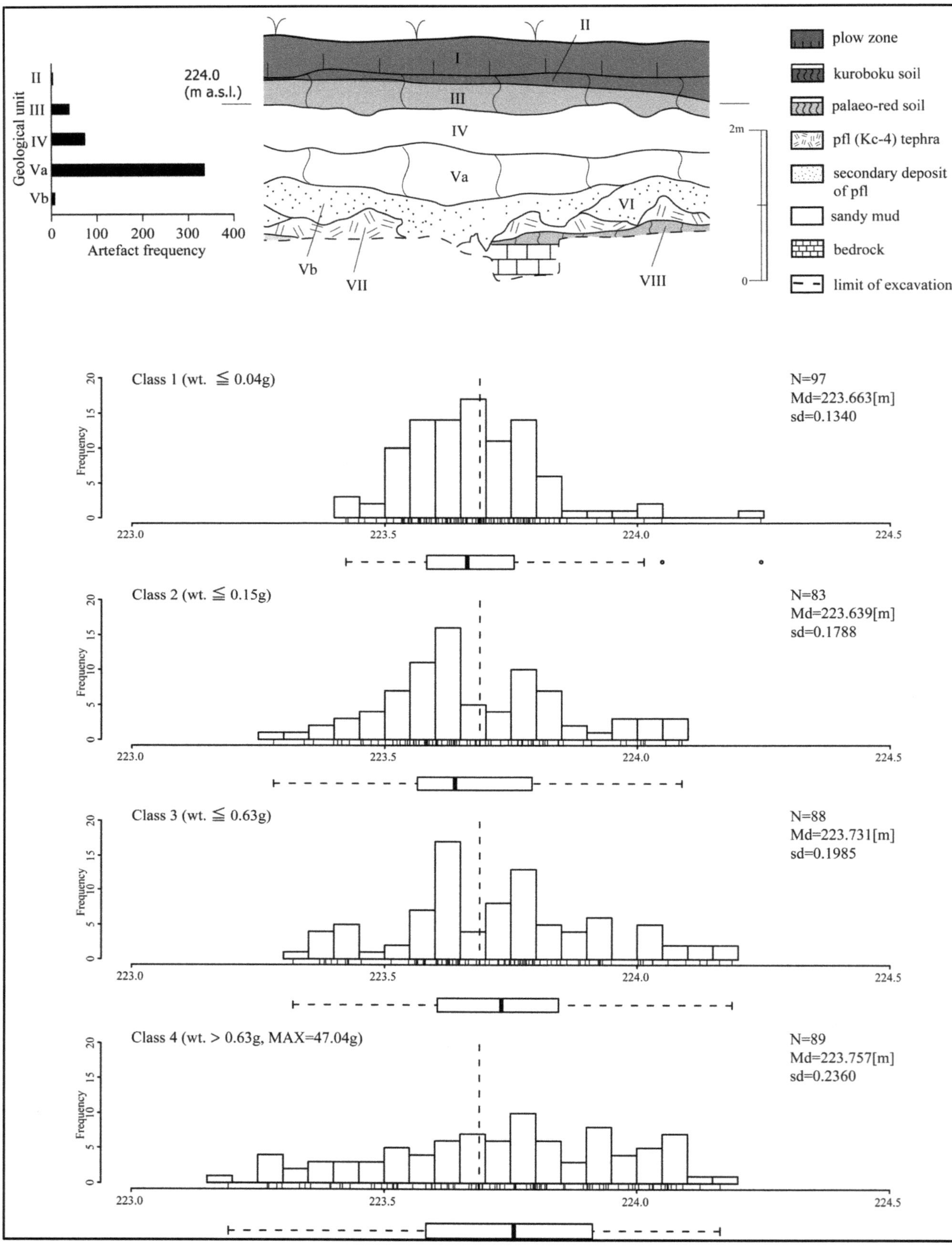

Figure 9.4. Stratigraphic profile and artefact frequencies at the Ogachi-Kato 2 site

Figure 9.5. The horizontal distribution extends 7.5 m east to west and 7.2 m north to south. The concentration extends up to 1.5 m in depth from the surface. Lithic artefacts were recovered in stratigraphic units Vb through II with the following counts per unit: Unit Vb (n = 5), Unit Va (n = 265), Unit IV (n = 48), Unit III (n = 30), and

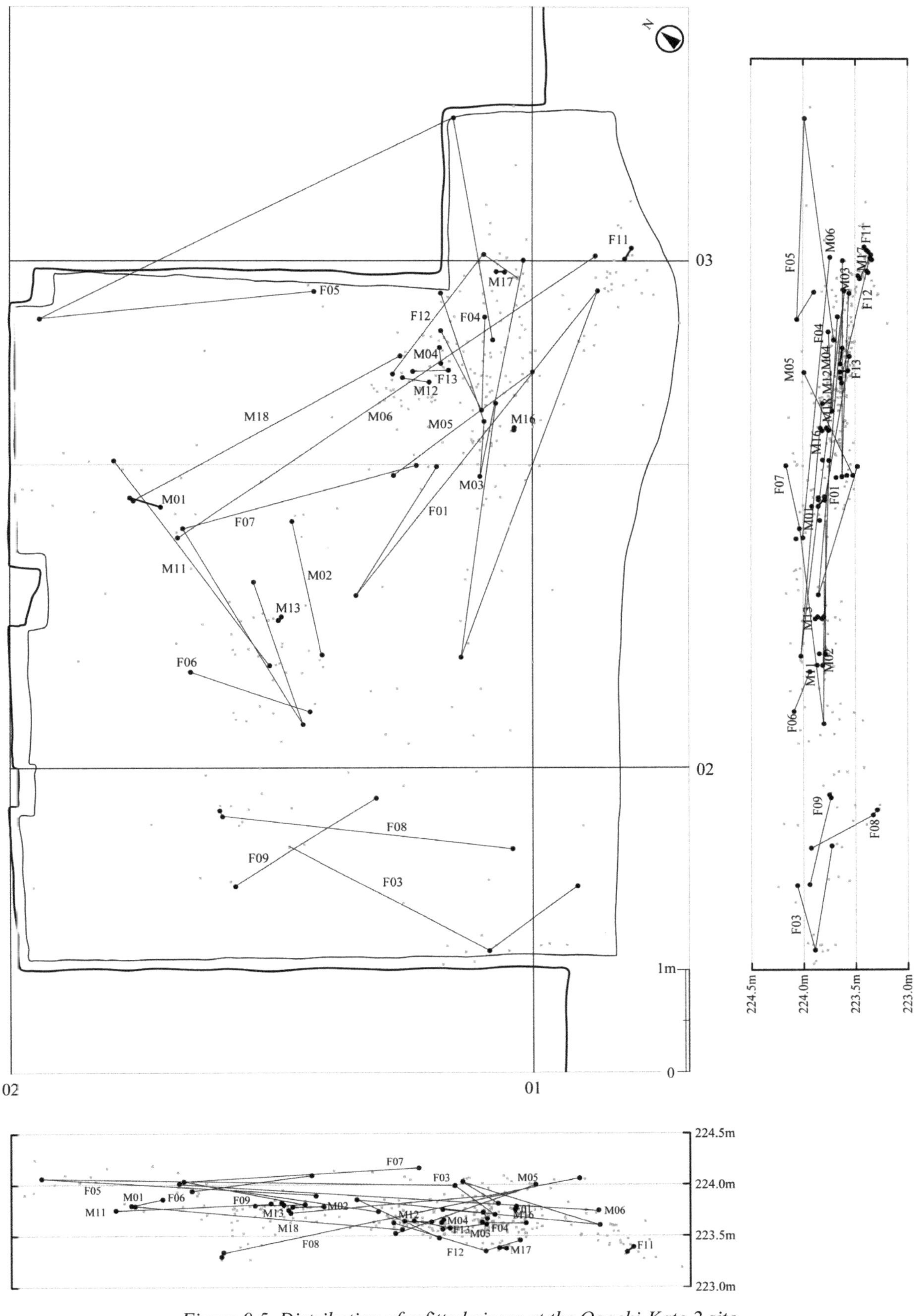

Figure 9.5. Distribution of refitted pieces at the Ogachi-Kato 2 site

Unit II (n = 4). The mean elevation of the piece-plotted artefacts is 223.701 m a.s.l. with a standard deviation of 19.4 cm. The median elevation is 223.686 m a.s.l., 1.5 cm lower than the mean. There is a clear bimodal distribution in a histogram of elevation by 5 cm increments with peaks at 223.640 m and 223.790 m a.s.l.

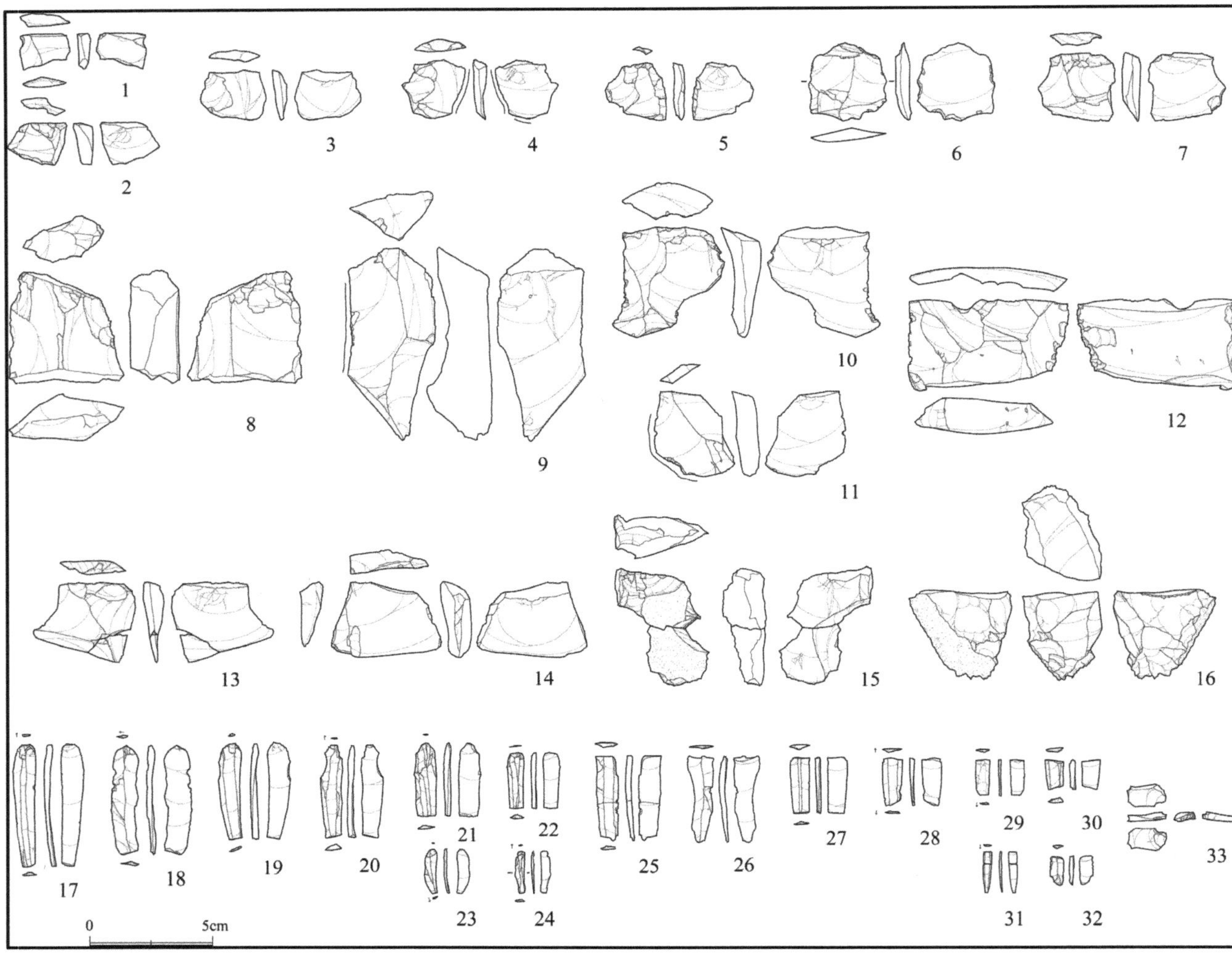

Figure 9.6. Lithic specimens of the Ogachi-Kato 2 site showing small flake reduction sequence (Nos. 1–7, 12), flake reduction sequence (8–11 and 13–16), and microblade reduction sequence (17–33). Tool categories: 1–7 – trapezoids; 8 – side scraper; 9 and 11 – flakes with micro-flaking; 10 – retouched flake; 17–32 – microblades; 33 – microblade core tablet; 13–15 – flakes; and 12 and 16 – cores

Contrary to the bimodal distribution by depth, the horizontal distribution is relatively tight. There are 82 refit examples that span as far as between units II and Vb (II–IV: 3; II–Va: 4; II–III: 2; III–IV: 7; III–Va: 7; III–Vb: 3; IV–IV: 5; IV–Va: 15; IV–Vb: 3; Va–Va: 31; and Va–Vb: 2) (refits shown in Figure 9.5). Based on both spatial distribution analysis (Izuho *et al.* 2012a) and geoarchaeological studies mentioned above, we suggest that the artefacts were originally deposited in Unit Va prior to transformation by cryogenic processes and mass movement. Unit IV is a redeposition of the upper portion of Unit Va (Izuho 2012a, 2012b, 2012c).

Interesting vertical distribution patterns by weight are apparent (Figure 9.4). The lower the mass, the tighter the vertical distribution. Significant statistical differences (Kruskal–Wallis test: $H = 8.27$, $df = 3$, $p < 0.041$) were observed between the medians of the four size classes: 1) > 0.04 g; 2) 0.04 ≤ 0.14 g; 3) 0.14 ≤ 0.62 g; and 4) > 0.62 g. The vertical distribution by weight is further evidence of size sorting within the lithic concentration. One radiocarbon date with calibrated range of 14,120–13,790 cal BP (± 2σ) is reported on charcoal from Unit Va. Based on the apparent stratigraphic mixing, combined with the geochronology of the Upper Palaeolithic assemblages in Hokkaido (Izuho *et al.* 2012b), we believe the human occupation predated the reported radiocarbon date.

Illustrations of representative lithic specimens are shown in Figure 9.6. The Ogachi-Kato 2 assemblage includes trapezoids (n = 7), microblades (n = 93), side scrapers (n = 3), retouched flakes (n = 4), flakes with microflaking (n = 6), debitage (n = 370), cores (n = 2), and ochre (n = 3) (Table 9.1). Lithic raw materials include obsidian (n = 470; 587 g total weight; 95.1% by count and 87.3% by weight), "hard-shale" (n = 6; 6.9 g; 1.2% by count and 1.0% by weight), and andesite (n = 9; 57.5 g; 1.8% by count and 8.5% by weight). The assemblage exhibits a microblade reduction sequence related to the Rankoshi-type microblade cores, a flake reduction sequence identified as a Shimaki assemblage, and a small flake reduction sequence. Based on the geochronology of the Upper Palaeolithic in Hokkaido (Izuho and Akai 2005; Izuho *et al.* 2012b), these assemblages are characteristic of a microblade, flake, and small flake industry below the Eniwa-a tephra (En-a).

Table 9.1. Lithic assemblage of the Ogachi-Kato 2 site

	Obsidian		"Hard-Shale"		Andesite		Chert		Ochre		Total	
	No.	*Weight (g)*	*No.*	*Weight (g)*	*No.*	*Weight (g)*	*No.*	*Weight (g)*	*No.*	*Weight (g)*	*No.*	*Weight (g)*
Microblade	91	26.28	2	0.42	–	–	–	–	–	–	93	26.70
Trapezoid	7	21.85	–	–	–	–	–	–	–	–	7	21.85
Side-scraper	2	38.72	–	–	1	46.48	–	–	–	–	3	85.20
Retouched flake	4	66.25	–	–	–	–	–	–	–	–	4	66.25
Flake with microflaking	6	23.17	–	–	–	–	–	–	–	–	6	23.17
Flake	358	321.42	4	6.46	8	10.97	–	–	–	–	370	338.85
Core	2	89.32	–	–	–	–	–	–	–	–	2	89.32
Orcher	–	–	–	–	–	–	–	–	3	5.3	3	5.30
Rock fragment	–	–	–	–	–	–	6	15.45	–	–	6	15.45
Total	470	587.01	6	6.88	9	57.45	6	15.45	3	5.3	494	672.09

Table 9.2. Artefact frequency by recovery methods at the Ogachi-Kato 2 site

	Piece-Plotted Specimens	Non-Piece-Plotted Specimens	Materials Collected from the Surface of the Site	Total
Ogachi-Kato 2	353	71	64	488

REFITTING ANALYSIS AND REDUCTION SEQUENCE

Materials and Methods

The assemblage of 488 stone artefacts from the single concentration at Ogachi-Kato 2 includes 353 piece-plotted specimens; 71 pieces obtained during screening with only level context; and 64 samples collected from the surface of the site (Table 9.2). Due to the uncertain context, the surface materials were not included in this analysis unless they were part of a refit with other piece-plotted artefacts. The detailed interpretation of the lithic assemblage involves three parts: 1) the identification of refits; 2) an analysis of the reduction sequences to determine the industries present; and 3) the analysis of the artefacts by X-ray Fluorescence (XRF) to determine the geologic sources.

M. Izuho, F. Akai, Y. Nakazawa, and N. Oda identified the individual refits within the different raw material types based on refitting analysis and interpretation of the reduction sequence. Refitting analysis of obsidian is generally difficult, but this study uses methods based on visual identification that have been developed for the study of the Upper Palaeolithic in Hokkaido (Akai 2005a, 2005b, 2008; Hokkaido Centre for Buried Cultural Property 2000, 2001, 2002, 2004, 2006, 2007a, 2007b, 2008, 2010, 2011, 2012; Izuho 1998; Oda 2009; Yamada 1999).

The artefacts were first grouped by raw material type, then by distinct material characteristics such as colour, transparency, flow structure, and cortex type (Izuho 1997). Artefacts exhibiting burning or weathering were compared across all groups. Based on our experience, we expect the refit rate to vary according to the excavation recovery methods and the skill of the analysts, but it is also correlated with distance from the raw material source(s) and the stage and type of reduction (Akai 2008). Table 9.3 lists the refit rates of Ogachi-Kato 2 along with those from other major Upper Palaeolithic sites in Hokkaido. The Ogachi-Kato 2 refit rate is 17.0%, and is in line with the other studies; therefore, we do not believe there is a significant sampling bias that would preclude comparison with other assemblages (Table 9.3, see also Figure 9.7).

The goal of the reduction sequence analysis is to reconstruct the lithic industries and raw material use of the Upper Palaeolithic hunter–gatherer groups in the Palaeo-Sakhalin–Hokkaido Peninsula. In particular, we are interested in a quantitative understanding of the flow of raw material in and out of an occupation site. The reconstruction of the reduction sequence is based on the removal order of the flakes as evidenced by the analysis of the refitted reduction sequences. This approach to analysis is common in studies of the Upper Palaeolithic in Hokkaido (Akai 2005a, 2005b, 2008; Izuho 1998; Naoe 2009; Naoe and Nagasaki 2005; Oda 2009; Suzuki 2007; Takakura 2001; Yamada 1999). Each of the refit reduction sequences is separated by a reduction stage, including primary and secondary reduction.

A total of 83 individual pieces are included in 31 separate refit sequences (Table 9.4) for an overall refit rate of 17.0%. Major refit groups are illustrated in Figure 9.8. One refit (T01) is a small flake reduction sequence; five

Table 9.3. Rate of refitted pieces of the Upper Palaeolithic sites on Hokkaido

Site	Distance from the Major Obsidian Primary Source	Rate of Refitted Pieces*	No. of Refitted Pieces	No. of Lithic Artefacts**	Reference
Ogachi-Kato 2	10 km	17.0	83	488	This study
Kamishirataki 2	5 km	24.6/2.8	12,301	50,085/432,429	Hokkaido Centre for Buried Cultural Property (2001)
Kamishirataki 6	5 km	29.6/9.4	477	1609/5062	Hokkaido Centre for Buried Cultural Property (2001)
Okushirataki 1	5 km	28.4/3.7	23,859	84,105/647,321	Hokkaido Centre for Buried Cultural Property (2002)
Kamishirataki 5	5 km	20.9/5.5	4699	22,441/86,034	Hokkaido Centre for Buried Cultural Property (2002)
Kamishirataki 8, East Area	5 km	25.8/3.3	23,052	89,310/702,618	Hokkaido Centre for Buried Cultural Property (2004)
Kamishirataki 8, West Area	5 km	30.9/6.2	39,046	126,475/626,305	Hokkaido Centre for Buried Cultural Property (2006)
Hattoridai 2	5 km	25.7/2.2	17,388	67,754/798,648	Hokkaido Centre for Buried Cultural Property (2007a)
Shirataki 8	5 km	36.9/15.8	636	1722/4030	Hokkaido Centre for Buried Cultural Property (2007b)
Shirataki 18	5 km	13.5/6.6	3162	23,355/47,762	Hokkaido Centre for Buried Cultural Property (2007b)
Kyushirataki 5	5 km	27.0/7.0	18,250	43,172/261,571	Hokkaido Centre for Buried Cultural Property (2008)
Kyushirataki 16	5 km	42.3/9.1	1653	3326/18,071	Hokkaido Centre for Buried Cultural Property (2010a)
Kyushirataki 1	5 km	34.7/0.7	504	920/71,243	Hokkaido Centre for Buried Cultural Property (2010a)
Horokazawa I	5 km	25.9/6.5	7561	22,722/115,574	Hokkaido Centre for Buried Cultural Property (2011)
Kyushirataki 15	5 km	43.7/12.5	14,543	35,541/115,965	Hokkaido Centre for Buried Cultural Property (2012)
Oruika 2	170 km	12.0/–	365	3301/–	Hokkaido Centre for Buried Cultural Property (2005)
Kamihoronai-Moi	125 km	10.2/–	143	1400/–	Atsuma Board of Education (2006)
Ankaritoh 7	75 km	23.3/–	484	2078/–	Hokkaido Centre for Buried Cultural Property (2010b)

* In percent; the left side: piece-plotted artefacts; the right side: non-piece-plotted artefacts.

** The left side: piece-plotted artefacts; the right side: non-piece-plotted artefacts.

refits (F01–05) are flake reduction sequences; 18 refits (M01–18) are microblade reduction sequences; one example (O01) is a bifacial reduction; and six refits (U01–06) are uncertain. Here we present a detailed discussion of the three dominant types: small flake, flake, and microblade reduction.

Small Flake Reduction Sequence

One refit (consisting of three samples) and five individual specimens are related to the small flake reduction sequence. Gravel cortex is present on a trapezoid (Figure 9.6, 4), suggesting raw material procurement from the river deposits. The small flake technology involves primary reduction. In T01, small flakes were removed from a discoidal core made of a large thick flake (Figure 9.8, 1). The detached flakes are approximately 2–3 cm in length, and typical forms are trapezoidal, triangular, or shell. They have multifaceted platforms, abrupt flaking angles (approximately 100–125°), and a salient bulb. Secondary reduction included snap fracture and retouch of the lateral edge and/or distal end of the small flake. A trapezoid represents a relatively informal tool form.

Cortex is present on only one trapezoid and large flakes, such as T01, which could serve as core blanks that were produced elsewhere and then transported to the site. The lack of cortex suggests that raw material procurement, cortex removal, and the production of the core blanks were undertaken elsewhere. The reduction apparent in the

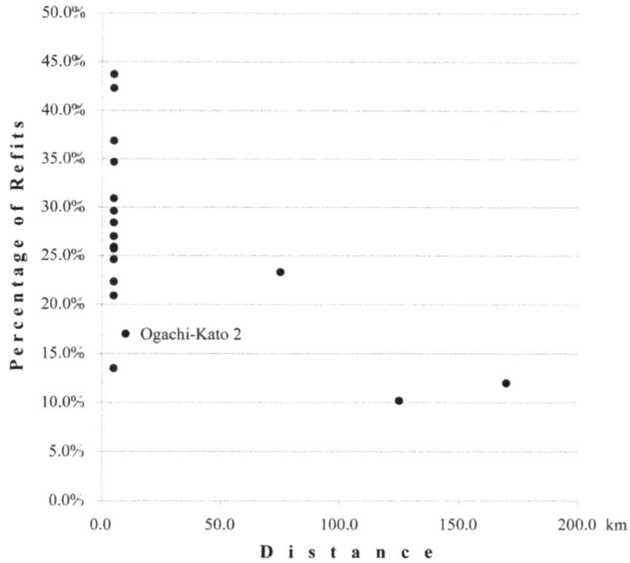

Figure 9.7. Scatter diagram of the percentage of refitted pieces and the distances from the major primary obsidian sources (see Table 9.3)

Ogachi-Kato 2 assemblage includes primary reduction of the core blank and further secondary reduction. Because several trapezoids (Figure 9.6, 5–6) were made on obsidian from apparently different sources (e.g., they have different macroscopic characteristics), they were brought in without further reduction, and used and discarded at the site. While trapezoids and a core were discarded, the parent core of the T01 trapezoid was not recovered, suggesting that it may have been transported to the next occupation.

The small trapezoidal pieces detached in the small flake reduction sequence are not expedient tools. This is in sharp contrast to the flake reduction sequence (Sato 2003; Izuho *et al.* 2012a).

Flake Reduction Sequence

Four refits (consisting of 11 total pieces) and six individual specimens are assigned to the flake reduction sequence. Based on refits such as F02 (Figure 9.8, 2), the obsidian gravel procured from the river deposits was at least 5 cm in diameter. The flake reduction sequence involved primary reduction. In F02, flakes were removed from a multifaceted core made on gravel or split-gravel, and the sizes of the detached pieces were approximately 3–5 cm. The typical shapes include rectangular and convergent forms. They exhibit single-faceted platforms, abrupt flaking angles (approximately 115–135°), a salient bulb, and a pronounced cone at the point of impact. Secondary reduction included flat flaking and marginal retouch on lateral edges and/or the distal end of the flakes. The specimens, such as retouched flakes, are relatively informal.

The primary and secondary reduction of gravel or split-gravel was performed at this site as seen in F02 and F01. Because several refits and flakes (Figure 9.6, 14) have

Table 9.4. Lithic refits of the Ogachi-Kato 2 site

Refit No.	Raw Material/Shape	Reffited Artefact Nos.
T01	Obsidian/Unknown	319, 325, 344
F01	Obsidian/Unknown	035, 045, 180
F02	Obsidian/Gravel	031, 043, 164, 181
F03	Obsidian/Unknown	313, 322
F04	Obsidian/Gravel	127, 137
F05	Andesite/Unknown	144, 157, 278
M01	Obsidian/Unknown	224, 281
M02	Obsidian/Unknown	211, 236
M03	Obsidian/Unknown	053, 068, 089, 250
M04	Obsidian/Unknown	252, 254, 426
M05	Obsidian/Unknown	014, 300
M06	Obsidian/Unknown	117, 189, 419
M07	Obsidian/Unknown	305, 420
M08	Obsidian/Unknown	216, 379
M09	Obsidian/Unknown	187, 395
M10	Obsidian/Unknown	216, 369
M11	Obsidian/Unknown	201, 207
M12	Obsidian/Unknown	261, 279
M13	Obsidian/Unknown	220, 235, 391
M14	Obsidian/Unknown	256, 453
M15	Obsidian/Unknown	093, 399
M16	Obsidian/Unknown	063,080
M17	Hard-Shale/Unknown	152,153
M18	Obsidian/Unknown	282, 294
O01	Obsidian/Unknown	012, 051, 139, 174, 302, 411, 446
U01	Obsidian/Unknown	170, 368, 380, 410
U02	Obsidian/Unknown	169, 191
U03	Obsidian/Unknown	162, 188, 198, 222, 370
U04	Obsidian/Unknown	312, 351, 352
U05	Obsidian/Unknown	241, 259
U06	Obsidian/Unknown	226, 425

different macroscopic characteristics of obsidian without a trace of primary reduction at the site, they were likely brought in without producing blanks, then used and discarded at this location. Although retouched flakes, flakes with microflaking, and cores were discarded, the parent core of F01 (Figure 9.8, 3) was not recovered, suggesting it was transported to the next occupation site. This reduction sequence is comparable to that of the Shimaki site (Izuho *et al.* 2012b).

Microblade Reduction Sequence

The microblade reduction assemblage consists of primary reduction and includes 18 refits (39 pieces) and 52 individual specimens of obsidian, and one refit (two pieces) of "hard-shale". Macroscopic visual source

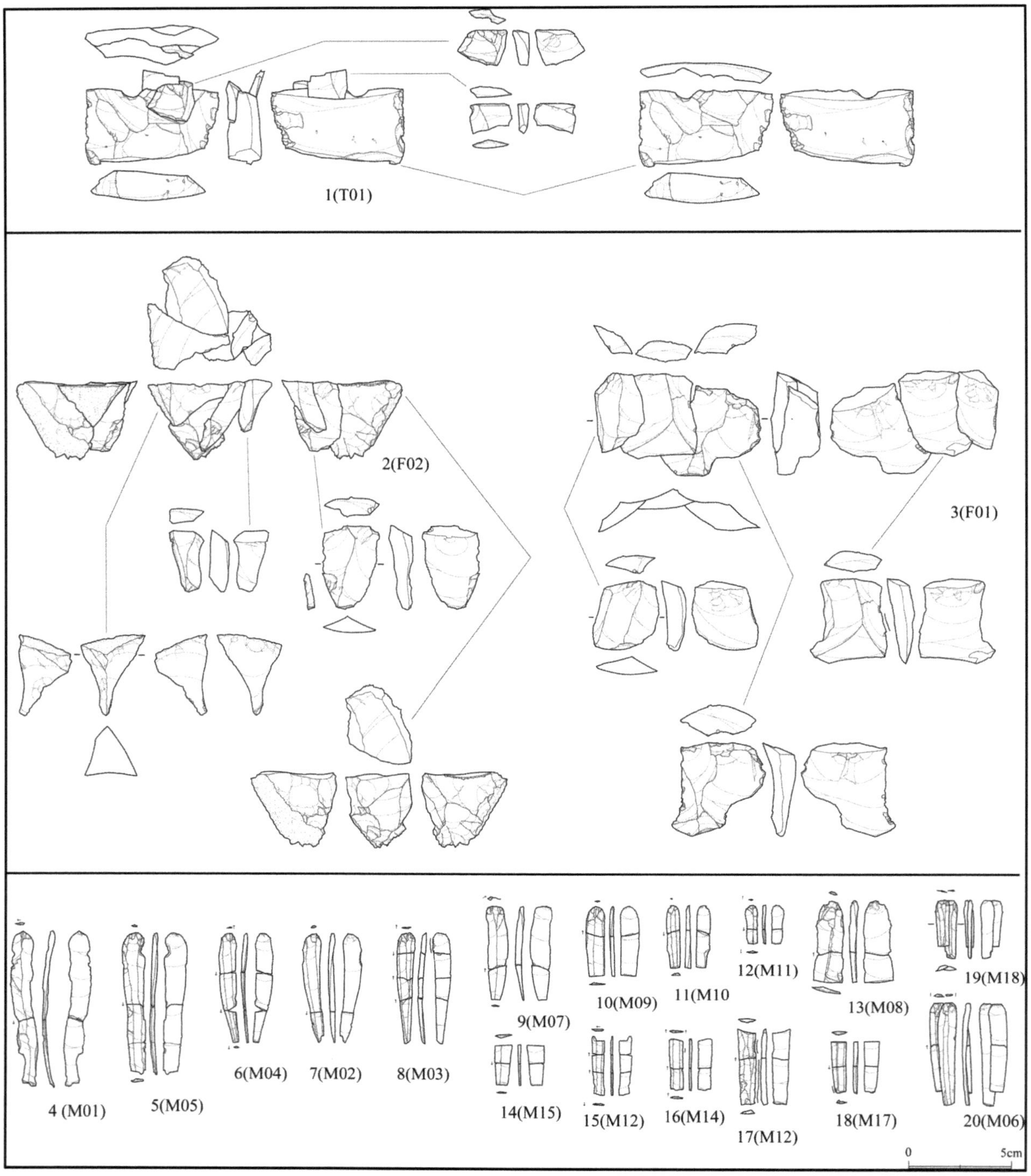

Figure 9.8. Major lithic refits of the Ogachi-Kato 2 site

classification of the obsidian yields five sub-groups (Oda and Izuho 2012). There is no cortex present on the artefacts; thus, there is no indication of the original form of the raw material.

The proximal portion of the microblades exhibits parallel margins with some tapering and gentle twisting (usually to the right) of the distal portion. Given these morphological characteristics, we believe that most of the microblades were removed from Rankoshi-type microblade cores (Kimura 1978; Tsurumaru 1981; Nakazawa *et al.* 2005).

Secondary reduction includes snap fractures and marginal retouch of the lateral edges of the microblades. The microblades vary considerably in size. Retouched microblades account for only 7.5% (n = 7) of the total microblade assemblage.

The refit sequences indicating primary reduction and rejuvenation of the obsidian microblade core tablet provide clear evidence of microblade production at the site. Because refit M06 is assigned to Visual Group 5, refit M18 to Visual Group 7 (Figure 9.8, 19–20), and a microblade core tablet rejuvenation to Visual Group 4

(Figure 9.6, 33), there are at least three separate cores with evidence of local microblade production. Interestingly, there are only a few refit sequences which suggest that many of the microblades recovered from the site were not made there. There is very little retouch present on the microblades, indicating non-intensive use.

There are no cores or other reduction debris of the "hard-shale" – only microblades. The lack of production evidence indicated import and discard, but not production, of the "hard-shale" microblades at Ogachi-Kato 2.

OBSIDIAN GEOCHEMICAL ANALYSIS

A total of 129 obsidian artefacts were submitted for analysis by XRF (Table 9.5), with a subset also subjected to Neutron Activation Analysis (NAA). M. Izuho and N. Oda selected the samples, J. Ferguson and M. Izuho conducted the XRF analysis, and M. Glascock performed the short-irradiation NAA. A more detailed discussion of the analytical methodology is presented in Ferguson *et al.* (this volume).

Methodology

The XRF analysis was conducted using a Bruker III-SD handheld XRF mounted in a Dewitt Systems sample changer. The XRF spectrometer has a Rh-based X-ray tube operated at 40 kV and uses a thermoelectrically-cooled silicon detector. The obsidian calibration incorporates a set of 40 well-characterised obsidian standards with elemental compositions based on previous Inductive Couple Plasma – Mass-Spectrometry (ICP–MS), XRF, and NAA measurements (Glascock and

Table 9.5. Samples for obsidian source assignment at the Ogachi-Kato 2 site

<table>
<tr><th colspan="2">Refits</th><th colspan="4">Individual Specimens</th></tr>
<tr><th>Refit No.</th><th>Artefact No(s).</th><th>Small Flake Reduction Sequence</th><th>Flake Reduction Sequence</th><th>Microblade Reduction Sequence</th><th>Others/Unknown</th></tr>
<tr><td>T01</td><td>319, 325</td><td rowspan="31">100, 353, 354</td><td rowspan="31">042, 045, 345</td><td rowspan="31">029, 037, 054, 082, 085, 090, 102, 109, 116, 126, 133, 134, 145, 150, 156, 165, 166, 185, 190, 192, 194, 195, 199, 202, 206, 219, 225, 228, 260, 264, 274, 276, 285, 286, 295, 334, 365, 375, 376, 389, 405, 424, 454, 456, 458</td><td rowspan="31">006, 008, 009, 056, 076, 077, 078, 084, 091, 128, 135, 141, 158, 173, 205, 210, 215, 217, 230, 234, 253, 259, 266, 280, 287, 298, 301, 309, 317, 333, 359, 361, 362, 367, 381, 382, 412, 422, 427, 428, 429, 442, 443, 447, 450, 455, 459, 474</td></tr>
<tr><td>F01</td><td>045</td></tr>
<tr><td>F02</td><td>031, 181</td></tr>
<tr><td>F03</td><td>313</td></tr>
<tr><td>F04</td><td>137</td></tr>
<tr><td>F05</td><td>–</td></tr>
<tr><td>M01</td><td>224</td></tr>
<tr><td>M02</td><td>236</td></tr>
<tr><td>M03</td><td>068</td></tr>
<tr><td>M04</td><td>254, 426</td></tr>
<tr><td>M05</td><td>014</td></tr>
<tr><td>M06</td><td>117</td></tr>
<tr><td>M07</td><td>305</td></tr>
<tr><td>M08</td><td>379</td></tr>
<tr><td>M09</td><td>395</td></tr>
<tr><td>M10</td><td>369</td></tr>
<tr><td>M11</td><td>201</td></tr>
<tr><td>M12</td><td>261</td></tr>
<tr><td>M13</td><td>391</td></tr>
<tr><td>M14</td><td>453</td></tr>
<tr><td>M15</td><td>399</td></tr>
<tr><td>M16</td><td>–</td></tr>
<tr><td>M17</td><td>–</td></tr>
<tr><td>M18</td><td>282, 294</td></tr>
<tr><td>O01</td><td>051</td></tr>
<tr><td>U01</td><td>170</td></tr>
<tr><td>U02</td><td>191</td></tr>
<tr><td>U03</td><td>188</td></tr>
<tr><td>U04</td><td>352</td></tr>
<tr><td>U05</td><td>259</td></tr>
<tr><td>U06</td><td>–</td></tr>
</table>

Ferguson 2012). The samples were counted for three minutes each to measure minor and trace elements present. The elements measured include Mn, Fe, Zn, Ga, Pb, Th, Sr, Y, Zr, and Nb; however, due in part to the small size of the artefacts only data for Rb, Sr, Y, Zr, and Nb are considered reliable.

A limited subset of the samples was analysed by short-irradiation NAA to clarify the small number of samples with inconclusive source assignment based on XRF data alone. The samples were plased into clean, high-density polyethylene vials used for short irradiations at the Missouri University Research Reactor Center (MURR). The samples were all small enough to fit into the vials without damage and were safely returned to the collections after a short decay period. Individual sample weights were recorded to the nearest 0.01 mg using an analytical balance. Along with the unknown samples, standards made from the National Institute of Standards and Technology (NIST) certified standard reference material of SRM-278 (obsidian rock) were used. Quality control also included samples of SRM-1633b (coal fly ash) and the Alca obsidian source (Peru, South America). Concentrations are determined for Al, Ba, Cl, Dy, K, Mn, and Na. NAA methods are discussed in detail by Glascock (1992) and Glascock *et al.* (1994).

Statistical analysis of the data was carried out on base-10 logarithms of concentrations for both the XRF and NAA datasets. The use of log concentrations rather than raw data compensates for differences in magnitude between the major, minor, and trace elements. Transformation to base-10 logarithms also yields a more normal distribution for many trace elements.

The interpretation of compositional data obtained from the analysis of archaeological materials is discussed in detail elsewhere (e.g., Baxter and Buck 2000; Bieber *et al.* 1976; Bishop and Neff 1989; Glascock 1992; Harbottle 1976; Neff 2000), and will only be summarised here. The main goal of data analysis is to identify distinct homogenous groups within the analytical Database and match these groups to the chemical signatures of known geologic sources.

Compositional groups can be viewed as "centres of mass" in the compositional hyperspace described by the measured elemental data. Groups are characterised by the locations of their centroids and the unique relationships (i.e., correlations) between the elements. Decisions about whether to assign a specimen to a particular compositional group are based on the overall probability that the measured concentrations for the specimen could have been obtained from that group.

The main complicating factor in this analysis is the size of the artefacts. Many XRF laboratories will not attempt to analyse artefacts smaller than 1 cm in diameter and 1 mm in thickness. The difficulty of small/thin artefacts is that the resulting X-rays from different elements in the sample that make it to the detector can originate from different depths in the sample (Ferguson 2012). Thus, a thin sample will show a difference in the relative size of the peaks within a specimen when compared to a thicker piece from the same source. Normalisation allows for some correction, but the changes are not consistent for each element because the calibration treats all areas of the spectra as the same, resulting in misleading concentrations for elements with peaks far from the portion of the spectra used for normalisation. Rb, Sr, Y, Zr, and Nb are all close to the normalisation region of the spectra, therefore providing the most reliable results for smaller flakes.

Results

The source characterisation for Hokkaido is a joint project of K. Wada, M. Mukai, M. Izuho, J. Ferguson, and M. Glascock, and a full description of these efforts is presented in Ferguson *et al.* (this volume). Most source groups are readily identified by XRF analysis alone, but a few sources — Akaigawa, Tokachi-Mitsumata, Monbetsu, and Kushiro (Shitakara) — require further efforts, such as short-irradiation NAA, to differentiate them. Fortunately, only three artefacts from this site matched this cluster of sources and were submitted for NAA.

The compositional data for the XRF analysis of the artefacts is presented in Ferguson *et al.* (this volume, Table 2.4), and the NAA data in Table 9.6. Figure 9.9 shows the separation of the relevant source groups and assigned artefacts. The separation of the Oketo groups requires a second plot (Figure 9.10). The artefacts do not always fit within the confidence ellipse for a given source, mostly due to the small size of them. As described, sample size/thickness has variable effects on the data and tends to spread the artefacts along a correlation line.

The three artefacts with questionable source assignment by XRF were subjected to short-irradiation NAA for further clarification. Sample MIH051 is confirmed to be from the Tokachi-Mitsumata source, although it is difficult to differentiate this source from Kushiro (Shitakara). Sample MIH086 seems closest to Tokachi-Mitsumata, but this is not clear. For now it is considered a provisional assignment to Tokachi-Mitsumata. Sample MIH111 is similar to the Oketo (Oketoyama) source, but there are sufficient differences to consider it unassigned. A long-irradiation NAA might help to resolve this artefact's source assignment, but that would require sample destruction.

DISCUSSION

Here we discuss the relationships between the obsidian sources and the three main reduction technologies (small flake, flake, and microblade), and reveal how the lithic technology of the hunter–gatherers was tied to the timing of raw material procurement, manufacture, and use. The results of the obsidian source assignments of every refit artefact are shown in Table 9.7, and the scheme of the

Table 9.6. Elements concentrations by short-irradiated NAA (in ppm) for artefacts of the Ogachi-Kato 2 site

MURR ID	Artefact No.	Al	Ba	Cl	Dy	K	Mn	Na
MIH051	199	80,019.4	873.4	544.9	5.1612	36,438.9	358.5	30,400.4
MIH086	313	69,042.9	894.2	485	4.0967	38,432.8	379.3	29,014.4
MIH111	405	74,259.1	788	524.6	3.592	32,734.6	372.5	32,209.9

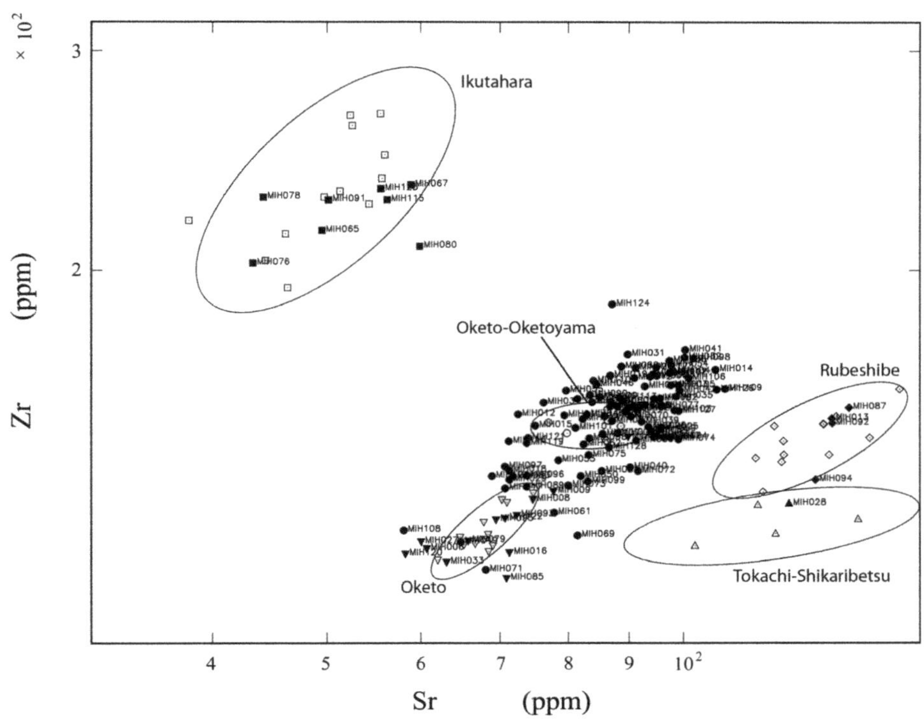

Figure 9.9. Bivariate plot of Sr vs. Zr concentrations (in ppm) showing all of the likely source reference groups (except for Tokachi-Mitsumata) along with all of the artefacts. Ellipses represent 90% confidence intervals for membership in the source group

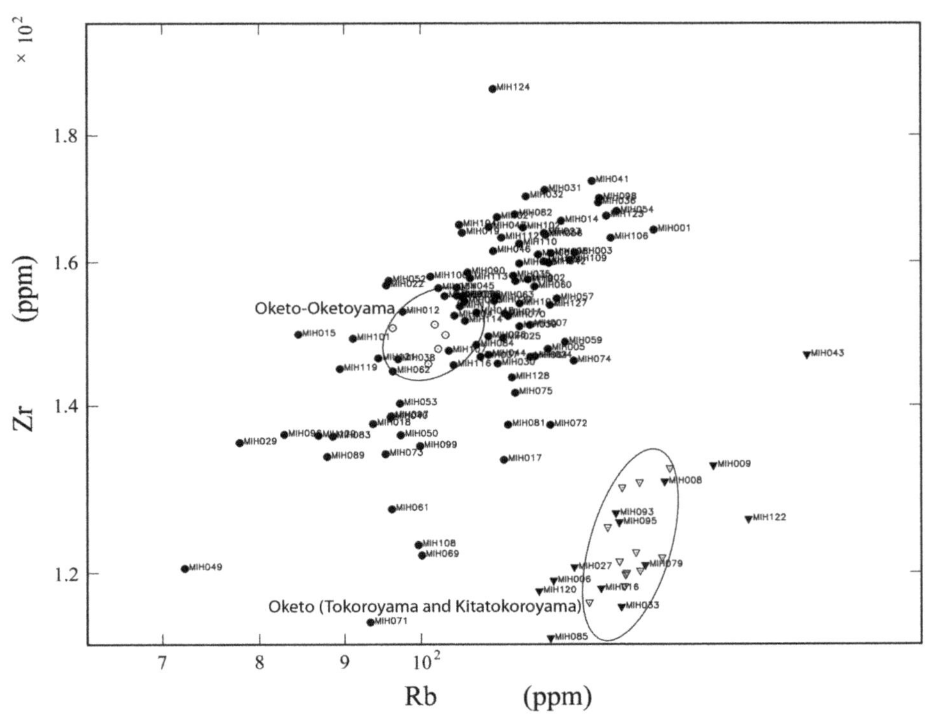

Figure 9.10. Bivariate plot of Rb vs. Zr concentrations (in ppm) showing the separation of the two Oketo source groups. Ellipses represent 90% confidence intervals for membership in the source group

Table 9.7. Results of obsidian source assignment at the Ogachi-Kato 2 site

Refit No.	Raw Material/Shape	Reffited Artefact Nos.	Analysed Artefact Nos.	Source
T01	Obsidian/Unknown	319, 325, 344	319, 325	Oketo (Oketoyama)
F01	Obsidian/Unknown	035, 045, 180	045	Oketo (Tokoroyama)
F02	Obsidian/Gravel	031,043, 164, 181	031, 181	Oketo (Tokoroyama)
F03	Obsidian/Unknown	313, 322	313	Tokachi-Mitsumata?
F04	Obsidian/Gravel	127, 137	137	Oketo (Tokoroyama)
F05	Andesite/Unknown	144, 157, 278	–	–
M01	Obsidian/Unknown	224, 281	224	Oketo (Oketoyama)
M02	Obsidian/Unknown	211, 236	236	Ikutahara
M03	Obsidian/Unknown	053, 068, 089, 250	068	Rubeshibe
M04	Obsidian/Unknown	252, 254, 426	254, 426	Ikutahara
M05	Obsidian/Unknown	014, 300	014	Oketo (Oketoyama)
M06	Obsidian/Unknown	117, 189, 419	117	Oketo (Oketoyama)
M07	Obsidian/Unknown	305, 420	305	Oketo (Oketoyama)
M08	Obsidian/Unknown	216, 379	369	Oketo (Oketoyama)
M09	Obsidian/Unknown	187, 395	395	Oketo (Oketoyama)
M10	Obsidian/Unknown	216, 369	369	Oketo (Oketoyama)
M11	Obsidian/Unknown	201, 207	201	Oketo (Oketoyama)
M12	Obsidian/Unknown	261, 279	261	Oketo (Oketoyama)
M13	Obsidian/Unknown	220, 235, 391	391	Oketo (Oketoyama)
M14	Obsidian/Unknown	256, 453	453	Oketo (Oketoyama)
M15	Obsidian/Unknown	093, 399	399	Oketo (Oketoyama)
M16	Obsidian/Unknown	063, 080	–	–
M17	Hard-Shale/Unknown	152, 153	–	–
M18	Obsidian/Unknown	282, 294	282, 294	Ikutahara

lithic reduction sequence at Ogachi-Kato 2 is shown in Figure 9.11.

Small Flake Reduction Sequence

The obsidian source of the refit T01 (Figure 9.8, 1) is Oketo (Oketoyama). The primary source is 10 km from the site, but the occupants most likely procured the cobbles from secondary deposits within 1 km of the settlement. Primary and secondary reductions were carried out at the site. Trapezoids were used and discarded at the site, and there is evidence that several cores were taken only when the locale was abandoned.

In addition to the use of the Oketo (Oketoyama) source, two other sources — Oketo (Tokoroyama) and Rubeshibe — were also exploited (Figure 9.6, 5–6). The tools made from raw material of these other sources were used and discarded at the site without further production of blanks. The primary source of Oketo (Tokoroyama) is approximately 15 km from the site, but secondary deposits occur within 5 km distance. The primary source of Rubeshibe is approximately 30 km from Ogachi-Kato 2, and secondary distribution extends to approximately 5 km from the site. The small flake reduction activities at the site exploited primarily high quality local obsidian.

Flake Reduction Sequence

All refits from the flake reduction sequence are assigned to the Oketo (Tokoroyama) source. The gravel cortex present on some specimens suggests procurement of the Oketo (Tokoroyama) obsidian from the local (< 5 km away) secondary deposits. Refit F02 (Figure 9.8, 2) is a good example of these primary and secondary reductions of the local gravel. Refit F01 (Figure 9.8, 3) shares the same reduction sequence as F02, but the core was not recovered from the site.

Among the individual tools, one sample (Figure 9.6, 14) is from Rubeshibe and another (Figure 9.6, 13) is from Tokachi-Mitsumata. These artefacts were used and discarded with no evidence of production or further reduction at this location. The use of Rubeshibe material might be considered local, assuming the closest secondary deposits (5 km away) were used. In contrast, the primary source of Tokachi-Mitsumata is 50 km from Ogachi-Kato 2 (Figure 9.1), and secondary deposits are at least 45 km from the site. The specimens belonging to the flake reduction sequence were seemingly expedient, yet they include a mix of local and non-local materials. Assuming direct procurement of the raw material, the presence of the

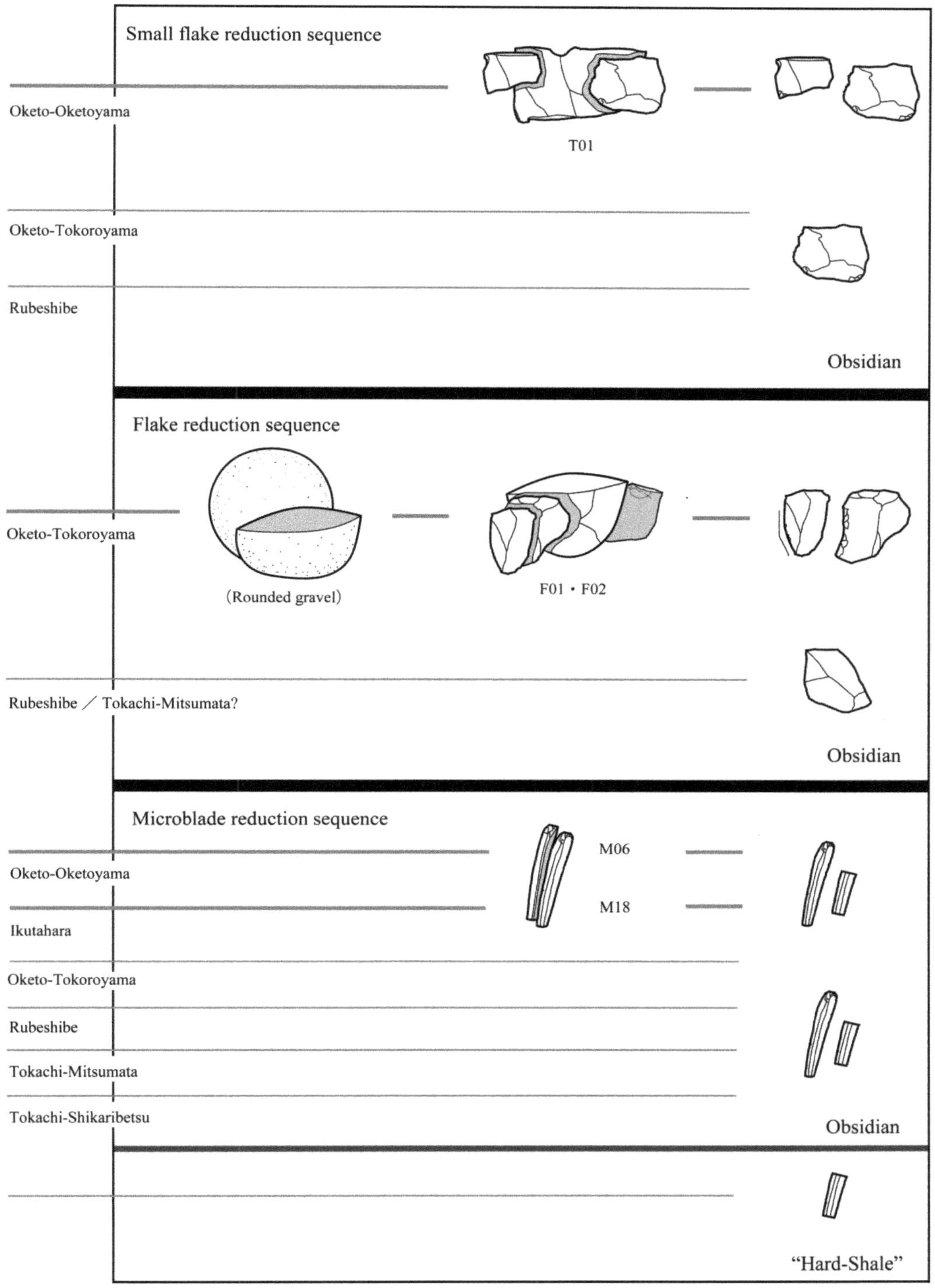

Figure 9.11. Lithic reduction sequences of the Ogachi-Kato 2 site; reduction flows are from left to right

Tokachi-Mitsumata source suggests a movement range of at least 50 km.

Microblade Reduction Sequence

A number of sources in the microblade reduction sequence assemblage include Oketo (Oketoyama) (Figure 9.6, Nos. 17–18, 20–21, 23, 25, 27, and 29; Figure 9.8, 4–5, 9–17, and 20); Oketo (Tokoroyama) (Figure 9.6, 24); Ikutahara (Figure 9.6, 19, 22, 26, and 28; Figure 9.8, 6–7 and 19); Rubeshibe (Figure 9.6, 30); Tokachi-Mitsumata (Figure 9.6, 31); and Tokachi (Shikaribetsu) (Figure 9.6, 32). At Ogachi-Kato 2, a few microblades were removed from microblade cores made of Oketo (Oketoyama) and Ikutahara obsidians, in addition to some secondary reduction of the microblades. There is no refit evidence of a complete sequence of microblade production from a single core, suggesting a lack of intensive microblade production at the site.

Because no artefacts in microblade assemblage are cortical, it is impossible to provide information as to the specific procurement location (primary or secondary). In addition to the source distances noted above, the Ikutahara primary deposits are 30 km away, and secondary deposits are as close as 25 km away; and the Tokachi (Shikaribetsu) primary source is 95 km away, and secondary deposits are as close as 90 km away (Figure 9.1). The microblade reduction sequence includes obsidian and "hard-shale" from a combination of local and non-local sources.

CONCLUSIONS

In this chapter, the following subjects were examined: 1) We identified and described refit sequences; 2) We separated and described the three main reduction sequence industries; 3) We assigned the obsidian artefacts to specific geologic sources using XRF and NAA; and 4) We discussed how the lithic technology of the Upper Palaeolithic hunter–gatherers was tied to the timing of raw material procurement, manufacture, and use.

In the small flake industry, three high quality local sources — Oketo (Oketoyama), Oketo (Tokoroyama), and Rubeshibe — were used for the manufacture of cores and tools. Primary and secondary small flake reductions were carried out at Ogachi-Kato 2, and several tools and cores were discarded. There is evidence of the transport of several cores from the site.

In the flake reduction industry, primary and secondary reductions were undertaken, and several cores and tools were discarded at the site. Nearby gravels from the local source of Oketo (Tokoroyama) were used for cores. Obsidian from the local Rubeshibe and non-local Tokachi-Mitsumata sources was used only for tools. Several cores were likely transported from the site.

The microblade industry at Ogachi-Kato 2 consists of the Rankoshi-type microblade detaching method. Oketo (Oketoyama), Oketo (Tokoroyama), Ikutahara, Rubeshibe, Tokachi-Mitsumata, and Tokachi (Shikaribetsu) obsidians were all present in the assemblage indicating procurement from a variety of local and non-local sources. A few microblades were removed at the site from cores made of Oketo (Oketoyama) and Ikutahara source materials, as well as some secondary microblade reduction. Microblades made from obsidian of Oketo (Tokoroyama), Rubeshibe, Tokachi-Mitsumata, and Tokachi (Shikaribetsu) sources have been recovered from the site, but there is no evidence of microblade production on obsidian from these sources at the site.

Although all three industries used high quality obsidian, we have demonstrated that the procurement and reduction strategies varied by industry. While we believe that the correlation between procurement strategy and industry is the result of patterns in human behaviour, it is possible that it is a result of the varying sample sizes. The smallest assemblage (small flake industry, n = 6) exhibits the least source variability (only two local sources); the next smallest industry (flake industry, n = 15) includes only three sources; and the largest assemblage (microblade industry, n = 91) has obsidian from six chemically-distinct sources.

In addition to this problem, we could not make a solid conclusion on following important questions. What does greater diversity of obsidian source utilisation through time mean? Why might people have increased distances to sources and why would they have transported either microblades or cores to and from the site, whereas those using the flake and small flake industries only transported cores away from the site? We believe that further research, combining technological analysis with compositional studies applied to the other sites and integrating the results between the sites, will solve the questions and will greatly improve our understanding of the lithic procurement and use strategies of Upper Palaeolithic hunter–gatherers in Hokkaido.

Acknowledgements

We would like to thank Mr Julian Pessarossi and Dr K. Terry for their help in carrying out this study. Financial support was provided by the Japan Society for the Promotion of Science, KAKENHI grants No. 21242026 (PI H. Sato) and No. 24320157 (PI M. Izuho). The compositional studies were funded in part by the US NSF, Grant 1110793 (PIs M. D. Glascock and J. R. Ferguson). We are grateful to Drs Y. V. Kuzmin and S. G. Keates for comments.

References

AKAI, F. 2005a. Ishikari-Teichitai-Nanbu niokeru Saise-kiji Sekkigun no Kenkyu [A Study of the Microblade Industry on the Southern Ishikari Lowland]. *Hokkaido Kyusekki Bunka Kenkyu* 10, 59–78.

AKAI, F. 2005b. Chitose-shi Marukoyama Iseki Eniwa-a Tephra joi Sekkigun no Saikento [Re-examination of the Lithic Industry above the Eniwa-a Tephra from Marukoyama Site, Chitose City, Hokkaido]. *Ronshu Oshorokko* I, 103–121.

AKAI, F. 2008. Kamihoronai-Moi Sekkigun no Hakurikatei [Lithic Reduction Processes of the Palaeolithic Assemblage of the Kamihoronai-Moi Site, Hokkaido]. *Ronshu Oshorokko* II, 83–97.

Atsuma Board of Education. 2006. *Kamihornai-Moi Iseki (1)* [Kamihoronai-Moi Site (1)]. Atsuma, Japan, Atsuma-cho Kyoiku Iinkai.

BAXTER, M. J., and C. E. BUCK. 2000. Data Handling and Statistical Analysis. In *Modern Analytical Methods in Art and Archaeology*, edited by E. Ciliberto and G. Spoto, 681–746. New York, Wiley.

BIEBER, A. M., JR., D. W. BROOKS, G. HARBOTTLE, and E. V. SAYRE. 1976. Application of Multivariate Techniques to Analytical Data on Aegean Ceramics. *Archaeometry* 18, 59–74.

BISHOP, R. L., and H. NEFF. 1989. Compositional Data Analysis in Archaeology. In *Archaeological Chemistry IV*, edited by R. O. Allen, 576–586. Washington, D.C., American Chemical Society.

BUTZER, K. W. 1982. *Archaeology as Human Ecology*. Cambridge, Cambridge University Press.

CLARKE, D. L. 1968. *Analytical Archaeology*. London, Methuen.

FERGUSON, J. R. 2012. X-ray Fluorescence of Obsidian: Approaches to Calibration and the Analysis of Small Samples. In *Handheld XRF for Art and Archaeology*, edited by A. N. Shugar and J. L. Mass, 401–422. Leuven, Leuven University Press.

FERGUSON, J. R., M. D. GLASCOCK, and M. IZUHO. 2012. Kokuyoseki Ibutsu no XRF Bunseki oyobi NAA Bunseki [X-Ray Fluorescence and Neutron Activation Analysis of Obsidian Artefacts from the Ogachi-Kato 2 Site, Hokkaido]. In *Kokuyoseki no Ryutsu to Shouhi Karamita Kan-Nihonkai Hokubu Chiiki Niokeru Koushinsei Shakai no Keisei to Henyou (I)*, edited by H. Sato, 125–131. Tokyo, University of Tokyo.

GLASCOCK, M. D. 1992. Characterization of Archaeological Ceramics at MURR by Neutron Activation Analysis and Multivariate Statistics. In *Chemical Characterization of Ceramic Pastes in Archaeology*, edited by H. Neff, 11–26. Madison, WI, Prehistory Press.

GLASCOCK, M. D., and J. R. FERGUSON. 2012. Report on the Analysis of Obsidian Source Samples by Multiple Analytical Methods. Report on File. Columbia, MO, University of Missouri Research Reactor Center, Archaeometry Laboratory.

GLASCOCK, M. D., H. NEFF, K. S. STRYKER, and T. N. JOHNSON. 1994. Sourcing Archaeological Obsidian by an Abbreviated NAA Procedure. *Journal of Radioanalytical and Nuclear Chemistry* 180, 29–35.

HARBOTTLE, G. 1976. Activation Analysis in Archaeology. *Radiochemistry* 3, 33–72.

Hokkaido Centre for Buried Cultural Property. 2000. *Shirataki Isekigun* [The Shirataki Site Group]. Sapporo, Hokkaido Maizo Bunkazai Centre.

Hokkaido Centre for Buried Cultural Property. 2001. *Shirataki Isekigun II* [The Shirataki Site Group II]. Sapporo, Hokkaido Maizo Bunkazai Centre.

Hokkaido Centre for Buried Cultural Property. 2002. *Shirataki Isekigun III* [The Shirataki Site Group III]. Sapporo, Hokkaido Maizo Bunkazai Centre.

Hokkaido Centre for Buried Cultural Property. 2004. *Shirataki Isekigun IV* [The Shirataki Site Group IV]. Sapporo, Hokkaido Maizo Bunkazai Centre.

Hokkaido Centre for Buried Cultural Property. 2006. *Shirataki Isekigun VI* [The Shirataki Site Group VI]. Sapporo, Hokkaido Maizo Bunkazai Centre.

Hokkaido Centre for Buried Cultural Property. 2007a. *Shirataki Isekigun VII* [The Shirataki Site Group VII]. Sapporo, Hokkaido Maizo Bunkazai Centre.

Hokkaido Centre for Buried Cultural Property. 2007b. *Shirataki Isekigun VIII* [The Shirataki Site Group VIII]. Sapporo, Hokkaido Maizo Bunkazai Centre.

Hokkaido Centre for Buried Cultural Property. 2008. *Shirataki Isekigun IX* [The Shirataki Site Group IX]. Sapporo, Hokkaido Maizo Bunkazai Centre.

Hokkaido Centre for Buried Cultural Property. 2010. *Shirataki Isekigun X* [The Shirataki Site Group X]. Sapporo, Hokkaido Maizo Bunkazai Centre.

Hokkaido Centre for Buried Cultural Property. 2011. *Shirataki Isekigun XI* [The Shirataki Site Group XI]. Sapporo, Hokkaido Maizo Bunkazai Centre.

Hokkaido Centre for Buried Cultural Property. 2012. *Shirataki Isekigun XII* [The Shirataki Site Group XII]. Sapporo, Hokkaido Maizo Bunkazai Centre.

IWASE, A., and K. TAKASE. 2012. Sekki no Shiyoukon Bunseki [Micro Use-Wear Analysis of Artefacts from the Ogachi-Kato 2 Site]. In *Kokuyoseki no Ryutsu to Shouhi Karamita Kan-Nihonkai Hokubu Chiiki Niokeru Koushinsei Shakai no Keisei to Henyou (I)*, edited by H. Sato, 152–163. Tokyo, University of Tokyo.

IZUHO, M. 1997. Tokorogawa Ryuiki niokeru Sekkisekizai no Kiso Kenkyu [A Fundamental Study of the Lithic Raw Material on the Tokoro River Basin]. *Hokkaido Kyusekki Bunka Kenkyu* 2, 1–14.

IZUHO, M. 1998. Hokkaido Tokoro-gun Oketo-cho Ogachi-Kato Iseki niokeru Saishu Shiryo [Collected Artefacts from the Surface of the Ogachi-Kato Site, Hokkaido]. *Hokkaido Kyusekki Bunka Kenkyu* 3, 13–28.

IZUHO, M. 2012a. Iseki Shuhen no Chikei to Chishitsu [Topography and Geology around the Ogachi-Kato 2 Site]. In *Kokuyoseki no Ryutsu to Shouhi Karamita Kan-Nihonkai Hokubu Chiiki Niokeru Koushinsei Shakai no Keisei to Henyou (I)*, edited by H. Sato, 43–45. Tokyo, University of Tokyo.

IZUHO, M. 2012b. Soujo Yunitto Kubun [Stratigraphic Divisions at the Ogachi-Kato 2 Site]. In *Kokuyoseki no Ryutsu to Shouhi Karamita Kan-Nihonkai Hokubu Chiiki Niokeru Koushinsei Shakai no Keisei to Henyou (I)*, edited by H. Sato, 46–52. Tokyo, University of Tokyo.

IZUHO, M. 2012c. Taisekibutsu no Ryudobunseki [Particle-Size Analysis at the Ogachi-Kato 2 Site]. In *Kokuyoseki no Ryutsu to Shouhi Karamita Kan-Nihonkai Hokubu Chiiki Niokeru Koushinsei Shakai no Keisei to Henyou (I)*, edited by H. Sato, 53–56. Tokyo, University of Tokyo.

IZUHO, M., and F. AKAI. 2005. Hokkaido no Kyusekki Hennen: Iseki Keisei Kateiron no Tekiyo [Geochronology of Palaeolithic Sites in Hokkaido]. *Kyusekki Kenkyu* 1, 39–55.

IZUHO, M., F. AKAI, Y. NAKAZAWA, and A. IWASE. 2012b. The Upper Paleolithic of Hokkaido: Current Evidence and Its Geochronological Framework. In *Environmental Changes and Human Occupation in East Asia during OIS3 and OIS2* (B.A.R. International Series 2352), edited by A. Ono and M. Izuho, 109–128. Oxford, BAR Publishing.

IZUHO, M., and K. HAYASHI. 2012. Dojoukagakusei Bunseki [Soil and Pedostratigraphic Analysis at the Ogachi-Kato 2 Site]. In *Kokuyoseki no Ryutsu to Shouhi Karamita Kan-Nihonkai Hokubu Chiiki Niokeru Koushinsei Shakai no Keisei to Henyou (I)*, edited by H. Sato, 57–61. Tokyo, University of Tokyo.

IZUHO, M., and W. HIROSE. 2010. A Review of Archaeological Obsidian Studies on Hokkaido Island (Japan). In *Crossing the Straits: Prehistoric Obsidian Source Exploitation in the North Pacific Rim* (B.A.R. International Series 2152), edited by Y. V. Kuzmin and M. D. Glascock, 9–25. Oxford, BAR Publishing.

IZUHO, M., N. ODA, and K. HAYASHI. 2012a. Kenshutsu Sareta Koukoshiryo to sono Kukanbunpu [Statistical Analysis of Artefact Spatial Distribution of the Ogachi-Kato 2 Site]. In *Kokuyoseki no Ryutsu to Shouhi Karamita Kan-Nihonkai Hokubu Chiiki Niokeru Koushinsei Shakai no Keisei to Henyou (I)*, edited by H. Sato, 70–86. Tokyo, University of Tokyo.

IZUHO, M., and T. TSURUMARU. 2012. Chosa Kenkyu no Gaiyo [A Summary of the Ogachi-Kato 2 Site]. In *Kokuyoseki no Ryutsu to Shouhi Karamita Kan-Nihonkai Hokubu Chiiki Niokeru Koushinsei Shakai no Keisei to Henyou (I)*, edited by H. Sato, 29–49. Tokyo, University of Tokyo.

KIMURA, H. 1978. Yoichigawa, Akaigawa Ryuiki no Sendoki Sekkigun Nitsuite [The Palaeolithic Sites on Yoichi and Akai River Basins]. *Hokkaido Kokogaku* 14, 23–48.

MACHIDA, H., and F. ARAI. 2003. *Shinpen Kazanbai Atlas: Nihon Retto to Sono Shuhen* [Atlas of Tephra in and Around Japan (Revised Edition)]. Tokyo, University of Tokyo Press.

NAKAZAWA, Y., M. IZUHO, J. TAKAKURA, and S. YAMADA. 2005. Toward an Understanding of Technological Variability in Microblade Assemblages in Hokkaido, Japan. *Asian Perspectives* 44, 276–292.

NAOE, Y. 2009. Hokkaido-Tobu niokeru Kokuyoseki Riyo [Obsidian Procurement and Consumption on Northeastern Hokkaido]. In *Kokuyoseki ga Hiraku Jinrui Shakai no Koryu*, edited by H. Sato, 14–31. Tokyo, University of Tokyo.

NAOE, Y., and J. NAGASAKI. 2005. Hokkaido Koki Kyusekki Jidai Zenhan Ki no Sekizai Shohi Senryaku [Raw Material Consumption Strategy at Late Upper Palaeolithic in Hokkaido]. *Hokkaido Kyusekki Bunka Kenkyu* 10, 45–58.

NEFF, H. 2000. Neutron Activation Analysis for Provenance Determination in Archaeology. In *Modern Analytical Methods in Art and Archaeology*, edited by E. Ciliberto and G. Spoto, 81–134. New York, Wiley.

ODA, N. 2009. Kitami-shi Momijiyama Iseki Shutsudo Sekkigun no Saikento [Re-examination of the Momijiyama Site, Hokkaido]. In *Nihonrettoh Hokubu no Koshinsei/Kanshinsei Ikoki niokeru Kyojukeitai to Bunkakeisei nikansuru Kenkyu*, edited by H. Sato, 139–221. Tokyo, University of Tokyo.

ODA, N. 2012. Hakurikatei [Reduction Sequences of the Ogachi-Kato 2 Site]. In *Kokuyoseki no Ryutsu to Shouhi Karamita Kan-Nihonkai Hokubu Chiiki Niokeru Koushinsei Shakai no Keisei to Henyou (I)*, edited by H. Sato, 132–141. Tokyo, University of Tokyo.

ODA, N., and M. IZUHO. 2012. Sekki no Kisai [Description of Artefacts from the Ogachi-Kato 2 Site]. In *Kokuyoseki no Ryutsu to Shouhi Karamita Kan-Nihonkai Hokubu Chiiki Niokeru Koushinsei Shakai no Keisei to Henyou (I)*, edited by H. Sato, 87–124. Tokyo, University of Tokyo.

OKUMURA, K. 1991. Quaternary Tephra Studies in the Hokkaido District, Northern Japan. *Daiyonki Kenkyu* 30, 379–390 (in Japanese with English Abstract).

SATO, H. 2003. Hokkaido no Koki Kyusekki Jidai Zenhanki no Yoso [The Early Upper Palaeolithic in Hokkaido]. *Kodai Bunka* 55, 181–194.

SODA, T. 2012. Tephra Bunseki [Tephra Analysis at the Ogachi-Kato 2 Site]. In *Kokuyoseki no Ryutsu to Shouhi Karamita Kan-Nihonkai Hokubu Chiiki Niokeru Koushinsei Shakai no Keisei to Henyou (I)*, edited by H. Sato, 62–67. Tokyo, University of Tokyo.

SUZUKI, H. 2007. Gensanchi niokeru Isekikanheni Kenkyu [A Study of Inter-Site Variability on the Shirataki Site Group]. In *Kokogaku Danso*, edited by Department of Archaeology, Graduate School of Arts and Letters, Tohoku University, 109–129. Sendai, Tohoku University.

TAKAKURA, J. 2001. Hokkaido Kitami-shi Yoshiizawa Iseki B chiten Shutudo Saisekijin Sekkigun no Saikento [Re-examination of the Yoshiizawa Site Location B, Hokkaido]. *Hokkaido Kyusekki Bunka Kenkyu* 5, 1–34.

TAKAKURA, J. 2012. Kyusekkisiryo nikansuru Hakurihoho no Dotei [Flaking Technique Determination Based on the Gull-wing Fracture Analysis of Artefacts from the Ogachi-Kato 2 Site]. In *Kokuyoseki no Ryutsu to Shouhi Karamita Kan-Nihonkai Hokubu Chiiki Niokeru Koushinsei Shakai no Keisei to Henyou (I)*, edited by H. Sato, 142–151. Tokyo, University of Tokyo.

TSURUMARU, T. 1981. Hokkaido Chihono Saisekijin Bunka [Microblade Culture in Hokkaido District]. *Sundai Shigaku* 47, 23–50.

YAMADA, S. 1999. Hokkaido no Zenhanki Saisekijin Sekkigun nituiteno Kenkyu [A Study of the Early Microblade Industry in Hokkaido]. *Senshi Kokogaku Ronshu* 8, 1–70.

WATERS, M. R. 1992. *Principles of Geoarchaeology: A North American Perspective.* Tucson, University of Arizona Press.

Chapter 10
GEOARCHAEOLOGICAL ASPECTS OF OBSIDIAN SOURCE STUDIES IN THE SOUTHERN RUSSIAN FAR EAST AND BRIEF COMPARISON WITH NEIGHBOURING REGIONS

Yaroslav V. KUZMIN

Abstract: *Obsidian provenance studies began in the southern Russian Far East (Primorye Province, Amur River basin, Sakhalin Island, and Kurile Islands) in the 1980s – early 1990s, and continue until now. These studies give archaeologists a powerful tool for independent evaluation of their empiric models of prehistoric migrations and exchange. Today, the most important sources of archaeological obsidian in the mainland (Primorye and the Amur River basin) and insular (Sakhalin Island and Kurile Islands) parts of the region are securely established. The exploitation of obsidian by the prehistoric people of the region began in the second half of the Upper Palaeolithic, ca. 19,000–18,000 BP, and continued until the Palaeometal/Early Iron Age (ca. 3000–1300 BP; in some places like the Kuriles until ca. 800 BP). Obsidian was most intensively used as a raw material in Upper Palaeolithic and Neolithic times. Several long-distance transport/exchange/trade networks have now been reconstructed for the southern Russian Far East and neighbouring Northeast Asia. The most important of them are: 1) a source on the Basaltic Plateau in southern Primorye, used by prehistoric people of Primorye, the Lower Amur River basin, and Northeast China; 2) a source at the Paektusan Volcano on the Chinese/North Korean border, exploited by people of the Korean Peninsula, Primorye, and Northeast China; 3) the Shirataki and Oketo sources in Hokkaido Island of Japan, utilised by inhabitants of Hokkaido, Sakhalin Island, Kurile Islands, and the Lower Amur River basin; and 4) a source on the Obluchie Plateau in the middle course of the Amur River, used by populations of the entire Amur River basin. Distances between sources and utilisation sites reach 1000–1200 km in a straight line during the Neolithic and Palaeometal times (ca. 8000–800 BP). It is not exactly clear how obsidian was transported from the sources to the archaeological sites in the southern Russian Far East; some studies suggest existence of exchange networks in the Neolithic of Primorye. Intensive transport of Hokkaido obsidian to Sakhalin and the Kuriles across sea straits allows the suggestion of the existence of watercraft since the early Holocene, ca. 10,000 BP, and possibly earlier. The northern part of the Russian Far East (Kamchatka Peninsula and Chukotka Region) should be the focus of obsidian provenance studies in the near future. Large number of prehistoric sites with obsidian artefacts made these territories perspective for in-depth research by international team.*

Keywords: *Obsidian, Provenance, Long-Distance Transport, Upper Palaeolithic, Neolithic, Palaeometal, Russian Far East, Northeast China, Korean Peninsula, Hokkaido Island, Northeast Asia*

INTRODUCTION

Since the early 1960s, investigations of obsidian provenance have been conducted in the Mediterranean and the Near East (Cann and Renfrew 1964; Renfrew *et al.* 1965, 1966, 1968). In Northeast Asia, the first research on obsidian sourcing quickly followed (Suzuki 1969, 1970a, 1970b, 1973a, 1973b; see also Ono 1976; 1984, 418–21). It was repeatedly demonstrated that this is an effective tool for studying human contacts and interactions (e.g., Renfrew and Dixon 1976; see also Cann 1983; Carlson 1994; Glascock 2002; Shackley 2005). Study of obsidian exploitation and sources of archaeological volcanic glasses in the southern Russian Far East (Primorye [Maritime] Province, Amur River basin, Sakhalin Island, and Kurile Islands) began in the 1980s, and has been most active since 1992. At that time, our Russian–US group (including scholars from the Institute of Geology and Mineralogy, Siberian Branch of the Russian Academy of Sciences, at Novosibirsk, and the Far Eastern Geological Institute, Far Eastern Branch of the Russian Academy of Sciences, at Vladivostok, both in Russia; and the Research Reactor Center, University of Missouri, Columbia, MO, in the USA) and other researchers initiated a long-term programme devoted to understanding the patterns of obsidian acquisition and transport in prehistoric time in the Russian Far East. Comprehensive summaries on obsidian from both 'geological' sources and archaeological sites were published in the 2000s (see major papers and edited volumes: Doelman 2008; Doelman *et al.* 2008, 2009, 2012; Glascock *et al.* 2011; Hall and Kimura 2002; Jia *et al.* 2010, 2013; Kim *et al.* 2007; Kluyev and Sleptsov 2007; Kuzmin 2006a, 2010, 2011; Kuzmin and Glascock 2007, 2010; Kuzmin and Popov 2000; Kuzmin *et al.* 2002a, 2002b; Obata 2009; Pantukhina 2007; Phillips 2010; Popov *et al.* 2005, 2008, 2009; Sato *et al.* 2002; Tomoda *et al.* 2004; Yoshitani *et al.* 2003). For obsidian sources from Hokkaido Island, recently some new data have been released (Izuho and Hirose 2010; Kuzmin *et al.* 2013; Watanabe and Suzuki 2009, 461–2; see also: Ferguson *et al.*, this volume; and Wada *et al.*, this volume).

In this overview, I would like to summarise the results accumulated as of late 2013 from the view of geoarchaeology, and briefly consider the future tasks for the Russian Far East and neighbouring Northeast Asia in terms of prehistoric obsidian source research. Some of these issues are already discussed in a recent review (see Kuzmin 2011), and here I will focus on the latest developments in Primorye, the Amur River basin, and the Kurile Islands, and on perspectives in obsidian sourcing for the rest of the Russian Far East (i.e., Kamchatka Peninsula) and neighbouring parts of Northeast Asia. Information on the age and environment of prehistoric cultural complexes of the Russian Far East is based on the author's results (see summaries: Kuzmin 2002, 2003, 2006b, 2008, 2012) and data generated by other scholars.

MATERIAL AND METHODS

Early Knowledge on Obsidian from Geological and Archaeological Contexts in Northeast Asia

Obsidian (i.e., high quality volcanic glass) in mainland Northeast Asia was initially studied in the late nineteenth century at the Paektusan Volcano, in the course of a survey related to the construction of the Chinese Eastern Railroad from Transbaikal to Vladivostok via Northeast China (Manchuria) (see Garin-Michailowski 2005 [1898], 153–6; Stephan 1994, 58–61). Earlier on, British explorers visited northern Korea and Manchuria in 1885, but they did not study the geology of the region and only described general topographic features (James 1888). In 1897, the Russian geologist Eduard E. Anert took a trip to the Paektusan Volcano. In the caldera of Paektusan, near the lake shore on the modern North Korean side, the black trachytic volcanic glass was detected with an indication that this was 'obsidian' (Anert 1904, 275). This is the first reliable information about this mineral in Northeast Asia, to the best of my knowledge, besides notes of the early Russian explorers of Kamchatka Peninsula (Krasheninnikov 1972 [1755]), Sea of Okhotsk coast (Fischer 1818, 278–9), and the Aleutian Islands (Veniaminov 1984 [1840], 203) in the eighteenth – nineteenth centuries AD.

After a long gap caused by political turmoil in Northeast Asia in the early-to-mid twentieth century, the Russian geologist Evgeny P. Denisov in 1958 studied the Korean side of the Paektusan Volcano and described volcanic rocks, including obsidian near the southern rim of the caldera (Denisov 1965, 65; Denisov and Ten 1966). It was confirmed that obsidian belongs to trachytes (Denisov and Ten 1966, 6). The latest study was conducted by our Russian–US group (see Popov *et al.* 2005, 2008); obsidian is a part of the trachyrhyolites and is dated to about 2.2 Ma ago. Sakhno (2007) mentioned black obsidians from deposits dated to ca. 650,000–950,000 years ago. Artefacts made of obsidian from regions neighbouring the Paektusan Volcano (the northern part of the Korean Peninsula and Manchuria) were initially found by Japanese scholars in the early twentieth century (see Matsushita 1998; see also Nelson 1993, 88–95; 1995, 22–33).

In Primorye Province, pillow lavas with volcanic glass on the Shkotovo Plateau (in our publications also Basaltic Plateau; e.g., Kuzmin *et al.* 2002a) were described in the late 1950s (Petrov and Zamurueva 1960). This source was later studied in detail (see summaries: Kuzmin and Popov 2000, 54; Popov *et al.* 2009, 2010). The famous Russian explorer Vladimir K. Arseniev (see Arseniev 2007 [1921], 220; see also Arsenjew 1924, 248) in 1906 described '...obsidian with prismatic cleavage' in the Zerkalnaya [Tadushi] River basin. Our studies in the 1990s showed that it was not a pure and good quality volcanic glass but perlite, unsuitable for tool-making (e.g., Kuzmin *et al.* 2002a, 509–10). Obsidian artefacts in Primorye were found for the first time by the Russian scholar and statesman Fedor F. Busse in the late nineteenth century (e.g., Okladnikov 1965, 14). In the 1950s, the Russian archaeologist Aleksei P. Okladnikov (see Ganeshin and Okladnikov 1956) and the geologist Viktor F. Petrun (see Petrun 1956) initiated the current stage of obsidian research in the Russian Far East. In the Amur River basin, volcanic glass-bearing rocks were studied by our group in the early 2000s (see Glascock *et al.* 2011). Obsidian artefacts were discovered in this region by Russian archaeologists in the 1960s (see Derevianko 1970).

On Sakhalin Island, which lacks primary obsidian sources (e.g., Kuzmin and Glascock 2007), artefacts made of obsidian were discovered by the Russian zoologist Ivan S. Polyakov in 1881 near the modern town of Aleksandovsk-Sakhalinsky (Polyakov 1883, 18). Obsidian artefacts from the southern Kurile Islands, which also lack obsidian sources, were initially found by Japanese and Swedish scholars in the early twentieth century (see, for example: Befu and Chard 1964; Schnell 1932, 52–3).

Obsidian in Prehistoric Complexes of the Southern Russian Far East and Their Chronology

As a result of geoarchaeological studies in Northeast Asia, the distribution of obsidian artefacts can be summarised (Table 10.1). Thus, obsidian can be used as a commodity for the study of transport/exchange patterns in prehistory. Basic information on the archaeology of the Stone Age and Palaeometal complexes of the southern Russian Far East can be found in Andreeva (2005) and Nelson *et al.* (2006). The chronology of these cultural complexes is briefly characterised below.

In Primorye, the Upper Palaeolithic sites are now dated to ca. 40,000–11,500 BP (Kuzmin 2006b). Obsidian was found at sites associated with the late Upper Palaeolithic, dated to ca. 13,000–11,500 BP and possibly up to ca. 20,000–18,000 BP (Ustinovka 1 site; see Kuzmin 1996). The Neolithic assemblages can be placed at ca. 10,800–3300 BP (Batarshev *et al.* 2010; Kuzmin 2006b, 2012); and Palaeometal complexes are dated to ca. 3300–1300 BP (Kuzmin *et al.* 2005). Obsidian is common in these cultural complexes, especially the Neolithic ones.

In the Amur River basin, obsidian artefacts were recorded only at a few sites of Upper Palaeolithic age (Table 10.1), despite the fact that numerous sites of this period were excavated (e.g., Derevianko 1998; Derevianko *et al.* 2006). In the Neolithic complexes (see Derevianko and Medvedev 2006), obsidian is quite rare, and only at some sites in the middle part of the Amur Basin (between the Zeya and Bureya rivers) does it sometimes reach 10–15% of the total raw material. The existence of the Neolithic in this region can be determined at ca. 13,300–3000 BP (Kuzmin 2003, 2006b, 2006c; Kuzmin and Nesterov 2010; Kuzmin and Shewkomud 2003; Shewkomud and Kuzmin 2009). The Palaeometal cultures of the Amur

Table 10.1. Obsidian in prehistoric cultural complexes of Northeast Asia (after Cho *et al.* 2010; Choi 2001; Doelman *et al.* 2008, 2009; Glascock *et al.* 2011; Hong and Kim 2008; Izuho and Hirose 2010; Jia *et al.* 2010, 2013; Kimura 1995; Kuzmin and Glascock 2007; Nelson 1993, 1995; Obata 2003; Phillips 2010, 2011; Seong 2007; Sohn 1993; Yanshina and Kuzmin 2010; Yonsei University Museum 2001)

Region	Periods		
	Upper Palaeolithic	*Neolithic*	*Bronze Age/Palaeometal*
Primorye Province	+ *	++	+
Amur River basin	+	+	+
Sakhalin Island	++	+ **	+ **
Kurile Islands	—	+	++
Hokkaido Island	++	++	+
Northeast China	+	+	+
Korean Peninsula	++ ***	++ ***	+

* Presence of obsidian in archaeological assemblages: "+" – present (less than ca. 10% of total stone artefacts); "++" – common (10–50%). Long cash means the absence of cultural complexes in given region.

** Obsidian is common in the southern part, and is rare in the central and northern parts.

*** Obsidian in common in the northern and central parts, and is relatively rare in southern part.

River basin with obsidian are dated to ca. 2900–1400 BP (Glascock *et al.* 2011; Kuzmin and Chernuk 2000, 36–9; Nesterov and Kuzmin 1999).

On Sakhalin Island, the earliest site with obsidian, Ogonki 5, corresponds to the Upper Palaeolithic and is dated to ca. 19,400–17,800 BP (Kuzmin 2006b); obsidian is common in the Upper – Final Palaeolithic assemblages (e.g., Vasilevsky 2006, 2008). The Neolithic complexes of Sakhalin are dated to ca. 8800–2800 BP (Kuzmin 2006b; Kuzmin and Glascock 2007; Kuzmin and Orlova 2000; Kuzmin *et al.* 2004; Vasilevsky *et al.* 2009, 2010); Palaeometal cultures can be placed at ca. 2800–800 BP (Kuzmin and Glascock 2007; Vasilevsky *et al.* 2003). Both cultural complexes contain obsidian, especially in the southern and central parts of Sakhalin (e.g., Vasilevski and Grishchenko 2011).

As for the Kurile Islands, reliable Palaeolithic sites are still unknown in this region (Table 10.1). The earliest Jomon/Neolithic locale, the Yankito cluster dated to ca. 7000 BP, has obsidian (Yanshina and Kuzmin 2010), but its source is not yet known. Generally, Jomon complexes in the Kuriles are dated to ca. 7000–2500 BP, with a gap at ca. 6800–4200 BP (see Kuzmin 2006b; Kuzmin *et al.* 2012); several of them contain obsidian artefacts. The Epi-Jomon [Zoku Jomon] archaeological complex is dated to ca. 2500–1400 BP (Fitzhugh 2012; Fitzhugh *et al.* 2011; Phillips 2010; Vasilevsky *et al.* 2010); and the Okhotsk cultural complex existed in the Kuriles at ca. 1400–800 BP (e.g., Fitzhugh 2012; Phillips 2010). The latter two archaeological complexes contain relatively large amounts of obsidian artefacts (e.g., Phillips 2010, 2011).

Methods of Obsidian Geochemical Study in Northeast Asia

In order to establish the geochemical signatures of known (and still unknown, see Shackley *et al.* 1996) 'geological' obsidian sources, different analytical methods were employed. Our Russian–US group used two techniques, Neutron Activation Analysis (NAA) and Energy Dispersive X-ray Fluorescence (EDXRF) analysis (see, for example: Kuzmin and Popov 2000). The results of analyses for individual specimens were combined into geochemical groups following the approach developed previously (see Glascock *et al.* 1998), and obsidian sources were revealed (see, for example: Kuzmin *et al.* 2002a, 2002b). The Japanese-led groups used both Wave-Dispersive X-ray Fluorescence (WDX) and EDXRF techniques in Primorye (Tomoda *et al.* 2004; Yoshitani *et al.* 2003), and EDXRF in Hokkaido Island (Hall and Kimura 2002). Australian-led groups applied several methods, including combined Proton-Induced X-ray Emission (PIXE) and Proton-Induced Gamma-ray Emission (PIGME), and relative density (Doelman *et al.* 2008); and NAA and portable XRF (pXRF) (Jia *et al.* 2010, 2013). Korean scholars used both NAA (Cho *et al.* 2010) and PIXE (Kim *et al.* 2007) methods for the determination of rare-earth elements. As for the Kurile Islands, Phillips and Speakman (2009) used pXRF equipment, and later a laser ablation version of the Inductively Coupled Plasma – Mass Spectrometry (ICP–MS) method; the abbreviated form is LA–ICP–MS. Details on these analytical techniques can be found in handbooks (e.g., Malainey 2011; Pollard *et al.* 2007).

It should be noted that at the initial stage of study, especially when the geochemical signatures of sources are unknown, the analysis which can measure the content of the maximal amount of the chemical elements should be chosen. In our case (e.g., Kuzmin 2010 and references therein), the NAA was employed, and this gave us secure data for the separation of different sources (see also Grebennikov *et al.* 2010). An example of insecure determination of an obsidian source by using a small number of elements is the study by Warashina (2004) who was unable to separate two sources in Hokkaido, Akaigawa and Tokachi-Mitsumata, using EDXRF

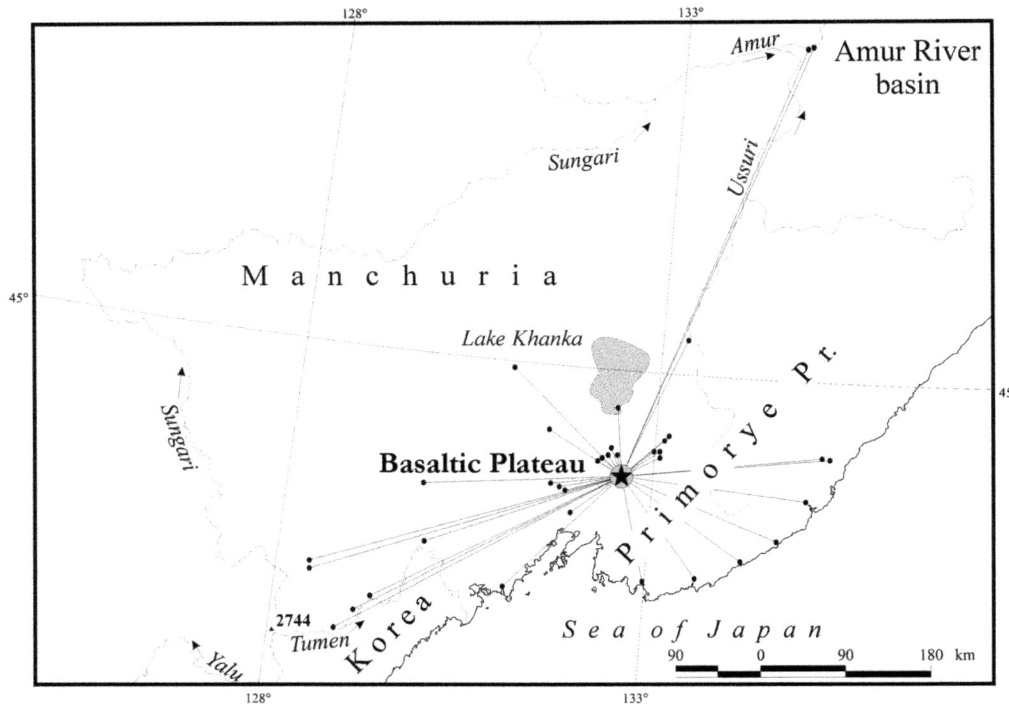

Figure 10.1. Distribution of obsidian from the Basaltic Plateau source in Northeast Asia (after Glascock et al. *2011; Jia* et al. *2010, 2013; Kuzmin 2011; Kuzmin* et al. *2002a, 2013). For Figures 10.1–10.5, sources are indicated by encircled stars*

analysis (12 elements) (see Izuho and Hirose 2010, 13–4). This discrimination became possible only with the help of the NAA method (28 elements) (Kuzmin and Glascock 2007; Kuzmin *et al.* 2013).

RESULTS

Sources of Archaeological Obsidian in the Southern Russian Far East: Spatiotemporal Patterns

After 20+ years of extensive work, the major sources for archaeological obsidian in the region under consideration can be established (Figures 10.1–10.5; Tables 10.2–10.5). A brief characterisation of major obsidian sources for the Upper Palaeolithic and Initial Neolithic cultural complexes (dated to ca. 19,000–10,000 BP) was recently published (Kuzmin 2011). Here I present a description of the obsidian distribution in all prehistoric complexes (Upper Palaeolithic – Palaeometal). The exact location for sites with obsidian analysed by geochemical methods (Tables 10.2–10.5) can be found in relevant publications; see the captions to these tables.

In Primorye Province (Figure 10.1), the Basaltic Plateau (a.k.a. the Shkotovo Plateau) is the main source of high quality volcanic glass, and it was investigated recently in detail (Doelman 2008; Doelman *et al.* 2008, 2009; Popov *et al.* 2010). However, the results of the first study published in the early 2000s (e.g., Kuzmin *et al.* 2002a) are still valid, and no other important sources of archaeological volcanic glass were found afterwards (e.g., Doelman *et al.* 2004, 2008). The obsidian from the

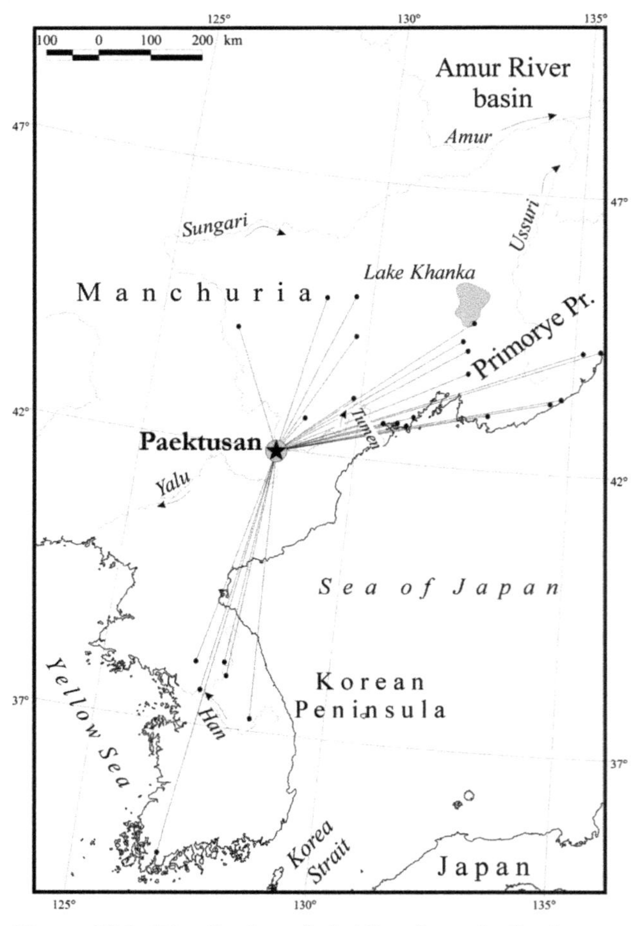

Figure 10.2. Distribution of obsidian from the Paektusan source in Northeast Asia (after Jia et al. *2010, 2013; Kim* et al. *2007; Kuzmin 2011; Kuzmin* et al. *2002a)*

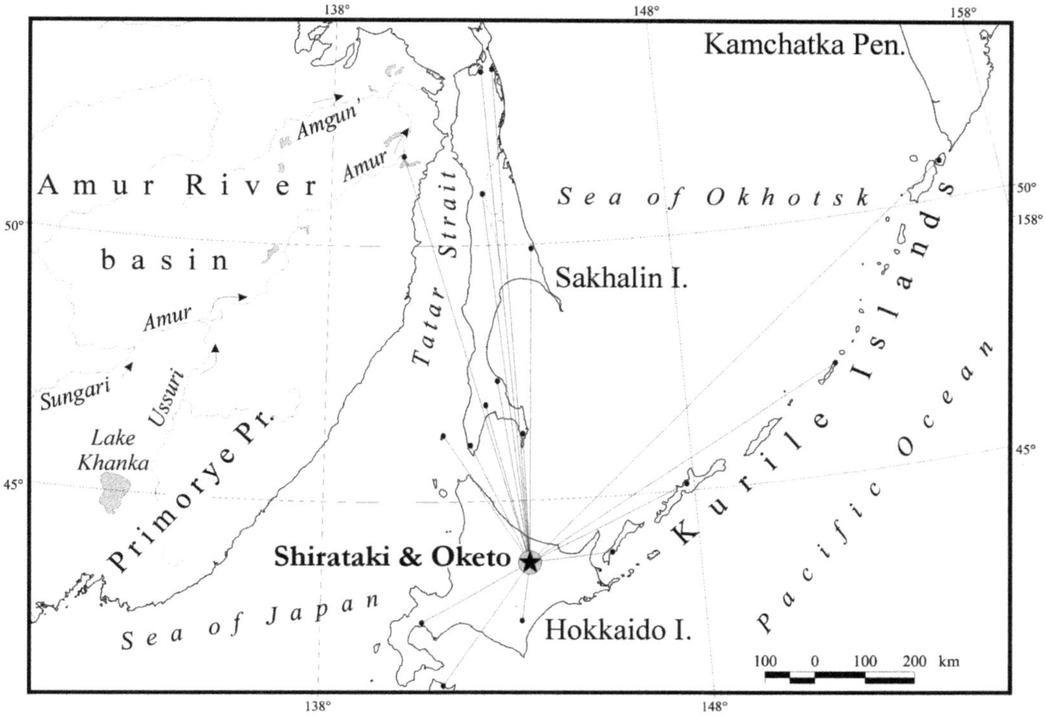

Figure 10.3. Distribution of obsidian from the Shirataki and Oketo sources in Northeast Asia (selective sites, mainly beyond Hokkaido Island) (after Glascock et al. *2011; Kimura 1995; Kuzmin and Glascock 2007; Phillips 2010)*

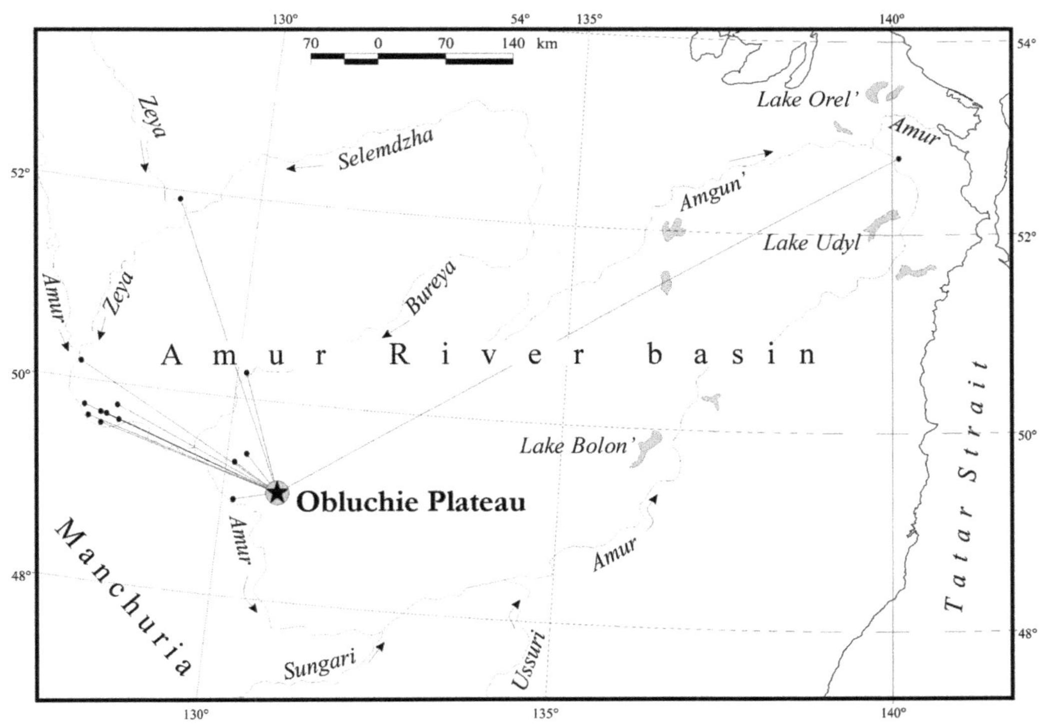

Figure 10.4. Distribution of obsidian from the Obluchie Plateau source in Northeast Asia (after Glascock et al. *2011)*

Basaltic Plateau spread beyond the Primorye region in prehistory, and reached the lower part of the Amur River basin and Manchuria (Figure 10.1; Tables 10.2–10.3); the distance from source to utilisation sites is from 20–300 to 660 km in a straight line.

The second most important source of archaeological obsidian in the Russian Far East and neighbouring Northeast Asia is the Paektusan Volcano. Its obsidian is now known not only in Primorye (Kuzmin *et al.* 2002a; Tabarev 2004; see Table 10.2) and the entire Korean

Table 10.2. Prehistoric sites in Primorye Province with obsidian artefacts and their sources (after Doelman 2008; Doelman *et al.* 2004; Komoto and Obata 2005; Kuzmin *et al.* 2002a; Obata 2007; Popov *et al.* 2006; see the precise location of sites in these references)

Site Name (No. of Samples)	Period*	Obsidian Sources			
		Basaltic Plateau	***Paektusan***	***Gladkaya River***	***Samarga***
Arizona 1 (1)	UP	+			
Borisovka (2)	UP	+			
Firsanova Sopka (5)	UP	+	+		
Gadyuchya Sopka (5)	UP	+			
Gorbatka 2 (4)	UP	+			
Gorbatka 3 (9)	UP	+	+		
Gorelaya Sopka (5)	UP				
Ilistaya 1 (5)	UP	+			
Ivanovka (5)	UP	+			
Ivanovka 1 (5)	UP	+	+		
Ivanovka 3 (4)	UP	+			
Kentsukhe (2)	UP	+			
Lesozavodsk (2)	UP	+			
Molodezhnaya 1 (63)	UP	+	+		
Novovarvarovka 1 (13)	UP	+	+		
Osinovka (8)	UP	+			
Razdolnoye (1)	UP			+	
Risovaya 1 (45)	UP	+	+		
Sheklyevo 6 (21)	UP	+	+		
Suvorovo 3 (2)	UP	+			
Tigrovy 2 (1)	UP	+			
Timofeevka 1 (3)	UP	+	+		
Ustinovka 1 (1)	UP	+			
Ustinovka 4 (18)	UP	+			
Ustinovka 6 (15)	UP	+	+		
Aleskee-Nikolskoe 1 (1)	N		+		
Boisman 2 (4)	N		+		
Chernaya Sopka (3)	N		+	+	
Eustaphy (1)	N		+		
Gladkaya (7)	N		+		
Gladkaya 4 (4)	N		+	+	
Gvozdevo 3 (4)	N		+	+	
Gvozdevo 4 (4)	N		+		
Khansi (4)	N		+		
Kievka (1)	N	+			
Klerk 5 (8)	N, Pal		+		
Luzanova Sopka 2 (6)	N	+	+		
Luzanova Sopka 3 (2)	N	+	+		
Luzanova Sopka 5 (2)	N	+			
Maikhe (3)	N		+		
Osinovka 1 (1)	N	+			
Pereval (1)	N		+		
Phusun (3)	N	+	+		
Sergeevka 1 (1)	N	+			

Site Name (No. of Samples)	Period*	Obsidian Sources			
		Basaltic Plateau	***Paektusan***	***Gladkaya River***	***Samarga***
Sinie Skaly (2)	N	+		+	
Troitsa (2)	N	+	+		
Ustinovka 3 (2)	N	+			+
Ustinovka 8 (1)	N		+		
Valentin-Peresheek (3)	N	+			
Zaisanovka 1 (15)	N		+		
Zaisanovka 7 (12)	N		+	+	
Zara 1A, 1C (5)	N	+			
Anuchino 1 (2)	Pal	+			
Anuchino 14 (5)	Pal	+			
Aurovka (2)	Pal	+			
Boisman 2 (1)	Pal		+		
Bukhta Uglovaya 1 (2)	Pal		+		
Bulochka (1)	Pal	+			
Kievka (3)	Pal	+			
Lebyazhye (1)	Pal	+			
Monastyrka 3 (1)	Pal		+		
Pavlova Pad 1 (1)	Pal	+			
Rybak (1)	Pal		+		
Samarga 2A (1)	Pal				+
Samarga 5 (2)	Pal				+
Senkina Shapka (1)	Pal	+			
Sheklyaevo 7 (1)	Pal	+			
Sheklyaevo 8 (1)	Pal	+			
Sheklyaevo 16 (1)	Pal	+			
Sinie Skaly (6)**	Pal	+	+	+	
Ust-Svetlaya (1)	Pal				+
Zara 3 (6)**	Pal	+	+		
Total sites/samples: 72/390					

* UP – Upper Palaeolithic; N – Neolithic; Pal – Palaeometal (Bronze Age/Early Iron Age).
** Obsidian from one unknown source was also identified (see Doelman *et al.* 2004, 119).

Peninsula (e.g., Hong 2012; Kim *et al.* 2007; Popov *et al.* 2005; see also Kim, this volume), but also in neighbouring Manchuria (Jia *et al.* 2010, 2013; Obata 2003). The distance from the Paektusan source to utilisation sites is 230–710 km (Figure 10.2). However, I do not include some dubious cases in the list of sites with obsidian from Paektusan (see Tables 10.2 and 10.4). For example, the identification of an obsidian source at the Khummi site in the lower part of the Amur River basin as Paektusan (Warashina *et al.* 1998; see also Sato 2011, 207), in my opinion need additional confirmation, because Warashina *et al.* (1998) did not use primary data (i.e., 'geological' samples) for the volcanic glasses from a source located at and near the Paektusan Volcano as it was done by our group (see Popov *et al.* 2005, 2008).

As for Sakhalin Island (Figure 10.3), sources at Shirataki and Oketo in Hokkaido Island were the main obsidian suppliers for prehistoric inhabitants (Table 10.3). The occasional use of another obsidian source, Akaigawa, located in the eastern part of Hokkaido (see Ferguson *et al.*, this volume), is also detected (Kuzmin and Glascock 2007). The distance between sources and archaeological sites is up to 1000 km in a straight line. Some data on XRF analysis of several sites from Sakhalin are available from a Database compiled by the Cultural Properties Research Laboratory, Meiji University, Tokyo (Keiko-Ekkususen 2009, 35–41). For the Upper Palaeolithic site of Ogonki 5, 28 obsidian specimens were studied, and all of them originated from the Shirataki area (Akaishiyama and Tokachi-Ishizawa locales). Eleven obsidian samples from four Palaeometal sites in southern Sakhalin, including the Susuya shellmidden, Kushunkotan (near the modern town of Korsakov [Otomari in 1905–45], and Tarabakotan (near the modern village of Starodubskoe [Sakaehama in 1905–45]) were analysed, and identified as derived from the Shirataki area (Akaishiyama and Tokachi-Ishizawa locales) and Oketo

Table 10.3. Prehistoric sites in Sakhalin Island with obsidian artefacts and their sources (after Keiko-Ekkususen 2009; Kuzmin and Glascock 2007; Kuzmin *et al.* 2002b; see the precise location of sites in these references)

Site Name (No. of Samples)	**Period***	**Obsidian Sources**			
		Shirataki-A	***Shirataki-B***	***Oketo-A***	***Akaigawa***
Bereznyaki 4 (2)**	UP	+			
Ogonki 5 (34)	UP	+	+		
Ogonki 6 (1)	UP	+			
Ogonki 7 (3)	UP	+	+		
Olimpiya 1 (1)	UP		+		
Ostantsevaya Cave (1)	UP	+			
Sennaya 2 (2)	UP	+			
Sokol (8)	UP	+	+		
Starorusskoe 3 (1)	UP		+		
Starorusskoe 5 (4)	UP		+		
Ado-Tymovo 4 (1)	N		+		
Blagodatny 3 (1)	N	+			
Bogataya 1 (5)	N	+		+	
Dolinsk 1 (4)	N	+	+		
Kirpichny 9 (1)	N			+	
Kuznetsovo 1 (1)	N	+			
Kuznetsovo 3 (1)	N				+
Lugovka (1)	N		+		
Moneron 5 (4)	N		+	+	
Naiba 6 (3)**	N	+			
Novoaleksandrovsk 2 (2)	N	+			
Novoaleksandrovsk 3 (6)	N	+	+		
Novoaleksandrovsk 6 (2)	N	+			
Novoaleksandrovsk 7 (2)**	N		+		
Odoptu (3)***	N	+		+	
Porechye 4 (1)	N	+			
Pugachevo 4 (1)	N	+			
Pugachevo 5 (3)	N	+			
Puzi 4 (3)**	N	+	+		
Sedykh 1 (1)	N		+		
Shebunino 1 (1)	N		+		
Slavnaya 2 (2)	N	+			
Starodubskoe (2)	N	+			
Vostochny 2 (2)	N	+	+		
Yasnoe 3 (1)	N		+		
Yuzhnaya 2 (2)	N		+	+	
Ado-Tymovo 33 (1)	Pal		+		
Arsentyevka (1)	Pal	+			
Astokh 1 (2)	Pal	+	+		
Astokh 6 (1)	Pal		+		
Bakhura (3)	Pal	+	+		
Baklan 1 (3)	Pal		+	+	
Belinskoe2 (2)	Pal			+	
Beregovoe 2 (5)	Pal	+		+	
Dolinsk 6 (2)	Pal	+	+		

Site Name (No. of Samples)	Period*	Obsidian Sources			
		Shirataki-A	*Shirataki-B*	*Oketo-A*	*Akaigawa*
Kalinino 1 (1)	Pal			+	
Lovetskoe 5 (1)	Pal			+	
Mys Krugly (1)	Pal		+		
Mys Ozerny 8 (1)	Pal		+		
Mys Peschany6 (1)	Pal		+		
Mys Svobodny (2)	Pal	+		+	
Mys Velikan (1)	Pal			+	
Ozersk 1 (1)	Pal	+			
Petropavlovskoe (1)	Pal	+			
Piltun 5 (3)	Pal		+	+	
Razmolovka (1)	Pal			+	
Russa (1)	Pal		+		
Sadovniki 1 (5)**	Pal	+	+	+	
Sadovniki 8 (1)	Pal			+	
Slavnaya 1 (1)	Pal	+			
Slavnaya 3 (1)	Pal	+			
Stary Nabil 5 (1)	Pal			+	
Tomari 2 (1)	Pal			+	
Tretya Pad 1 (3)	Pal	+	+		
Urozhainoe 3 (1)	Pal	+			
Ust-Ainskoe (1)	Pal		+		
Vostochny 1 (1)	Pal			+	
Vzmorye 2 (1)	Pal			+	
Yasnomorsk 3 (12)**	Pal	+	+	+	
Zapadnoe 2 (1)	Pal			+	
Zapadnoe 4 (2)	Pal		+	+	
Zyrianskoe 3 (1)	Pal				+
Zarechye (2)	Pal			+	
Total sites/samples: 76/193****					

* Abbreviations see in Table 10.2.

** These sites contain obsidian unassigned to sources known to us as of late 2011.

*** Previously this site was associated with the Upper Palaeolithic (see Kuzmin and Glascock 2007, 106), but now it is considered as the Early Neolithic (Vasilevsky *et al.* 2010, 18).

**** Due to uncertainty in the number of samples from Palaeometal sites found on southern Sakhalin before the 1945 (Keiko-Ekkususen 2009, 39–41; see also text), these sites are not included in this table but counted for total amount of sites/samples.

area (Tokoroyama locale) (Keiko-Ekkususen 2009, 39–41).

The major source of archaeological volcanic glass in the Amur River basin was the Obluchie Plateau (Table 10.4), and the distance from source to sites is 65–700 km (Figure 10.4). The limited amount of obsidian pebbles in the river channels which drain the source (see Glascock *et al.* 2011) was perhaps one of the factors responsible for the relatively small amount of obsidian in the prehistoric assemblages of the Amur River basin. One of the most important discoveries in recent years was the identification of Hokkaido obsidian in the lower course of the Amur River basin (Figure 10.4). At the Suchu Island, obsidian from the Shirataki-A source was securely established (Table 10.4; see Glascock *et al.* 2011). Before that, I was quite skeptical about the possibility of obsidian to be transported from Hokkaido to the Amur River basin (e.g., Kuzmin 2006a, 68; 2010, 148), although leaving the opportunity for such a route when it would be properly proven. The reason for that was that the data available on the geochemistry of obsidian artefacts from the Lower Amur River region obtained prior to 2011 were of dubious quality (see Kuzmin 2006a, 68; 2010, 149; Kuzmin and Popov 2000, 158), which was recently highlighted again (see Sato 2011, 209). Nowadays, we finally have solid evidence for obsidian transport from Hokkaido (presumably via Sakhalin Island) to the lower part of the Amur River basin (Glascock *et al.* 2011).

Table 10.4. Prehistoric sites in the Amur River basin with obsidian artefacts and their sources (after Glascock *et al.* 2011; see the precise location of sites in this reference)

Site Name (No. of Samples)	Period*	Obsidian Sources			
		Obluchie Plateau	***Basaltic Plateau***	***Samarga***	***Shirataki-A***
Amur 2 (2)	N			+	
Arkhara (3)	N	+			
Goncharka 1 (1)	N			+	
Gromatukha (1)	N	+			
Dim (3)	N	+			
Konstantinovka (1)	N				
Lake Beloberezovoe (2)	N	+			
Lake Dubovoe (1)	N	+			
Lake Peschanoe (3)	N	+			
Malaya Gavan (2)	N	+			
Novopetrovka (1)	N	+			
Novopokrovka (1)	N	+			
Novotroitskoe 10 (2)	N		+		
Orlovka (2)	N	+			
Osinovaya Rechka 10 (1)	N		+		
Osinovoe Ozero (2)	N	+			
Suchu Island (5)	N				+
Sukhie Protoki 2 (1)	Pal	+			
Total sites/samples: 18/39					

* Abbreviations see in Table 10.2.

The existence of a still unlocated source of volcanic glass called 'Samarga' was initially recognised in the early 2000s (Kuzmin *et al.* 2002a). It supplied the raw material for some sites in the Amur River basin and Primorye (Glascock *et al.* 2011; see also Tables 10.2 and 10.4). Recently, Budnitsky (2013) determined the absolute geological age of obsidian used for artefacts of the Ust-Svetlaya and Samarga 2A sites as around 32.2–35.1 Ma. This allows to suggest that the geological source for this volcanic glass is situated in the Samarga River basin in northern Primorye (Budnitsky 2013, 13), as it was originally proposed (Kuzmin *et al.* 2002a).

Recent progress is evident for the Kurile Islands research (Figure 10.5) where obsidian provenance studies have been conducted since the mid-2000s (Phillips 2010; Phillips and Speakman 2009). At some sites there are different cultural components with obsidian artefacts, but they are not described separately (Phillips 2011; Phillips and Speakman 2009; see Table 10.5). According to the most complete study (Phillips 2010), prehistoric people of the Kuriles received obsidian raw material from two regions neighbouring the island chain, that is, Hokkaido and Kamchatka (Figure 10.5). At least seven obsidian sources were used throughout the later prehistory of the Kuriles, represented by Epi-Jomon and Okhotsk complexes dated to about 2900–800 BP (see Phillips 2011). The association of some older sites (Early–Middle Jomon, Kuibyushevskaya 1 and Sernovodsk 1 sites; and Late Jomon, Berezovka 2 site; see Table 10.5) with particular cultural complexes is uncertain because obsidian artefacts come from surface collections. The distance between the Hokkaido sources (Shirataki and Oketo) and archaeological sites in the Kurile Islands where these obsidians were identified is ca. 160–1200 km; and between the Kamchatkan sources and utilisation sites it is ca. 400–1500 km (Figure 10.5).

Despite the progress described above, a discrepancy exists between the identification of obsidian sources in Phillips (2010) and Phillips (2011). In the latter study, the wide use of the Tokachimisumata source [Tokachi-Mitsumata in Kuzmin *et al.* (2002b) and Kuzmin and Glascock (2007)] in the Epi-Jomon and Okhotsk cultural complexes of the Kuriles is detected. This is in contrast with earlier work (see Phillips 2010) where the same dataset was used, and the Tokachi-Mitsumata source (situated in central Hokkaido; see Ferguson *et al.*, this volume) was not identified at all. Unfortunately, Phillips (2011) does not provide original geochemical data for the obsidian groups representing the different sources (besides the content of only three elements: Rb, Sr, and Zr; see Phillips 2011, 127–33), as was done in Phillips (2010, 128) where the composition of 29 elements is given. As a result, it is not possible to understand the cause of this controversy. It may be due to a situation when Phillips (2011) used XRF and LA–ICP–MS methods, and no comparison with NAA results was

Table 10.5. Prehistoric sites on the Kurile Islands with obsidian artefacts and their sources (after Phillips 2010, 2011; Phillips and Speakman 2009; see the precise location of sites in these references)

Site Name (No. of Samples)	**Period***	**Obsidian Sources (Hokkaido)**			
		Shirataki-A	***Shirataki-B***	***Oketo-A***	***Oketo-B***
Kuibyushevskaya 1 (10)	E–MJ	+	+		
Sernovodsk 1 (11)**	E–MJ		+	+	+
Berezovka 2 (4)	LJ	+	+	+	
Ainu Creek 1 (341)**, ***	E-J, OK	+	+	+	+
Alekhina (1)	E-J (?)		+		
Baikova 1 (63)**	E-J			+	+
Drobnyye 1 (79)***	E-J, OK				
Peschanaya 2 (4)	E-J (?)	+	+		+
Rasshua 1 (2)**	E-J				
Rikorda 1 (85)***	E-J, OK	+	+	+	+
Savushkina 1 (60)**	E-J				+
Vodopadnaya 2 (89)**, ***	E-J, OK	+		+	
Ekarma 1 (3)**	OK				
Bolshoi 1 (2)	?				
Peschanaya Bay 1 (11)**	?				
Ryponkicha 1 (6)**	?				
Tikharka River 1 (1)	?				
Tikhaya 1 (1)	?			+	

Site Name (No. of Samples)	**Period***	**Obsidian Sources (Kamchatka)**				
		KAM-01	***KAM-02***	***KAM-04***	***KAM-05***	***KAM-07***
Kuibyushevskaya 1 (10)	E–MJ					
Sernovodsk 1 (11)**	E–MJ					
Berezovka 2 (4)	LJ					
Ainu Creek 1 (341)**, ***	E-J, OK		+		+	
Alekhina (1)	E-J (?)					
Baikova 1 (63)**	E-J	+	+	+		
Drobnyye 1 (79)***	E-J, OK	+	+	+	+	+
Peschanaya 2 (4)	E-J (?)					
Rasshua 1 (2)**	E-J				+	
Rikorda 1 (85)***	E-J, OK	+			+	
Savushkina 1 (60)**	E-J	+	+	+		
Vodopadnaya 2 (89)**, ***	E-J, OK	+	+	+	+	+
Ekarma 1 (3)**	OK				+	
Bolshoi 1 (2)	?		+			
Peschanaya Bay 1 (11)**	?	+				
Ryponkicha 1 (6)**	?		+			
Tikharka River 1 (1)	?	+				
Tikhaya 1 (1)	?					
Total sites/samples: 18/773						

* E–MJ – Early–Middle Jomon; LJ – Late Jomon; E-J – Epi-Jomon [Zoku Jomon]; OK – Okhotsk Culture; ? – unknown cultural period.

** At these sites, obsidian from some unidentified sources is also detected (see Phillips 2010, 129–30).

*** No data are provided for each cultural complex (Epi-Jomon and Okhotsk) separately (see Phillips 2011, 115–56), and combined data are used (Phillips 2010).

conducted by analysing the samples from the dataset of Kuzmin and Glascock (2007). This was suggested by M. D. Glascock (personal communication 2011); therefore, the NAA data for the Hokkaido sources (see Kuzmin and

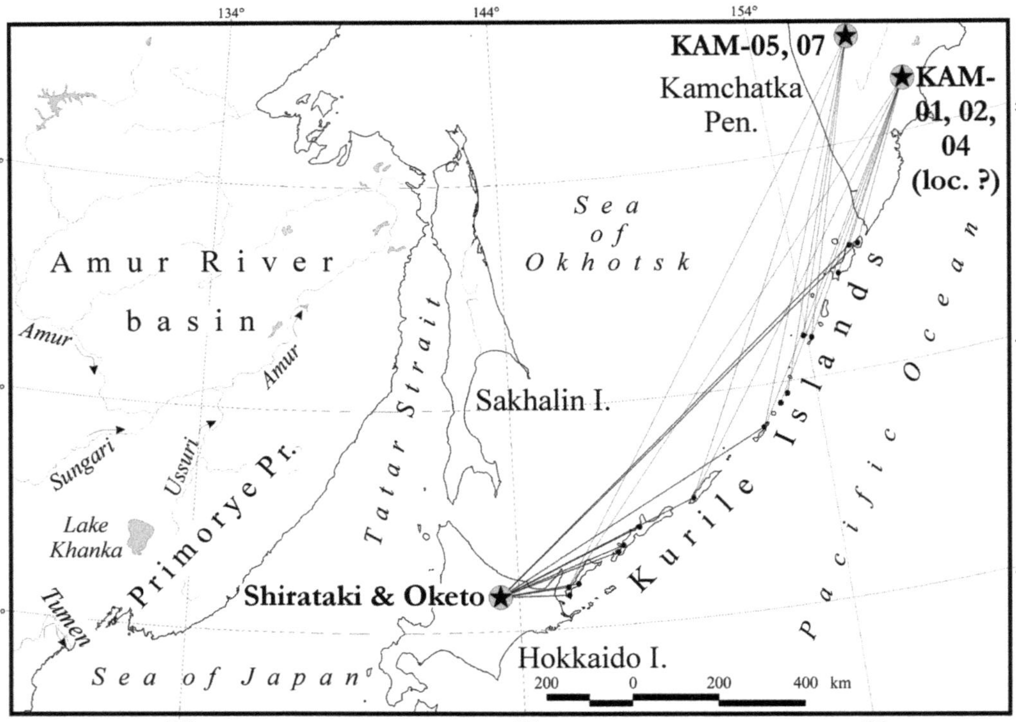

Figure 10.5. Sources of archaeological obsidian for the Kurile Islands (original data are after Phillips 2010)

Glascock 2007; Kuzmin *et al.* 2002b, 2013) are not directly compatible with Phillips' (2011) geochemical signatures. It would perhaps be methodologically more correct to analyse samples from the Hokkaido obsidian sources previously tested by NAA method (see Kuzmin and Glascock 2007) also by XRF and LA–ICP–MS techniques, in order to establish secure geochemical 'fingerprints' and make comparative analysis, but to the best of my knowledge this was not done by Phillips (2011).

Before the original geochemical data used by Phillips (2011) becomes available, the conclusion about the intensive use of the Tokachi-Mitsumata source by inhabitants of the Kurile Islands should be treated with great caution. It is improbable that this obsidian was widely exploited in the Kuriles while on Sakhalin Island no obsidian from the Tokachi-Mitsumata source was detected (Kuzmin and Glascock 2007; see also Table 10.3). In Hokkaido itself, the use of this source was quite limited despite the fact that obsidian from it has been identified at several sites (e.g., Kimura 1995, 11).

Concerning the degree of investigation of archaeological obsidian from the southern Russian Far East (Figure 10.6), Primorye and Sakhalin Island are the best-studied regions in terms of the number of sites; both the Amur River basin and Kurile Islands have not been adequately investigated so far. In terms of the number of specimens analysed, the Kurile Islands have the largest amount (little more than half of the total number of artefacts), and Primorye region includes more than a quarter of the overall sample size. For both the Amur River basin and Sakhalin Island, much less obsidian specimens have been analysed. Generally, the Amur River basin is the most poorly investigated part of the Russian Far East for obsidian sourcing (Figure 10.6).

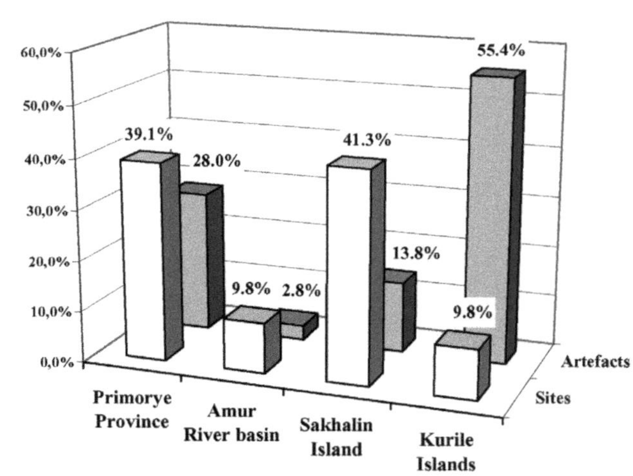

Figure 10.6. Percentage of sites/artefacts in the Russian Far East studied in terms of obsidian geochemistry (see Tables 10.2–10.5)

DISCUSSION

Long-Distance Obsidian Exchange Networks in the Southern Russian Far East

After the geochemical characterisation of obsidian artefacts and their sources, several large-scale networks in Primorye, Amur River basin, Sakhalin Island, Kurile

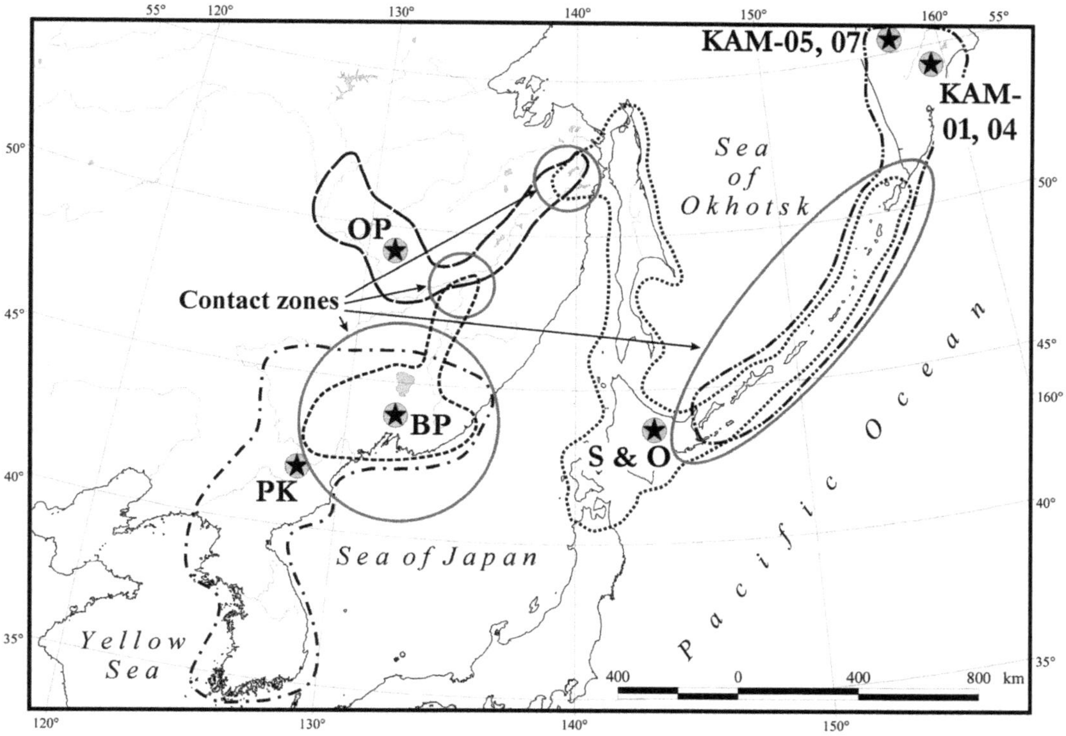

Figure 10.7. Contact zones of the obsidian networks in Northeast Asia. Source abbreviations: BP – Basaltic Plateau; PK – Paektusan Volcano; OP – Obluchie Plateau; S & O – Shirataki and Oketo; KAM-01, -04, -05, and -07 – different Kamchatkan sources (after Grebennikov et al. *2010)*

Islands, and adjacent Northeast Asia have been established (see, for example: Kuzmin 2010, 2011; Kuzmin and Popov 2000; Kuzmin *et al.* 2002a, 2002b). Below a brief description of the networks is presented, using information available as of late 2013.

The Basaltic Plateau source in southern Primorye was actively used by prehistoric people of this region, and obsidian was occasionally transported further away, to the lower reaches of the Amur River and Manchuria (Figure 10.1). Another important source is Paektusan (Figure 10.2); this network covers the entire Korean Peninsula, part of Manchuria adjacent to Paektusan, and Primorye. The third crucial obsidian dispersal sphere centred around two sources of Hokkaido Island, Shirataki and Oketo (Figure 10.3). Obsidian from these locales was widely transported and traded in Northeast Asia; the range of distribution covers Hokkaido, neighbouring Sakhalin Island and Kurile Islands, and the lower reaches of the Amur River. The fourth obsidian network existed in the middle and lower courses of the Amur River, with the Obluchie Plateau as a primary source of high quality volcanic glass (Figure 10.4).

In some parts of the Russian Far East and neighbouring Northeast Asia, the distribution networks of different obsidian sources overlap (Figure 10.7), and here so-called 'contact zones' appeared. In these regions we can expect the prehistoric use of several obsidian sources at the same time, as it is known in Primorye and Manchuria (e.g., Doelman *et al.* 2008; Jia *et al.* 2010, 2013; Kuzmin *et al.* 2002a). However, in some regions (like the Amur River basin, for example) the limited amount of artefacts analysed so far prevents an understanding of the complexity involved in obsidian acquisition. In order to achieve this, the application of 'exhaustive analysis' (*sensu* Tsutsumi 2010, 37) — in other words, examination of as many artefacts as possible — would be a perspective, and this will allow us to determine the exact number of obsidian sources used in prehistory.

Sato (2011) reviewed the obsidian distribution in Northeast Asia, with a particular focus on the possibility of finding volcanic glass from the Japanese Islands in mainland Russian Far East. He suggests that because the size of obsidian pebbles in the Ilistaya River channel near the Ilistaya and Gorbatka sites is small (no more than 5 cm; see Sato 2011, 206), Palaeolithic people used larger pieces of raw material acquired from the Paektusan source. Sato (2011, 206) determined two phases of obsidian exploitation in Primorye, with procurement of raw material from 'local' (i.e., Basaltic Plateau) and 'remote' (i.e., Paektusan) sources. However, in a recent study of volcanic glass debris eroded from the Basaltic Plateau it is indicated that the size of pebbles near the volcanic glass-containing outcrops is in the 15–20 cm range, and only about 50 km downstream are the pebbles less than 5 cm in diameter (see Doelman *et al.* 2008, 254–5). It is impossible to establish two phases of obsidian exploitation with different sources for each phase, because there is no stratigraphic data from late Upper

Palaeolithic sites in the Ilistaya River basin with cultural layers containing obsidian from either the Basaltic Plateau or Paektusan, to the best of my knowledge. Also, there is no reliable chronology for late Upper Palaeolithic sites in the Ilistaya River basin (e.g., Doelman *et al.* 2008; Kuznetsov 1996; Pantukhina 2007), and it is not possible to separate sites with obsidian from each major source mentioned above. It seems that the strategy of obsidian acquisition in southern Primorye by late Upper Palaeolithic inhabitants was quite complex (see Kuzmin *et al.* 2002a), and its patterns are not yet fully understood. More work is certainly needed to get closer to answer the questions related to the ways of obsidian procurement and transport in southern Primorye, even though this is the best-studied region of the whole Russian Far East in terms of both archaeology and obsidian provenance (see Andreeva 2005; Kluyev 2003; Kuzmin 2010; Nelson *et al.* 2006; see also Table 10.2).

Modes of Obsidian Acquisition: Some Thoughts

The use of multiple sources of high quality volcanic glass by the same populations in the southern Russian Far East is an important pattern, originally recognised in the early 2000s (see Kuzmin *et al.* 2002a, 2002b) and confirmed later (see Dolman 2008; Doelman *et al.* 2004, 2008, 2009). It is unclear why prehistoric people in southern Primorye were bringing obsidian from the 'remote' Paektusan source located more than 300 km away (Kuzmin *et al.* 2002a; Popov *et al.* 2005), given the fact that obsidian pebbles eroded from the 'local' Basaltic Plateau source were readily available from the nearby streams. In the Kurile Islands, obsidian from at least seven different sources located more than 1000 km apart was used (Phillips 2010; see Figure 10.5). This is especially true for the Kamchatka Peninsula where the use of a multitude of obsidian sources is a common feature (Grebennikov *et al.* 2010; Kuzmin *et al.* 2008). This issue is related to some extent to the diversity of human behaviour, but it also shows that the strategy of raw material acquisition was quite complex in the late Upper Palaeolithic (ca. 12,000 BP) and afterwards (see Kuzmin 2011).

As for the issues of obsidian raw material acquisition and movement, they are not well-studied in the southern Russian Far East. At some sites situated at the sources or near them, like in the Ilistaya River basin of Primorye (see Doelman *et al.* 2008, 2009, 2012; Kluyev and Sleptsov 2007; Kuznetsov 1996), direct procurement of volcanic glass from primary outcrops is obvious. In terms of sites located further away from the source (see Pantukhina 2007), the mechanism of obsidian transport is not yet clear; it could be either direct visits to the source or down-the-line exchange (*sensu* Renfrew 1969; see also: Renfrew and Dixon 1976, 145–9). Doelman *et al.* (2008, 264; 2012, 110–1) suggested that obsidian from the Paektusan source was transported to sites in southern Primorye in the Upper Palaeolithic as prepared blade cores and blades, without a definite conclusion how it was moved: either directly by the inhabitants of Primorye or by means of exchange/trade with people who occupied the vicinities of the source. In the Neolithic, Doelman *et al.* (2008, 266) assumed the existence of obsidian exchange networks in the northern Korean Peninsula and Primorye. When distance between source(s) and site(s) is significant, exceeding about 100–200 km or so, direct visits to the source were less likely, in my opinion. However, more research is necessary in order to understand the ways of obsidian transport in the southern Russian Far East.

A study of volcanic glass transport from the Basaltic Plateau source by the Ilistaya River was conducted by Doelman *et al.* (2008, 2012). It was found that after about 20 km from the source, the amount of obsidian pebbles in the channel decreased significantly, and a few rounded volcanic glass pieces (with nodule size of 2–4 cm only) were found 30–35 km from the source (see Doelman *et al.* 2008, 255; 2012, 103). This distance is much smaller than in other parts of the world, where obsidian pebbles are sometimes situated in alluvium up to 500 km and even further away from the primary source, as data from the American Southwest testify (see Shackley 2005, 63–4).

The existence of water space between archaeological sites and obsidian sources in insular regions of Northeast Asia, notably between Hokkaido, Sakhalin, and the Kuriles, allows us to assume that prehistoric people were able to use some kind of watercraft (rafts or primitive dugout boats) to move raw material across the wide straits between Hokkaido, Sakhalin Island, Kurile Islands, and the Kamchatka Peninsula. The earliest actual finds of boats in Northeast Asia are currently known from sites dated to ca. 7100–6800 BP (Jiang and Liu 2005; Park *et al.* 2010; see also Habu 2010). It is possible that people built dugout canoes even before that, because the La Pérouse [Soya] Strait between Hokkaido and Sakhalin islands after the Last Glacial Maximum "closure" was reopened at ca. 11,000–10,000 BP (Korotkii 1985). Some islands of the Kurile Archipelago remained disconnected from neighbouring landmasses for even longer times, and the only way people could get there was by sea using boats (e.g., Erlandson and Braje 2011).

Comparison with Neighbouring Northeast Asia: Exchange Networks and Watercraft

After several decades of research in the field of obsidian sourcing, it became clear that large-scale raw material exchange networks existed in prehistory in several regions of Northeast Asia, notably the main islands of Japan (see, for example: Izuho and Sato 2007; Ono *et al.* 1992, 35–6, 79–80; Tsutsumi 2010), the Ryukyu Archipelago (Obata *et al.* 2010), and the Korean Peninsula (Kim *et al.* 2007). In some parts of Northeast Asia, use of obsidian as a raw material for tool manufacture began earlier than in the Russian Far East. In Japan it can be dated to ca. 35,000 BP (e.g., Shimada 2009), and in Korea to ca. 25,500 BP (Kim *et al.* 2007). Along with the evidence from the southern Russian Far East, it shows that human contacts were very active in the entire region of Northeast Asia since the second part of the Upper Palaeolithic, ca. 25,000–20,000 BP. The issue of human contacts and migrations in Northeast Asia, based on knowledge of obsidian transport and/or exchange, deserves special attention. Some models have already been put forward (see, for example: Ambiru 2003, 2009), but in my opinion more research is still needed to obtain reliable information about human movements in this region in the Upper Palaeolithic and Neolithic/Jomon.

Discoveries of obsidian from remote sources in insular Northeast Asia, separated from utilisation sites by wide water space, gave some scholars the idea about seafaring in the Upper Palaeolithic (see, for example: Erlandson 2001, Oda 1990). The latest data (although still indirect!) on seafaring in the Pacific, based on finds of fish hooks and pelagic fish exploitation (such as tuna) at the Jerimalai site in Timor-Leste [East Timor], on the eastern tip of Timor Island in Southeast Asia (O'Connor *et al.* 2011), goes back to ca. 41,800 cal BP (or ca. 36,700 BP; see conversion to ^{14}C ages: Reimer *et al.* 2009) in terms of pelagic fish, and to older than ca. 23,200 cal BP (or ca. 19,300 BP) in relation to fish hooks. Due to the disturbance of cultural layers, it is impossible to pinpoint the age of the earliest fish hook precisely (see O'Connor *et al.* 2011, 118). There is also doubt about offshore fishing in the Pacific prior to the Holocene (e.g., Anderson 2013), and discussion on this issue is ongoing (e.g., O'Connor and Ono 2013). On the North Pacific coast of North America, the earliest sites on islands which could have been settled with the presumed help of boats are now dated to ca. 10,600 BP on Haida Gwaii [Queen Charlotte Islands], British Columbia (Fedje *et al.* 2011, 457); and to ca. 10,400 BP on the Channel Islands off California (Erlandson *et al.* 2011). These sites can also be used as indirect evidence for the existence of watercraft.

Additional data in favour of the early existence of watercraft in Northeast Asia is the presence of obsidian from the Koshidake source on Kyushu Island in the southern part of the Korean Peninsula, at the Shinbuk [Sinbuk] site (Kim *et al.* 2007; see also Kim, this volume). This Upper Palaeolithic site is dated to ca. 25,500–18,500 BP (see Kim *et al.* 2007; Seong 2007, 2011). If the source identification is correct, the transport of obsidian from Kyushu Island to Korea across the Korea [Tsushima] Strait, which still existed at this time and was probably ca. 15–20 km wide (see, for example: Lee and Nam 2003; Lee *et al.* 2008), can be established. Cultural contacts between the Korean Peninsula and Kyushu Island were suggested earlier based on archaeological data (see Matsufuji 2003). On the other hand, the content of four elements only was measured by PIXE (see Kim *et al.* 2007, 125), and perhaps application of NAA, XRF, or LA–ICP–MS methods is necessary to confirm the finding (see Kim, this volume).

The phenomenon of obsidian source exploitation at the Kozu-shima [Kozushima] Islet in the open sea off central Honshu Island (see Tsutsumi 2010, 2012) is important and still enigmatic because several sites in the Kanto Plain (Honshu Island), dated to ca. 34,000–30,000 BP, have obsidian taken from Kozu-shima (e.g., Tsutsumi 2010, 33–6; see also Ikeya, this volume). This conclusion was initially based on early research when the content of less than ten elements was measured (see Suzuki 1973b; Suzuki and Tomura 1983). Later this was supplemented by the application of the EDXRF method (see Tsutsumi 2010), and was confirmed by the results of a NAA study of the Ide-Maruyama and Doteue sites (Ikeya *et al.* 2005; Ikeya and Glascock 2013). As a result of these new studies, we are now much closer to a valid identification of the Kozu-shima obsidian in Upper Palaeolithic contexts of the Kanto Plain. This is directly related to the possibility of watercraft use dated to at least ca. 30,000 BP.

Dennell (2013) completely misunderstood the information on obsidian transport from the Kozu-shima source to the neighbouring region, by saying that '... boats were being used routinely ... to obsidian trade between Japan and Kamchatka after 30,000 BP (Kuzmin 2006)' (Dennell 2013, 1227). In the cited source (see Kuzmin 2006a), there is no any information about the obsidian exchange between Japan and Kamchatka, especially in such old times, after 30,000 BP. What exists is the indication about the find of Kozu-shima obsidian at the Musashidai site dated to ca. 30,000 BP (Kuzmin 2006a, 66). Today, more data on the use of Kozu-shima source in the early prehistory of Japan are available (see Ikeya, this volume).

Perspectives: Northern Russian Far East

Upon the completion of the first stage of obsidian provenance studies in the southern Russian Far East in the late 1990s, an edited volume was published (Kuzmin and Popov 2000; see also Kuzmin *et al.* 2002a, 2002b). In the 2000s, special attention was given to the more northern territories, notably Kamchatka Peninsula and Chukotka Region. Fieldwork on Kamchatka in 2004–5 and geochemical analysis of several hundred samples allowed us to establish the main sources of archaeological obsidian (see the latest summary: Grebennikov *et al.* 2010), and to conduct an in-depth study of the late Upper Palaeolithic cluster of Ushki (Kuzmin *et al.* 2008). Nowadays, at least 14 sources of archaeological obsidian have been detected on Kamchatka, and the localisation of seven sources is securely pinpointed (Grebennikov *et al.* 2010; see also Grebennikov *et al.*, this volume).

At the Ushki cluster, in cultural layers 5–7 dated to ca. 11,500–9000 BP (according to the latest data, up to ca. 11,900 BP; see Krenke *et al.* 2011), and possibly up to ca. 14,000 BP (see discussion: Goebel *et al.* 2010; Kuzmin *et al.* 2010), obsidian from three to six different sources was identified. The distance between the Ushki cluster and obsidian sources is ca. 140–260 km in a straight line; and sources are also hundreds of kilometres apart from each other. This shows that acquisition of good quality raw material was quite sophisticated in the Upper Palaeolithic in this remote part of Northeast Asia with its harsh terrain and climate. Kamchatka is one of the key regions for future detailed study of obsidian sources used by ancient people.

CONCLUSIONS

The results of obsidian source studies in different parts of Northeast Asia (Russian Far East, Hokkaido and Honshu islands, Ryukyu Archipelago, and Korean Peninsula) in the last two decades are summarised in Kuzmin and Glascock (2010). The primary geochemical data presented in this collection create a solid background for the continuation of this kind of research in the near future, based on cooperation between scientists from Russia, Japan, Republic of Korea, and USA.

The exploitation of obsidian by prehistoric people of the southern Russian Far East began in the Upper Palaeolithic, at ca. 19,000–18,000 BP, and continued until the Palaeometal/Early Iron Age (ca. 3000–1300 BP, and in some regions like the Kurile Islands until ca. 800 BP). Obsidian was most intensively used as a raw material in Upper Palaeolithic and Neolithic times. Distances between sources and utilisation sites reach 1000–1200 km in a straight line during the Neolithic and Palaeometal times (ca. 8000–800 BP). Intensive transport of Hokkaido obsidian to Sakhalin and the Kuriles across the sea straits allows us to suggest the existence of watercraft since the early Holocene, ca. 10,000 BP, and possibly even earlier.

The beginning of obsidian exploitation in Northeast Asia is dated to ca. 35,000–30,000 BP. Obsidian was used as a raw material until the end of the Palaeometal, generally about ca. 2000–1500 BP. Several long-distance transport and/or exchange networks existed in Northeast Asia in prehistory, and human contacts were very active since the early Upper Palaeolithic. Additional studies of important obsidian sources and archaeological assemblages in Northeast Asia are now necessary using advanced analytical methods, which provide minimal (if any) damage to valuable artefacts and can establish the source of archaeological obsidian with a high degree of confidence. A good possibility exists now for international collaboration in the investigation of obsidian sources and human movements, migrations, and contacts in this part of the North Pacific Rim.

Acknowledgements

This review is the result of long-term studies of archaeological obsidian sources in the Russian Far East and neighbouring Northeast Asia, conducted since 1992 by the author together with numerous colleagues from Russia, USA, Japan, and Republic of Korea. I am grateful to all of them (see relevant references where their names are indicated) for fruitful collaboration and support. My thanks also go to Prof. A. Ono and the Centre for Obsidian and Lithic Studies of Meiji University (Tokyo) for invitation to visit Japan in September 2010 and October–November 2011 and financial support, provided within the activity of the Project "Historical Variation in Interaction between Humans and Natural Resources: Towards the Construction of a Prehistoric Anthropography". Profs A. Ono and S. Sugihara kindly helped with the translation of information about obsidian sources from sites in southern Sakhalin, and A. Ono made me aware of the early papers on obsidian sourcing in Japan conducted by Prof. M. Suzuki in the 1960s – early 1970s. I am also thankful for the possibility to participate in the Workshop "Methodological Issues of Obsidian Provenance Studies and the Standardisation of Geologic Obsidian", held at Centre for Obsidian and Lithic Studies, Nagawa Town (Nagano Pref.), on 5–6 November 2012. This paper is an extended version of my presentation delivered at the Workshop. Finally, I am very grateful to Prof. R. E. Ackerman and Dr N. A. Klyuev for helpful comments and suggestions on the earlier version of this chapter, and to Dr S. G. Keates for grammar corrections.

References

AMBIRU, M. 2003. Approach to the Stone Age in the Japanese Archipelago from the Standpoint of Obsidian Archaeology. In *Palaeolithic Men's Lives and their Sites*, edited by Y.-J. Lee, 661–681. Seoul, Hakyeon Press.

AMBIRU, M. 2009. The Paleolithic Cultural Corridor around the Sea of Japan and Obsidian Road. *Sundai Shigaku* 135, 147–168 (in Japanese with English Abstract).

ANDERSON, A. 2013. The Antiquity of Sustained Offshore Fishing. *Antiquity* 87, 879–895.

ANDREEVA, Z. V. (ed.). 2005. *Rossiisky Dalny Vostok v Drevnosti i Srednevekovye: Otkrytiya, Problemy, Gipotezy* [The Russian Far East in Prehistory and the Mediaeval Times: Discoveries, Problems, and Hypotheses]. Vladivostok, Dalnauka Publishers.

ANERT, E. E. 1904. *Puteshestvie po Manchzurii* [À Travers la Mandjourie] (Memoirs of the Imperial Russian Geographic Society on General Geography. Vol. XXXV). St.-Petersburg, Imperatorskoe Russkoe Geograficheskoe Obshchestvo.

ARSENIEV, V. K. 2007 [1921]. *Po Ussuriiskomu Krayu. Puteshestvie v Gornuyu Oblast Sikhote-Alin v 1906 G.* [Around the Ussuri Region. Travel in Mountainous District of Sikhote-Alin in 1906]. Vladivostok, Rubezh Press.

ARSENJEW, W. K. 1924. *In der Wildnis Ostsibiriens: Forschungsreisen im Ussurigebiet. Erster Band.* Berlin, August Scherl G.m.b.H. [originally published in Russian in 1921].

BATARSHEV, S. V., O. L. MOREVA, and Y. V. KUZMIN. 2010. The Rudnaya Neolithic Culture in Primorye: Issues of Periodization and Chronology. *Rossiiskaya Arkheo-logiya* 4, 5–17 (in Russian with English Abstract).

BEFU, H., and C. S. CHARD. 1964. A Prehistoric Maritime Culture of the Okhotsk Sea. *American Antiquity* 30, 1–18.

BUDNITSKY, S. Y. 2013. *Novaya Kalii–Argonovaya Geokhronologiya Redkikh i Unikalnykh Geologicheskikh i Arkheologicheskikh Obyektov* [The New Potassium–Argon Geochronology of Rare and Unique Geological and Archaeological Objects]. Synopsis of the Candidate of Geological–Mineralogical Sciences (PhD-equivalent) Dissertation. Vladivostok, Dalnevostochny Geologichesky Institut.

CANN, J. R. 1983. Petrology of Obsidian Artefacts. In *The Petrology of Archaeological Artefacts*, edited by D. R. C. Kempe and A. P. Harvey, 227–255. Oxford, Clarendon Press.

CANN, J. R., and C. RENFREW. 1964. The Characterization of Obsidian and Its Application to the Mediterranean Region. *Proceedings of the Prehistoric Society* 30, 111–133.

CARLSON, R. L. 1994. Trade and Exchange in Prehistoric British Columbia. In *Prehistoric Exchange Systems in North America*, edited by T. G. Baugh and J. E. Ericson, 307–361. New York, Plenum Press.

CHO, N.-C., J.-C. KIM, and H.-T. KANG. 2010. Provenance Study of Obsidian Artefacts Excavated from Palaeolithic Sites on the Korean Peninsula. In *Crossing the Straits: Prehistoric Obsidian Source Exploitation in the North Pacific Rim* (B.A.R. International Series 2152), edited by Y. V. Kuzmin and M. D. Glascock, 73–87. Oxford, BAR Publishing.

CHOI, B.-K. 2001. *The Janghung-ri Palaeolithic Site*. Chuncheon, Institute of Kangwon Archaeology (in Korean with English Title).

DENISOV, E. P. 1965. *Noveishaya Tektonika i Podznekainozoisky Vulkanizm Yuzhnogo Primorya i Prilegayushchikh Oblastei* [The Modern Tectonics and Late Cenozoic Volcanism of Southern Primorye and Adjacent Regions]. Vladivostok, Dalnevostochnoe Knizhnoe Izdatelstvo.

DENISOV, E. P., and H. C. TEN. 1966. *Kratkaya Geologicheskaya Kharakteristika Vulkana Pektusan (Baitoushan)* [The Short Geological Characteristic of Paektusan (Baitoushan) Volcano] (Supplement Brochure to: Khudyakov, G. I. (ed.). 1965. *Voprosy Geomorfologii i Morfotektoniki Yuzhnoi Chasti Dalnego Vostoka* [Questions on Geomorphology and Morphotectonics of the Southern Part of the Far East]. Vladivostok, Dalnevostochnoe Knizhnoe Izdatelstvo). Vladivostok, Dalnevostochnoe Knizhnoe Izdatelstvo.

DENNELL, R. 2013. Review of Dennis J. Stanford and Bruce A. Bradley. *Across Atlantic Ice: The Origins of America's Clovis Culture*. Berkeley and Los Angeles: University of California Press. *Antiquity* 87, 1226–1227.

DEREVIANKO, A. P. 1970. Gromatukhinskaya Kultura [The Gromatukha Culture]. In *Sibir i Ee Sosedi v Drevnosti*, edited by V. E. Larichev, 195–209. Novosibirsk, Nauka Publishers.

DEREVIANKO, A. 1998. Sites of the Selemdzha Type. In *The Paleolithic of Siberia: New Discoveries and Interpretations*, edited by A. Derevianko, D. B. Shimkin and W. R. Powers, 281–285. Urbana & Chicago, University of Illinois Press.

DEREVIANKO, A. P., and V. E. MEDVEDEV. 2006. Neolithic of the Nizhnee Priamurye (Lower Amur River Basin). In *Archaeology of the Russian Far East: Essays in Stone Age Prehistory* (B.A.R. International Series 1540), edited by S. M. Nelson, A. P. Derevianko, Y. V. Kuzmin and R. L. Bland, 123–149. Oxford, BAR Publishing.

DEREVIANKO, A. P., V. N. ZENIN, and I. Y. SHEWKOMUD. 2006. Palaeolithic of the Priamurye (Amur River Basin). In *Archaeology of the Russian Far East: Essays in Stone Age Prehistory* (B.A.R. International Series 1540), edited by S. M. Nelson, A. P. Derevianko, Y. V. Kuzmin and R. L. Bland, 53–73. Oxford, BAR Publishing.

DOELMAN, T. 2008. Flexibility and Creativity in Microblade Core Manufacture in Southern Primorye, Far East Russia. *Asian Perspectives* 47, 352–370.

DOELMAN, T., N. KONONENKO, B. POPOV, G. SUMMERHAYES, R. TORRENCE, A. BONETTI, A. GUGLIELMETTI, A. MANZONI, and M. ODDONE. 2004. Acquisition and Movement of Volcanic Glass in the Primorye Region of Far Eastern Russia. *Rossiya i ATR* 4(46), 112–125.

DOELMAN, T., R. TORRENCE, N. KLUYEV, I. SLEPTSOV, and V. POPOV. 2009. Innovations in Microblade Core Production at the Tigrovy-8 Late Palaeolithic Quarry in Eastern Russia. *Journal of Field Archaeology* 34, 367–384.

DOELMAN, T., R. TORRENCE, V. POPOV, M. IONESCU, N. KLUYEV, I. SLEPTSOV, I. PANTYUKHINA, P. WHITE, and M. CLEMENS. 2008. Source Selectivity: An Assessment of Volcanic Glass Sources in the Southern Primorye, Far East Russia. *Geoarchaeology* 23, 243–273.

DOELMAN, T., R. TORRENCE, V. POPOV, N. KLUYEV, and I. SLEPTSOV. 2012. Volcanic Glass Procurement and Use in the Late Paleolithic, Central Primorye, Far East Russia. In *Obsidian and Ancient Manufactured Glass*, edited by J. Liritzis and C. M. Stevenson, 97–114. Albuquerque, University of New Mexico Press.

ERLANDSON, J. M. 2001. The Archaeology of Aquatic Adaptations: Paradigms for a New Millennium. *Journal of Archaeological Research* 9, 287–350.

ERLANDSON, J. M., and T. J. BRAJE. 2011. From Asia to the Americas by Boat? Paleogeography, Paleoecology, and Stemmed Points of the Northwest Pacific. *Quaternary International* 239, 28–37.

ERLANDSON, J. M., T. C. RICK, T. J. BRAJE, M. CASPERSON, B. CULLETON, B. FULFROST, T. GARCIA, D. A. GUTHRIE, N. JEW, D. J. KENNETT, M. L. MOSS, L. REEDER, C. SKINNER, J. WATTS, and L. WILLIS. 2011. Paleoindian Seafaring, Maritime Technologies, and Coastal Foraging on California's Channel Islands. *Science* 331, 1181–1185.

FEDJE, D., Q. MACKIE, T. LACOURSE, and D. MCLAREN. 2011. Younger Dryas Environments and Archaeology on the Northwest Coast of North America. *Quaternary International* 242, 452–462.

FISCHER, G. 1818. *Oriknognoziya, ili Kratkoe Opisanie Vsekh Iskopaemykh Veshchestv, s Izyasneniyami Terminov. Chast 1* [The Orictognozy, or Short Description of All Fossil Compounds, with Explanation of Terms. Part 1]. Moscow, Tipografiya Imperatorskoi Mediko-Khirurgicheskoi Akademii.

FITZHUGH, B. 2012. Hazards, Impacts, and Resilience among Hunter–Gatherers of the Kuril Islands. In *Surviving Sudden Environmental Changes: Answers from Archaeology*, edited by J. Cooper and P. Sheets, 19–42. Boulder, University Press of Colorado.

FITZHUGH, B., S. C. PHILLIPS, and E. GJESFJELD. 2011. Modeling Variability in Hunter–Gatherer Information Networks: An Archaeological Case Study from the Kuril Islands. In *Information and Its Role in Hunter–Gatherer Bands*, edited by R. Whallon, W. A. Lovis and R. K. Hitchcock, 85–115. Los Angeles, Cotsen Institute for Archaeology, University of California at Los Angeles.

GANESHIN, G. S., and A. P. OKLADNIKOV. 1956. O Nekotorykh Arkheologicheskikh Pamyatnikakh Primorya i Ikh Geologicheskom Znachenii [About Some Archaeological Sites in Primorye and Their Geological Significance]. In *Materialy po Geologii i Poleznym Iskopaemym Vostochnoi Sibiri i Dalnego Vostoka*, edited by Y. F. Chemekov, 50–57. Moscow, Gosgeoltekhizdat Publishers.

GARIN-MICHAILOWSKI, N. G. 2005. *Koreas Schlafender Vulkan. Die Reise zum Pektusan.* Lenningen, Edition Erdmann [originally published in Russian in 1898].

GLASCOCK, M. D. (ed.). 2002. *Geochemical Evidence for Long-Distance Exchange.* Westport, CT, Bergin & Garvey.

GLASCOCK, M. D., G. E. BRASWELL, and R. H. COBEAN. 1998. A Systematic Approach to Obsidian Source Characterization. In *Archaeological Obsidian Studies: Method and Theory*, edited by M. S. Shackley, 15–65. New York & London, Plenum Press.

GLASCOCK, M. D., Y. V. KUZMIN, A. V. GREBENNIKOV, V. K. POPOV, V. E. MEDVEDEV, I. Y. SHEWKOMUD, and N. N. ZAITSEV. 2011. Obsidian Provenance for Prehistoric Complexes in the Amur River Basin (Russian Far East). *Journal of Archaeological Science* 38, 1832–1841.

GOEBEL, T., S. B. SLOBODIN, and M. R. WATERS. 2010. New Dates from Ushki-1, Kamchatka, Confirm 13,000-cal-BP Age for Earliest Paleolithic Occupation. *Journal of Archaeological Science* 37, 2640–2649.

GREBENNIKOV, A. V., V. K. POPOV, M. D. GLASCOCK, R. J. SPEAKMAN, Y. V. KUZMIN, and A. V. PTASHINSKY. 2010. Obsidian Provenance Studies on Kamchatka Peninsula (Far Eastern Russia): 2003–9 Results. In *Crossing the Straits: Prehistoric Obsidian Source Exploitation in the North Pacific Rim* (B.A.R. International Series 2152), edited by Y. V. Kuzmin and M. D. Glascock, 89–120. Oxford, BAR Publishing.

HABU, J. 2010. Seafaring and the Development of Cultural Complexity in Northeast Asia: Evidence from the Japanese Archipelago. In *The Global Origins and Development of Seafaring*, edited by A. Anderson, J. H. Barrett and K. V. Boyle, 159–170. Cambridge, McDonald Institute for Archaeological Research.

HALL, M. E., and H. KIMURA. 2002. Quantitative EDXRF Studies of Obsidian Sources in Northern Hokkaido. *Journal of Archaeological Science* 29, 259–266.

HONG, M.-Y. 2012. Obsidian Appeared in Paleolithic Industries on the Korean Peninsula. In *International Symposium "Lithic Raw Material Exploitation and Circulation in Prehistory: A Comparative Perspective in Diverse Palaeoenviornment"*, edited by A. Ono, 17–18. Tokyo, Centre for Obsidian and Lithic Studies, Meiji University.

HONG, M.-Y., and J.-H. KIM. 2008. *Hopyeong-dong Paleolithic Site (Namyangju, Gyeonggi Province, Korea). Volumes I–II.* Seoul, Gijeon Institute of Cultural Properties (in Korean with English Title).

IKEYA, N., and M. D. GLASCOCK. 2013. Maritime Transportation of Obsidian Across the Pacific during the Early Upper Palaeolithic Japan. In *5-th Arheoinvest Symposium 'Stories Written in Stone'. International Symposium on Chert and Other Knappable Material. Iaşi, Romania, 20–24 August 2013. Programme and Abstracts*, edited by O. N. Crandell and V. Cotiugă, 89–90. Iaşi, Editura Universităţii "Alexandru Ioan Cuza" din Iaşi.

IKEYA, N., K. WATANABE, and M. SUZUKI. 2005. Kyusekki Jidai no Kozushima Kokuyouseki to Kaijyoukoutsuu [Kozu-shima Island Obsidian and a Maritime Voyage of the Palaeolithic Period]. *Kokogaku Janaru* 525, 12–14.

IZUHO, M., and W. HIROSE. 2010. A Review of Archaeological Obsidian Studies on Hokkaido Island (Japan). In *Crossing the Straits: Prehistoric Obsidian Source Exploitation in the North Pacific Rim* (B.A.R. International Series 2152), edited by Y. V. Kuzmin and M. D. Glascock, 9–25. Oxford, BAR Publishing.

IZUHO, M., and H. SATO. 2007. Archaeological Obsidian Studies in Hokkaido, Japan: Retrospect and Prospects. *Bulletin of the Indo-Pacific Prehistory Association* 27, 114–121.

JAMES, H. E. M. 1888. *The Long White Mountain, or Journey in Manchuria: with Some Account of the History, People, Administration and Religion of that Country.* London, Longmans, Green & Co.

JIA, P. W., T. DOELMAN, C. CHEN, H. ZHAO, S. LIN, R. TORRENCE, and M. D. GLASCOCK. 2010. Moving Sources: A Preliminary Study of Volcanic Glass Artifact Distributions in Northeast China Using pXRF. *Journal of Archaeological Science* 37, 1670–1677.

JIA, P. W., T. DOELMAN, R. TORRENCE, and M. D. GLASCOCK. 2013. New Pieces: The Acquisition and Distribution of Volcanic Glass Sources in Northeast

China during the Holocene. *Journal of Archaeological Science* 40, 971–982.

JIANG, L., and L. LUI. 2005. The Discovery of an 8000-Year-Old Dugout Canoe at Kuahuqiao in the Lower Yangzi River, China. *Antiquity* 79(305), Project Gallery (available at: http://www.antiquity.ac.uk/projgall/liu/index.html; accessed 13 October 2013).

Keiko-Ekkususen. 2009. *Keiko-Ekkususen Bunseki Sochi ni Yori Kokuyisekisei Ibitsu no Gensanchi Suitei. Kiso Data 1* [Geologic Source Identification of Obsidian Artefacts by X-ray Fluorescence Analysis: A Database. Book 1]. Tokyo, Meiji Daigaku Kobunkazai Kenkyu Sho.

KIM, J.-C., D. K. KIM, M. YOON, C. C. YUN, G. PARK, H. J. WOO, M.-Y. HONG, and G. K. LEE. 2007. PIXE Provenancing of Obsidian Artefacts from Paleolithic Sites in Korea. *Bulletin of the Indo-Pacific Prehistory Association* 27, 122–128.

KIMURA, H. 1995. Kokuyouseki, Hito, Gijyutsu [Obsidian, Humans, and Technology]. *Hokkaido Kokogaku* 31, 3–63.

KLUYEV, N. A. 2003. *Arkheologiya Yuga Dalnego Vostoka Rossii v XIX – XX Vekakh. Biobibliografichesky Ukazatel* [Archaeology of the Southern Russian Far East in the XIXth – XXth Centuries. Biobibliographic Directory]. Vladivostok, Dalnauka Publishers.

KLUYEV, N. A., and I. Y. SLEPTSOV. 2007. Late Pleistocene and Early Holocene Uses of Basaltic Glass in Primorye, Far East Russia: A New Perspective Based on Sites near the Sources. *Bulletin of the Indo-Pacific Prehistory Association* 27, 129–134.

KOMOTO, M., and H. OBATA (eds). 2005. *Zaisanovka 7 Site, in Primorsky, Russia. Preliminary Results of Excavation in 2004.* Kumamoto, Department of Archaeology, Kumamoto University.

KOROTKII, A. M. 1985. Quaternary Sea-Level Fluctuations on the Northwestern Shelf of the Japan Sea. *Journal of Coastal Research* 1, 293–298.

KRASHENINNIKOV, S. P. 1972. *Explorations of Kamchatka, North Pacific Scimitar; Report of a Journey Made to Explore Eastern Siberia in 1735–1741, by Order of the Russian Imperial Government.* Portland, Oregon Historical Society [originally published in Russian in 1755, reprinted in 1994].

KRENKE, N. A., E. V. LEONOVA, I. V. MELEKESTSEV, and M. M. PEVZNER. 2011. New Data on the Stratigraphy of the Ushki Sites in the Valley of the River Kamchatka. *Rossiiskaya Arkheologiya* 3, 14–24 (in Russian with English Abstract).

KUZMIN, Y. V. 1996. Palaeoecology of the Palaeolithic of the Russian Far East. In *American Beginnings: The Prehistory and Palaeoecology of Beringia*, edited by F. H. West, 136–146. Chicago & London, University of Chicago Press.

KUZMIN, Y. V. 2002. Chronology and Environment of the Palaeolithic and Neolithic Cultures of the Southern Russian Far East. *The Korean Journal of Quaternary Research* 16(2), 39–56.

KUZMIN, Y. V. 2003. The Paleolithic-to-Neolithic Transition and the Origin of Pottery Production in the Russian Far East: A Geoarchaeological Approach. *Archaeology, Ethnology & Anthropology of Eurasia* 3(15), 16–26.

KUZMIN, Y. V. 2006a. Recent Studies of Obsidian Exchange Networks in Prehistoric Northeast Asia. In *Archaeology in Northeast Asia: on the Pathway to Bering Strait*, edited by D. E. Dumond and R. L. Bland, 61–71. Eugene, University of Oregon Museum of Natural and Cultural History.

KUZMIN, Y. V., 2006b. Palaeoenvironment and Chronology. In *Archaeology of the Russian Far East: Essays in Stone Age Prehistory* (B.A.R. International Series 1540), edited by S. M. Nelson, A. P. Derevianko, Y. V. Kuzmin and R. L. Bland, 13–40. Oxford, BAR Publishing.

KUZMIN, Y. V. 2006c. Chronology of the Earliest Pottery in East Asia: Progress and Pitfalls. *Antiquity* 80, 362–371.

KUZMIN, Y. V. 2008. Geoarchaeology of Prehistoric Cultural Complexes in the Russian Far East: Recent Progress and Problems. *Bulletin of the Indo-Pacific Prehistory Association* 28, 3–10.

KUZMIN, Y. V. 2010. Crossing Mountains, Rivers, and Straits: A Review of the Current Evidence for Prehistoric Obsidian Exchange in Northeast Asia. In *Crossing the Straits: Prehistoric Obsidian Source Exploitation in the North Pacific Rim* (B.A.R. International Series 2152), edited by Y. V. Kuzmin and M. D. Glascock, 137–153. Oxford, BAR Publishing.

KUZMIN, Y. V. 2011. The Patterns of Obsidian Exploitation in the Late Upper Pleistocene of the Russian Far East and Neighbouring Northeast Asia. In *Natural Resource, Environment and Humans* (Proceedings of the Meiji University Centre for Obsidian and Lithic Studies No. 1), edited by A. Ono, 67–82. Tokyo, Centre for Obsidian and Lithic Studies, Meiji University.

KUZMIN, Y. V. 2012. Radiocarbon Chronology for Prehistoric Complexes of the Russian Far East: 15 Years Later. *Radiocarbon* 54, 727–736.

KUZMIN, Y. V., V. I. BOLDIN, and Y. G. NIKITIN. 2005. Chronology of Cultures of the Early Iron Age and the Mediaeval Period of Primorye. *Rossiya i ATR* 4(50), 44–55 (in Russian with English Abstract).

KUZMIN, Y. V., and A. V. CHERNUK. 2000. Paleogeografiya Drevnikh Poseleniy Doliny Reki Burei [Palaeogeography of Ancient Settlements in the Bureya River Basin]. In *Drevnosti Burei*, edited by E. I. Derevianko, 33–58. Novosibirsk, Izdatelstvo Instituta Arkheologii i Etnografii.

KUZMIN, Y. V., M. A. DIKOVA, and R. J. CRUZ. 2010. Radiocarbon Age of the Paleolithic Layers at Ushki 1 Site, Kamchatka (Northeastern Siberia): New Dates from the N. N. Dikov Excavations in 1989. *Current Research in the Pleistocene* 27, 44–45.

KUZMIN, Y. V., and M. D. GLASCOCK. 2007. Two Islands in the Ocean: Prehistoric Obsidian Exchange between Sakhalin and Hokkaido, Northeast Asia. *Journal of Island & Coastal Archaeology* 2, 99–120.

KUZMIN, Y. V., and M. D. GLASCOCK (eds). 2010. *Crossing the Straits: Prehistoric Obsidian Source Exploitation in the North Pacific Rim* (B.A.R. International Series 2152). Oxford, BAR Publishing.

KUZMIN, Y. V., M. D. GLASCOCK, and M. IZUHO. 2013. The Geochemistry of the Major Sources of Archaeological Obsidian on Hokkaido Island (Japan): Shirataki and Oketo. *Archaeometry* 55, 355–369.

KUZMIN, Y. V., M. D. GLASCOCK, and H. SATO. 2002b. Sources of Archaeological Obsidian on Sakhalin Island (Russian Far East). *Journal of Archaeological Science* 29, 741–749.

KUZMIN, Y. V., and S. P. NESTEROV. 2010. Khronologiya Neoliticheskikh Kultur Zapadnogo Priamurya [Chronology of the Neolithic Cultures of Western Amur River Basin]. In *Traditsionnaya Kultura Vostoka Azii. Vypusk 6*, edited by A. P. Zabiyako, 103–110. Blagoveshchensk, Izdatelstvo Amurskogo Gosudarstvennogo Universiteta.

KUZMIN, Y. V., and L. A. ORLOVA. 2000. The Neolithization of Siberia and the Russian Far East: Radiocarbon Evidence. *Antiquity* 74, 356–365.

KUZMIN, Y. V., and V. K. POPOV (eds). 2000. *Vulkanicheskie Stekla Dalnego Vostoka Rossii: Geologicheskie i Arkheologicheskie Aspekty* [Volcanic Glasses of the Russian Far East: Geological and Archaeological Aspects]. Vladivostok, Dalnevostochny Geologichesky Institut.

KUZMIN, Y. V., V. K. POPOV, M. D. GLASCOCK, and M. S. SHACKLEY. 2002a. Sources of Archaeological Volcanic Glass in the Primorye (Maritime) Province, Russian Far East. *Archaeometry* 44, 505–515.

KUZMIN, Y. V., and I. Y. SHEWKOMUD. 2003. The Palaeolithic–Neolithic Transition in the Russian Far East. *The Review of Archaeology* 24(2), 37–45.

KUZMIN, Y. V., R. J. SPEAKMAN, M. D. GLASCOCK, V. K. POPOV, A. V. GREBENNIKOV, M. A. DIKOVA, and A. V. PTASHINSKY. 2008. Obsidian Use at the Ushki Lake Complex, Kamchatka Peninsula (Northeastern Siberia): Implications for Terminal Pleistocene and Early Holocene Human Migrations in Beringia. *Journal of Archaeological Science* 35, 2179–2187.

KUZMIN, Y. V., A. A. VASILEVSKI, S. V. GORBUNOV, G. S. BURR, A. J. T. JULL, L. A. ORLOVA, and O. A. SHUBINA. 2004. Chronology of Prehistoric Cultural Complexes of Sakhalin Island (Russian Far East). *Radiocarbon* 46, 353–362.

KUZMIN, Y. V., O. V. YANSHINA, S. M. FITZPATRICK, and O. A. SHUBINA. 2012. The Neolithic of the Kurile Islands (Russian Far East): Current State and Future Prospects. *Journal of Island & Coastal Archaeology* 7, 234–254.

KUZNETSOV, A. M. 1996. Late Palaeolithic Sites of the Russian Maritime Province Primorye. In *American Beginnings: The Prehistory and Palaeoecology of Beringia*, edited by F. H. West, 267–282. Chicago & London, University of Chicago Press.

LEE, E., S. KIM, and S. NAM. 2008. Paleo-Tsushima Water and Its Effect on Surface Water Properties in the East Sea during the Last Glacial Maximum: Revisited. *Quaternary International* 176–177, 3–12.

LEE, E., and S. NAM. 2003. Freshwater Supply by Korean Rivers to the East Sea during the Last Glacial Maximum: a Review and New Evidence from the Korea Strait Region. *Geo-Marine Letters* 23, 1–6.

MALAINEY, M. E. 2011. *A Consumer's Guide to Archaeological Science: Analytical Techniques*. New York, Springer.

MATSUFUJI, K. 2003. A Cultural Exchange between Kyushu Island and the Korean Peninsula in the Upper Palaeolithic Age. In *Palaeolithic Men's Lives and their Sites*, edited by Y.-J. Lee, 641–659. Seoul, Hakyeon Press.

MATSUSHITA, H. 1998. Chosen Ikyo Hokudo Kainei Jyo Fukin no Sekki Nitsuite [Stone Tools near Kainei Castle in North Korea]. In *Toki no Kizuna Michi wo Tadoru*, edited by Ishizuki Kimio Sensei wo Shinobu, 439–464. Sapporo, Hon Kanko Iinkai Publishers.

NELSON, S. M. 1993. *The Archaeology of Korea.* Cambridge, Cambridge University Press.

NELSON, S. M. (ed.). 1995. *The Archaeology of Northeast China: Beyond the Great Wall.* London & New York, Routledge.

NELSON, S. M., A. P. DEREVIANKO, Y. V. KUZMIN, and R. L. BLAND (eds). 2006. *Archaeology of the Russian Far East: Essays in Stone Age Prehistory* (B.A.R. International Series 1540). Oxford, BAR Publishing.

NESTEROV, S. P., and Y. V. KUZMIN. 1999. Periodizatsia Kultur Rannego Zheleznogo Veka i Srednevekovya Priamurya [Periodisation of the Early Iron Age and Mediaeval Age of the Amur River Basin]. In *Traditsionnaya Kultura Vostoka Azii. Vypusk 2*, edited by D. P. Bolotin and A. P. Zabiyako, 140–152. Blagoveshchensk, Izdatelstvo Amurskogo Gosudarstvennogo Universiteta.

O'CONNOR, S., and R. ONO. 2013. The Case for Complex Fishing Technologies: A Response to Anderson. *Antiquity* 87, 885–888.

O'CONNOR, S., R. ONO, and C. CLARKSON. 2011. Pelagic Fishing at 43,000 Years before the Present and the Maritime Skills of Modern Humans. *Science* 334, 1117–1121.

OBATA, H. 2003. Kyoukutoni Okeru Kokuyouseki Shutsudoisekito Gensannchi Kenkyu [Study on the Prehistoric Obsidian Utilisation in Far East Asia – Review and Perspective]. *Sekki Gensanchi* 2, 67–88.

OBATA, H. (ed.). 2007. *Klerk 5 Site, in Primorsky, Russia. Preliminary Result of Excavation in 2005.* Kumamoto, Department of Archaeology, Kumamoto University.

OBATA, H. 2009. Obsidian Archaeology in Sakhalin and Siberia. *Sundai Shigaku* 135, 1–23 (in Japanese with English Abstract).

OBATA, H., I. MORIMOTO, and S. KAKUBUCHI. 2010. Obsidian Trade between Sources on Northwestern Kyushu Island and the Ryukyu Archipelago (Japan) during the Jomon Period. In *Crossing the Straits: Prehistoric Obsidian Source Exploitation in the North Pacific Rim* (B.A.R. International Series 2152), edited by Y. V. Kuzmin and M. D. Glascock, 57–71. Oxford, BAR Publishing.

ODA, S. 1990. A Review of Archaeological Research in the Izu and Ogasawara Islands. *Man and Culture in Oceania* 6, 53–79.

OKLADNIKOV, A. P. 1965. *The Soviet Far East in Antiquity: An Archaeological and Historical Study of the Maritime Region of the USSR.* Toronto, University of Toronto Press.

ONO, A. 1976. Koki Kyusekki Jidai no Shudan Kankei [The Group Relationship in the Late Palaeolithic Japan]. *Kokogaku Kenkyu* 23(1), 9–22.

ONO, A. 1984. East Asia. In *Neue Forschungen zur Altsteinzeit* (Forschungen zur Allgemeinen und Vergleichenden Archäologie, Band 4), edited by O. Bar-Yosef, G. Corvinus and J. Hahn, 405–423. München, C. H. Beck.

ONO, A., H. HARUNARI, and S. ODA (eds). 1992. *Atlas of Japanese Archaeology.* Tokyo, University of Tokyo Press (in Japanese with English Title).

PANTUKHINA, I. 2007. The Role of Raw Material in Microblade Technology at Three Late Palaeolithic Sites, Russian Far East. *Bulletin of the Indo-Pacific Prehistory Association* 27, 144–153.

PARK, G., J.-C. KIM, M. YOUN, C. YUN, J. KANG, Y.-M. SONG, S.-J. SONG, H.-J. NOH, D.-K. KIM, and H.-J. IM. 2010. Dating the Bibong-ri Neolithic Site in Korea: Excavating the Oldest Ancient Boat. *Nuclear Instruments and Methods in Physics Research B* 268, 1003–1007.

PETROV, V. P., and M. G. ZAMURUEVA. 1960. O Steklovatykh Sharovykh Lavakh r. Levaya Lefu na Dalnem Vostoke [On the Vitreous Spheroidal Lavas of Levaya Lefu River in the Far East]. *Izvestiya Akademii Nauk SSSR. Seriya Geologicheskaya* 11, 69–75.

PETRUN, V. F. 1956. K Voprosu o Vozraste Rechnykh Terras Yuzhnogo Primoria (Nakhodka Kamennykh Orudyi na Poberezhye Yaponskogo Morya) [To the Question of the Age of River Terraces in Southern Primorye (The Finding of Stone Tools on the Sea of Japan Coast)]. In *Materialy po Geologii i Poleznym Iskopaemym Vostochnoi Sibiri i Dalnego Vostoka*, edited by Y. F. Chemekov, 58–73. Moscow, Gosgeoltekhizdat Publishers.

PHILLIPS, S. C. 2010. Bridging the Gap between Two Obsidian Source Areas in Northeast Asia: LA–ICP–MS Analysis of Obsidian Artefacts from the Kurile Islands of the Russian Far East. In *Crossing the Straits: Prehistoric Obsidian Source Exploitation in the North Pacific Rim* (B.A.R. International Series 2152), edited by Y. V. Kuzmin and M. D. Glascock, 121–136. Oxford, BAR Publishing.

PHILLIPS, S. C. 2011. Networked Glass: Lithic Raw Material Consumption and Social Networks in the Kuril Islands, Far Eastern Russia. Unpublished PhD Dissertation. Department of Anthropology, University of Washington, Seattle, WA, USA.

PHILLIPS, S. C., and R. J. SPEAKMAN. 2009. Initial Source Evaluation of Archaeological Obsidian from the Kurile Islands of the Russian Far East Using Portable XRF. *Journal of Archaeological Science* 36, 1256–1263.

POLLARD, M., C. BATT, B. STERN, and S. M. M. YOUNG. 2007. *Analytical Chemistry in Archaeology.* Cambridge, Cambridge University Press.

POLYAKOV, I. S. 1883. *Puteshestvie na Ostrov Sakhalin v 1881–1882 GG.* [Journey to the Sakhalin Island in 1881–1882] (Supplement to *Izvestia Imperatorsogo Russkogo Geograficheskogo Obshchestva*, Vol. XIX). St.-Petersburg, Tipografiya A. S. Suvorina.

POPOV, V. K., N. A. KLYUEV, I. Y. SLEPTSOV, T. DOELMAN, R. TORRENCE, N. A. KONONENKO, and P. WHITE. 2010. Gialoslastity Shkotovskogo Bazaltovogo Plato (Primorye) – Vazhneishyi Istochnik Arkheologicheskogo Obsidiana na Yuge Dalnego Vostoka Rossii [The Hyaloclastites of the Basaltic Shkotovo Plateau (Primorye) – the Most Important Source of Archaeological Obsidian in the Southern Russian Far East]. In *Priotkryvaya Zavesu Tysyacheletyi*, edited by N. A. Kluyev and Y. E. Vostretsov, 295–314. Vladivostok, Reya Publishers.

POPOV, V. K., V. G. SAKHNO, Y. V. KUZMIN, M. D. GLASCOCK, and B.-K. CHOI. 2005. Geochemistry of Volcanic Glasses of the Paektusan Volcano. *Doklady Earth Sciences* 403, 254–259.

POPOV, V. K., G. P. SANDIMIROVA, and T. A. VELIVETSKAYA. 2008. Strontium, Neodymium, and Oxygen Isotopic Variations in the Alkali Basalt–Trachyte–Pantellerite–Comendite Series of Paektusan Volcano. *Doklady Earth Sciences* 419, 329–334.

POPOV, V. K., V. A. SOLYANIK, and D. G. FEDOSEEV. 2009. Decorative Volcanic Glasses from Hyaloclastites of the Shkotovo Basaltic Plateau (Primorye, Russia). *The Journal of the Gemmological Association of Hong Kong* 30, 51–56.

REIMER, P. J., M. G. L. BAILLIE, E. BARD, A. BAYLISS, J. W. BECK, P. G. BLACKWELL, C. BRONK RAMSEY, C. E. BUCK, G. S. BURR, R. L. EDWARDS, M. FRIEDRICH, P. M. GROOTES, T. P. GUILDERSON, I. HAJDAS, T. J. HEATON, A. G. HOGG, K. A. HUGHEN, K. F. KAISER, B. KROMER, F. G. MCCORMAC, S. W. MANNING, R. W. REIMER, D. A. RICHARDS, J. R. SOUTHON, S. TALAMO, C. S. M. TURNEY, J. VAN DER PLICHT, and C. E. WEYHENMEYER. 2009. IntCal09 and Marine09 Radiocarbon Age Calibration Curves, 0–50,000 Years Cal BP. *Radiocarbon* 51, 1111–1150.

RENFREW, C. 1969. Trade and Culture Process in European Prehistory. *Current Anthropology* 10, 151–169.

RENFREW, C., J. R. CANN, and J. E. DIXON. 1965. Obsidian in the Aegean. *The Annual of the British School at Athens* 60, 225–247.

RENFREW, C., and J. DIXON. 1976. Obsidian in Western Asia: a Review. In *Problems in Economic and Social Archaeology*, edited by G. de G. Sieveking, I. H. Longworth and K. E. Wilson, 137–150. London, Duckworth.

RENFREW, C., J. E. DIXON, and J. R. CANN. 1966. Obsidian and Early Cultural Contact in the Near East. *Proceedings of the Prehistoric Society* 32, 30–72.

RENFREW, C., J. E. DIXON, and J. R. CANN. 1968. Further Analysis of Near Eastern Obsidian. *Proceedings of the Prehistoric Society* 34, 319–331.

SAKHNO, V. G. 2007. Chronology of Eruptions, Composition, and Magmatic Evolution of the Paektusan Volcano: Evidence from K–Ar, $^{87}Sr/^{86}Sr$, and $\delta^{18}O$ Isotope Data. *Doklady Earth Sciences* 412, 22–28.

SATO, H. 2011. Did the Japanese Obsidian Reach the Continental Russian Far East in the Upper Paleolithic? In *Emlékkönyv Violának: Papers in Honour of Viola T. Dobosi*, edited by K. T. Biró and M. András, 206–223. Budapest, Hungarian National Museum.

SATO, H., Y. V. KUZMIN, and M. D. GLASCOCK. 2002. Source Analysis of Obsidian in Prehistoric Sakhalin and the Assessment of Its Distribution in Northeast Asia. *Hokkaido Kokogaku* 38, 1–13 (in Japanese with English Title).

SCHNELL, I. 1932. Prehistoric Finds from the Island World of the Far East. *Bulletin of the Museum of the Far Eastern Antiquities* 4, 15–104.

SEONG, C. 2007. Late Pleistocene Microlithic Assemblages in Korea. In *Origin and Spread of Microblade Technology in Northern Asia and North America*, edited by Y. V. Kuzmin, S. G. Keates and C. Shen, 103–114. Burnaby, B.C., Archaeology Press.

SEONG, C. 2011. Evaluating Radiocarbon Dates and Late Paleolithic Chronology in Korea. *Arctic Anthropology* 48(1), 93–112.

SHACKLEY, M. S. 2005. *Obsidian: Geology and Archaeology in the North American Southwest.* Tucson, University of Arizona Press.

SHACKLEY, M. S., M. D. GLASCOCK, Y. V. KUZMIN, and A. V. TABAREV. 1996. Geochemical Characterization of Archaeological Obsidian from the Russian Far East: a Pilot Study. *International Association for Obsidian Studies Bulletin* 17, 16–19.

SHEWKOMUD, I. Y., and Y. V. KUZMIN. 2009. Khronologiya Kamennogo Veka Nizhnego Priamurya (Dalny Vostok Rossii) [The Chronology of the Stone Age of the Lower Amur River Basin (Russian Far East)]. In *Kulturnaya Khronologiya i Drugie Problemy v Issledovaniyakh Drevnostei Vostoka Azii*, edited by I. Y. Shewkomud, 7–46. Khabarovsk, Khabarovsky Kraevoi Muzei.

SHIMADA, K. 2009. Pioneer Phase of Obsidian Use and the Human Culture Origin in the Japanese Islands. *Sundai Shigaku* 135, 51–70 (in Japanese with English Abstract).

SOHN, P. 1993. *Sokchang-ni Prehistoric Site*. Seoul, Yonsei University Press (in Korean with English Abstract).

STEPHAN, J. J. 1994. *The Russian Far East: A History*. Stanford, CA, Stanford University Press.

SUZUKI, M. 1969. Fission Track Dating and Uranium Contents of Obsidian (I). *Daiyonki Kenkyu* 8, 123–130 (in Japanese with English Abstract).

SUZUKI, M. 1970a. Fission Track Dating and Uranium Contents of Obsidian (II). *Daiyonki Kenkyu* 9, 1–6 (in Japanese with English Abstract).

SUZUKI, M. 1970b. Fission Track Ages and Uranium Content of Obsidians. *Zinruigaku Zassi* 78(1), 50–58.

SUZUKI, M. 1973a. Chronology of Prehistoric Human Activity in Kanto, Japan. Part 1: Framework for Reconstructing Prehistoric Human Activity in Obsidian. *Journal of the Faculty of Science, University of Tokyo. Section 5 (Anthropology)* 4(3), 241–318.

SUZUKI, M. 1973b. Chronology of Prehistoric Human Activity in Kanto, Japan. Part 2: Time-Space Analysis of Obsidian Transportation. *Journal of the Faculty of Science, University of Tokyo. Section 5 (Anthropology)* 4(4), 396–469.

SUZUKI, M., and K. TOMURA. 1983. Basic Data for Identifying the Source of Archaeological Obsidian by Activation Analysis and Discriminant Analysis. *St. Paul's Review of Science* 4(4), 99–110.

TABAREV, A. V. 2004. Obsidian in the Neolithic of the Maritime Region. *Archaeology, Ethnology & Anthropology of Eurasia* 1(17), 2–6.

TOMODA, T., N. A. KONONENKO, A. YOSHITANI, and I. U. SLEPTSOV. 2004. An Investigation of the Origin of Paleolithic Obsidian Implements Excavated in Central Primorye, Far East Russia. *Kokogaku Kenkyu* 51(3), 93–102 (in Japanese with English Abstract).

TSUTSUMI, T. 2010. Prehistoric Procurement of Obsidian from Sources on Honshu Island (Japan). In *Crossing the Straits: Prehistoric Obsidian Source Exploitation in the North Pacific Rim* (B.A.R. International Series 2152), edited by Y. V. Kuzmin and M. D. Glascock, 27–55. Oxford, BAR Publishing.

TSUTSUMI, T. 2012. MIS 3 Edge–Ground Axes and the Arrival of the First *Homo sapiens* in the Japanese Archipelago. *Quaternary International* 248, 70–78.

VASILEVSKI [VASILEVSKY], A. A., and V. A. GRISHCHENKO. 2011. The Definition of Raw–Material Centers during the Late Paleolithic, Neolithic, and Paleometal Ages of Sakhalin Island, Eastern Russia. *Current Research in the Pleistocene* 28, 11–14.

VASILEVSKY, A. A. 2006. The Upper Palaeolithic of Sakhalin Island. In *Archaeology of the Russian Far East: Essays in Stone Age Prehistory* (B.A.R. International Series 1540), edited by S. M. Nelson, A. P. Derevianko, Y. V. Kuzmin and R. L. Bland, 75–100. Oxford, BAR Publishing.

VASILEVSKY, A. A. 2008. *Kamenny Vek Ostrova Sakhalin* [The Stone Age of Sakhalin Island]. Yuzhno-Sakhalinsk, Sakhalinskoe Knizhnoe Izdatelstvo.

VASILEVSKY, A. A., V. A. GRISCHENKO, Y. V. KUZMIN, and L. A. ORLOVA. 2009. Khronologiya i Periodizatsiya Epokhi Neolita na Sakhaline i Kurilskikh Ostrovakh (po Dannym Radiouglerodnogo Datirovaniya) [Chronology and Periodisation of the Neolithic of Sakhalin Island and Kurile Islands (Based on Data from Radiocarbon Dating)]. In *Kulturnaya Khronologiya i Drugie Problemy v Issledovaniyakh Drevnostei Vostoka Azii*, edited by I. Y. Shewkomud, 74–82. Khabarovsk, Khabarovsky Kraevoi Muzei.

VASILEVSKY, A. A., V. A. GRISCHENKO, and L. A. ORLOVA. 2010. Periods, Boundaries, and Contact Zones in the Far Eastern Insular World of the Neolithic (Based on the Radiocarbon Chronology of Sites on the Sakhalin and Kuril Islands). *Archaeology, Ethnology & Anthropology of Eurasia* 38/1, 10–25.

VASILEVSKY, A. A., Y. V. KUZMIN, S. V. GORBUNOV, and O. A. SHUBINA. 2003. Chronology of the 1-st Millennium B.C. – 2-nd Millennium A.D. Archaeological Sites at the Sakhalin Island and Kuril Islands. In *Amuro-Sakhalinsky Region v Epokhu Paleometalla i Srednevekovya. 1*, edited by V. A. Deryugin, 37–46. Khabarovsk, Khabarovsky Kraevoi Muzei (in Russian with English Abstract).

VENIAMINOV, I. (INNOKENTII). 1984. *Notes on the Islands of the Unalashka District*. Kingston, ONT, Limestone Press [originally published in Russian in 1840, reprinted in 2011].

WARASHINA, T. 2004. Kyu Shirataki, Shimo Shirataki Iseki Shutsudo no Kokuyousekisei-Sekki no Genzai Sanchi Bunseki, Suiwasou Sokutei [Source Identification Analysis and Hydration Dating for Obsidian Lithic Artefacts Recovered from Kyu Shirataki and Shimo Shirataki Sites]. In *Shirataki Isekigun V*, edited by Hokkaido Maizo Bunkazai Centre, 188–205. Sapporo, Hokkaido Maizo Bunkazai Centre.

WARASHINA, T., T. HIGASHIMURA, H. SATO, and Z. S. LAPSHINA. 1998. Sekki Genzai no Sanchi Bunseki [Analysis of Lithic Artefact Sources]. In *Nihon Bunkazai Kagaku-kai dai 15 kai Taikai Kenkyu Happyo Yoshi*, edited by Nihon Bunkazai Kagaku-kai, 138–139. Nara, Nihon Bunkazai Kagaku-kai.

WATANABE, K., and M. SUZUKI. 2009. Analytical Data of Geologic Obsidians from Japan by INAA Using Middle- and Long-Lived Nuclides. *Journal of Radioanalytical and Nuclear Chemistry* 279, 459–473.

YANSHINA, O. V., and Y. V. KUZMIN. 2010. The Earliest Evidence of Human Settlement in the Kurile Islands (Russian Far East): the Yankito Site Cluster, Iturup Island. *Journal of Island & Coastal Archaeology* 5, 179–184.

Yonsei University Museum (ed.). 2001. *Old Stone Age Relics of Korea.* Seoul, Yonsei University Press (in Korean with English Title).

YOSHITANI, A., N. A. KONONENKO, T. TOMODA, V. K. POPOV, and I. U. SLEPTSOV. 2003. On the Sources of the Obsidian Flakes from Some Late Palaeolithic Sites in the Southern Part of Central Primorye, Far East Russia. *Kokogaku to Shizen Kagaku* 47, 1–12.

Chapter 11
THE PAEKTUSAN VOLCANO SOURCE AND GEOCHEMICAL ANALYSIS OF ARCHAEOLOGICAL OBSIDIANS IN KOREA

Jong-Chan KIM

Abstract: *The current status of provenance studies of the Paektusan Volcano obsidian in terms of both source material and artefacts is presented in this chapter. Sources of Type 2 (PNK2) and Type 3 (PNK3) Paektusan obsidian are well-established, including the analysis of geological samples collected during the joint Korean–Russian fieldtrip in 2007. However, the primary source of the Type 1 (PNK1) obsidian has not yet been located, and this remains an outstanding problem although there is circumstantial evidence that the PNK1 also derives from the Paektusan Volcano. Statistical analyses of archaeological obsidians from Korea were conducted, using geochemical data from Neutron Activation Analysis and Inductively Couple Plasma – Mass Spectrometry. From these analyses, it is quite possible to draw the conclusion that in general most obsidian from the Palaeolithic sites in Korea originated from the Paektusan Volcano source, while obsidian from Neolithic sites in the coastal area of the Korean Peninsula came from Japanese sources.*

Keywords: *Obsidian, Provenance, Paektusan Volcano, Korean Peninsula, Palaeolithic, Neolithic, Consistency Check*

INTRODUCTION

The study of prehistoric obsidian sources in Korea is rapidly developing (e.g., Cho *et al.* 2010). In the absence of common knowledge that a single volcanic area can have a multitude of eruptions, and thereby multiple obsidian sources with different geochemical compositions, the provenancing of Korean obsidian artefacts has been confusing until now. On Hokkaido Island of Japan, for example, 15 source areas and 21 composition groups of volcanic glass have been identified (see Ferguson *et al.*, this volume; see also Izuho and Hirose 2010). In Central Mexico, at least 20 obsidian sources have been identified (Glascock *et al.* 1998). This shows that the initial efforts to search for a single source of Korean archaeological obsidian (e.g., Lee 1998; Cho 2005) were not so fruitful.

According to Ri (1996), more than 380 volcanoes and craters, which belong mainly to the Cenozoic era, were identified in and around Mount Paektu [Baekdu] (a.k.a. Paektusan in Korea, and the Baitoushan and Changbaishan in China) on the modern Chinese – North Korean border (Figure 11.1). Therefore, it is reasonable to expect that the Paektusan Volcano will also have several obsidian sources. Popov *et al.* (2005) describe three different geochemical groups of Paektusan obsidian, namely PNK1, PNK2, and PNK3. This work has been a turning point in terms of deciphering the provenance puzzle for Korean archaeological obsidian.

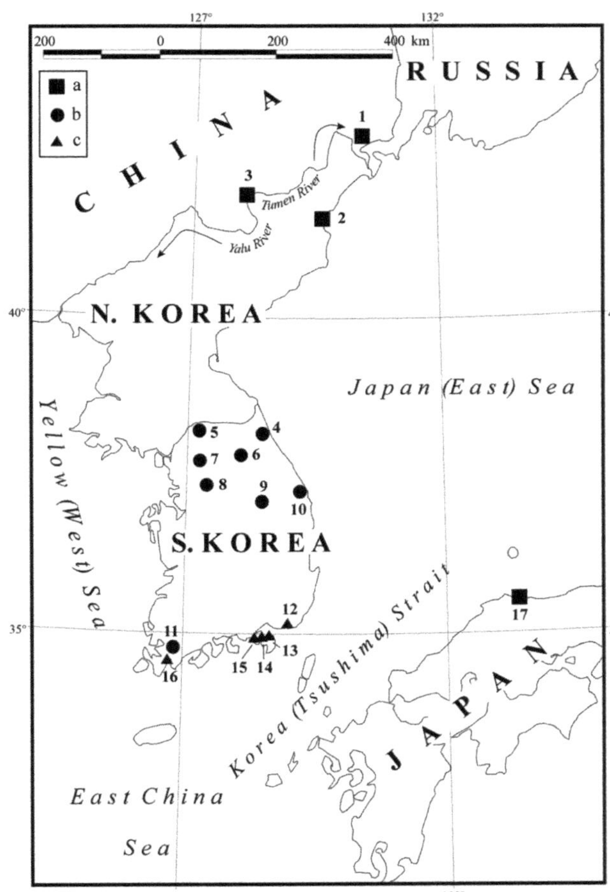

Figure 11.1. The position of archaeological sites and other places of interest: a (squares) – geographical localities and the Early Iron Age site; b (circles) – Palaeolithic sites; and c (triangles) – Neolithic sites. 1 – Hunchun; 2 – Chongjin; 3 – Paektusan Volcano (obsidian source); 4 – Sangmuyongri; 5 – Janghungri; 6 – Hahwageri; 7 – Hopyung; 8 – Samri; 9 – Suyanggae; 10 – Kigok; 11 – Shinbuk; 12 – Dongsamdong; 13 – Yokjido; 14 – Yondaedo; 15 – Sangnodaedo; 16 – Yeosu; and 17 – Sakaiminato (Tottori Pref., Japan)

Previously, there was a general belief in Korea that since there are no obsidian-producing explosive volcanoes other than Paektusan, and that obsidian found at numerous Korean archaeological sites is either from Paektusan or the Japanese Archipelago (e.g., Nishitani 1982, Sohn 1993). This approach was challenged when the actual results of chemical analyses became available, and comparisons were made with what was believed to be Paektusan obsidian. In retrospect, the obsidian of

Paektusan origin, acquired for analysis from occasional individual fieldtrips to the Chinese side of the Paektusan Volcano, was always identified as the PNK3 type *sensu* Popov *et al.* (2005); its quality is not suitable for tool-making due to numerous perlitic inclusions. Also, this type of obsidian was regarded as the only kind originating from the Paektusan Volcano (e.g., Lee 1998).

This chapter reviews the current data from the Paektusan source. It also takes a critical look at the paper by Cho and Choi (2010) on obsidian from the Kigok [Gigok] site, which reflects the present situation of Palaeolithic obsidian studies in Korea based on the firm belief of a single Paektusan obsidian source. Also, the status of provenance studies for the Neolithic obsidian in Korea is presented. Finally, measurements of obsidian source samples from Little Glass Buttes (Oregon, USA) and Sierra de Pachuca (Hidalgo, Mexico) by means of Laser Ablation Inductively Couple Plasma – Mass Spectrometry (hereafter LA–ICP–MS) are given for comparison with the previous results (Glascock 1999). This is in accord with the main topic of the 2011 Obsidian Workshop, namely the standardisation of obsidian analyses.

RESULTS OF RECENT FIELDWORK AT THE PAEKTUSAN OBSIDIAN SOURCE

To confirm the PNK geochemical groups of obsidian first reported by Popov *et al.* (2005), a joint Korean–Russian fieldtrip to the Paektusan Volcano was conducted on 14–17 August 2007, with the participation of this paper's author and Drs V. K. Popov and Y. V. Kuzmin. About 30 geological samples were collected from several localities, mostly from slope debris (Figure 11.2). The description of some samples provided by V. K. Popov is given in the Appendix (Table 1).

The Proton-Induced X-ray Emission (hereafter – PIXE) analysis was performed at the external PIXE beam line in the Accelerator Mass Spectrometry (AMS) machine of the Seoul National University (Seoul, Republic of Korea). The PIXE results on element concentrations, such as Rb, Sr, Zr, Mn, Fe, and Zn, were subjected to cluster analysis; three to four cluster groups were generated. However, the subsequent Inductively Couple Plasma – Mass Spectrometry (henceforth – ICP–MS) measurements reduced it to two groups, namely PNK2 and PNK3 (see Appendix, Table 1). Below is an explanation of this analysis.

The 1–2 MeV protons used for the PIXE analysis penetrate only tens of micron in depth to the sample exposed to the beam. Therefore, the PIXE method can measure only the chemical composition of the surface of a sample's material. Though this causes no problem when obsidian with a freshly-broken shiny surface is examined, considerable variation in trace element concentration will occur when the weathered surface of a sample is irradiated (Summerhayes *et. al.* 1998). For the ICP–MS analysis, the raw samples were ground using a mortar, and then selected obsidian-like glassy grains were

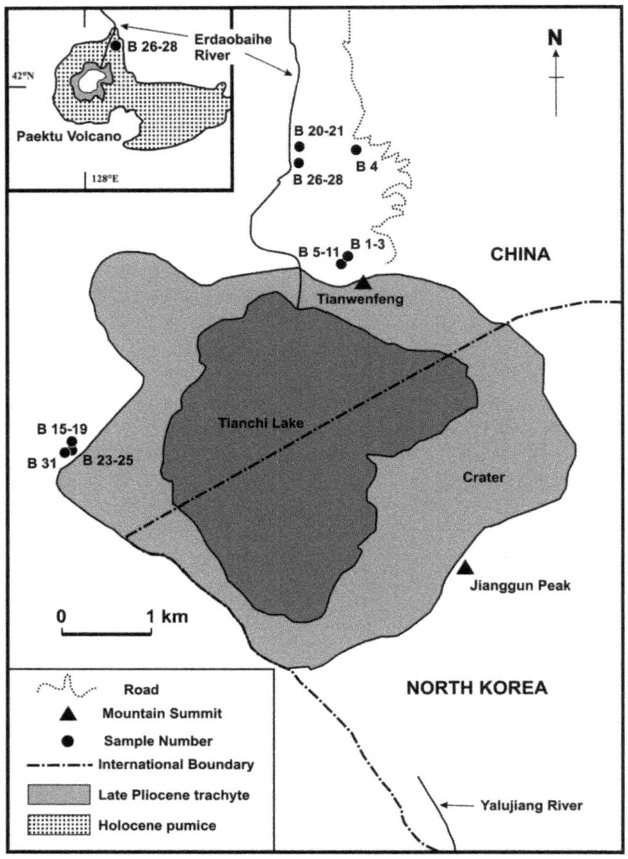

Figure 11.2. Locations sampled during the 2007 field trip to Paektusan Volcano (see Appendix) (after Jia et al. *2013, modified)*

segregated by sieving out the soft powdery material which is presumably perlite.

The PIXE method detects X-rays, and therefore is as sensitive to surface morphology and matrix effect as other related measurement techniques such as X-ray Fluorescence (henceforth – XRF), and its results are affected by absorption factors (Summerhayes *et. al.* 1998, Davis *et al.* 1998). Furthermore, some precaution is required when PIXE data are compared with Neutron Activation Analysis (hereafter – NAA) or ICP–MS results. Persistent discrepancies between the data obtained by NAA and PIXE methods have been observed (Doelman *et al.* 2004). Considering that NAA and ICP–MS results agree with each other, one has to apply this factor also when comparing PIXE and ICP–MS data. It was found that this discrepancy factor is largely caused by the matrix effect.

In PIXE analysis, the standard sample used is the NBS-278 obsidian available commercially from the National Institute of Standards and Technology (NIST, USA), which is a powder, while unknown samples are solid flakes of volcanic glass. This difference in matrices between the standard and the obsidian samples may be the cause of discrepancies in the measured concentration of elements. Table 11.1 shows the results of PIXE analysis of the Paektusan obsidians, using two different

Table 11.1. The matrix dependence of Fe concentration (in %) in PIXE measurement

Matrix Sample	Powder (NIST Standard Sample)	Solid Flake (Little Glass Buttes)
PNK1 obsidian	1.39–1.52	1.10–1.24
PNK2 obsidian	3.4–4.4	3.15–3.19

standard samples, flake-shaped obsidian from the Little Glass Buttes source and NBS-278. The result shows that the above mentioned discrepancies almost disappear when the flaked obsidian is used as a standard.

Summarising briefly the results of the 2007 Paektusan fieldtrip, one can say that the sources of PNK2 and PNK3 type obsidians are confirmed, but the attempt to verify the location of the PNK1 group source was unsuccessful. Results of the ICP–MS analysis for PNK2 and PNK3 'geological' samples are shown in Table 11.2, and they are compared with the previous results obtained by Popov *et al.* (2005). Although the source of PNK1 group has not yet been identified and this remains an outstanding problem for obsidian provenance studies in Korea and neighbouring regions, the following circumstantial evidence strongly suggests that the PNK1 obsidian originates from the Paektusan Volcano.

First of all, the PNK1 obsidians are frequently identified at archaeological sites near the Paektusan source, such as in the Primorye Province of Russia (Kuzmin *et al.* 2002), Jilin Province of China (Jia *et al.* 2010, 2013), and the central part of the Korean Peninsula (Kim *et al.* 2007). Further, an Early Iron Age site at Hunchun (Jilin Province, China; see Figure 11.1), which is very near Paektusan, contains the PNK1 type obsidian (Kim *et al.* 2007).

Secondly, Figure 11.3 shows rare-earth elements (hereafter – REE) concentrations of three PNK obsidian groups normalised to the chondritic abundances given by Thompson *et al.* (1983). The solid lines are from Dunlap (1996) and correspond to those of comendite and trachyte from the "Millennium Eruption" of Paektusan (see Discussion). As shown in Figure 11.3, the REE concentrations of PNK1 are identical to those of PNK3 which is definitely of Paektusan origin.

Thirdly, recently so-called Chongjin obsidian boulders have become available for analysis, with the results showing that they are definitely of PNK1 type (Kim *et al.* 2007). This obsidian was handed over to a Korean archaeologist from a Japanese colleague together with a message which reads: "A Japanese trade company named Materisk Ltd. at Sakaiminato, Tottori-ken [see Figure 11.1], found these obsidian boulders in their storehouse where they keep construction materials imported from North Korea. These materials were collected from the river bed in the Chongjin area, North Korea. Since these obsidian boulders were of no use as construction material, they gave the boulders away to local archaeologists who were practicing making stone tools." (M.-Y. Hong, personal communication 2010). The city of Chongjin, which is about 100 km southeast of the Paektusan source, lies near the tributary of the Tumen River [Tumangan] which originates close to the Paektusan Volcano, and this tributary river may be a secondary source of the Paektusan obsidian.

Table 11.2. The ICP–MS results (in ppm) for geological samples collected during the 2007 Paektusan fieldtrip (labelled as "present study") compared with the NAA results of Popov *et al.* (2005)

Groups Element	PNK1	PNK2	PNK2	PNK3	PNK3
	Popov et al. *(2005)*	*Popov* et al. *(2005)*	*Present Study*	*Popov* et al. *(2005)*	*Present Study*
Sc	1.25	1.55	0.98	5.45	4.62
La	65.81	145.45	157.8	75.69	80.29
Ce	135.47	273.45	313.06	148.38	154.82
Nd	62.28	111.87	126.7	62.55	64.31
Sm	10.41	21.05	26	11.5	12.09
Eu	0.37	0.32	0.28	0.64	0.53
Tb	1.54	2.85	4.09	1.47	1.6
Dy	10.38	16.55	22.43	7.62	8.01
Yb	4.01	8.97	9.67	3.49	3.42
Lu	0.7	1.28	1.36	0.52	0.52
Rb	226	300	351.97	127	132.5
Sr	28	—	3.8	—	16
Cs	3.71	2.85	5.3	1.37	1.42
Ba	143	7.00	9.1	79	65.6
Zr	250	1467	1961	484	600.2
Hf	9.98	40.11	42.3	14.23	13.8

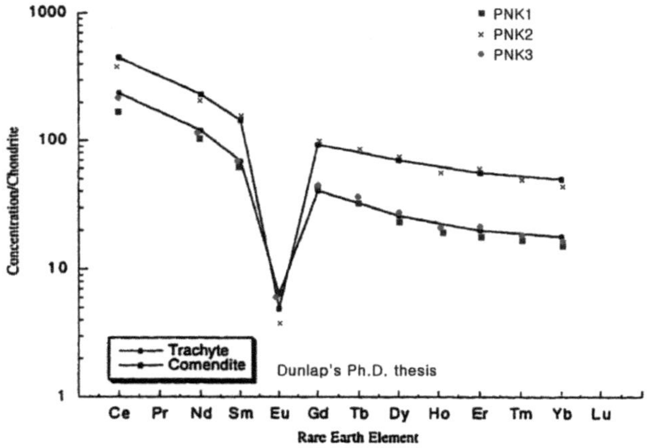

Figure 11.3. The REE concentrations of PNK obsidians normalised to the chondritic abundances (Thompson et al. *1983); the solid lines are from Dunlap (1996) and correspond to those of comendite and trachyte from the Millennium Eruption of Paektusan*

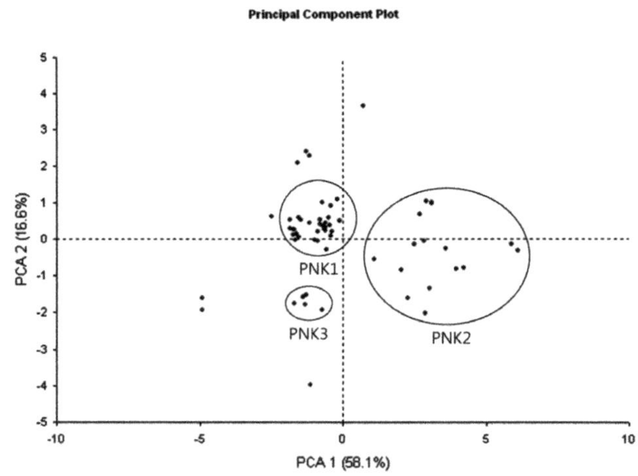

Figure 11.4. The result of PCA analysis using data given by Cho and Choi (2010) for the obsidians from four Palaeolithic sites (including the Kigok site)

THE PAEKTUSAN OBSIDIAN AT ARCHAEOLOGICAL SITES IN KOREA

The Palaeolithic Sites

The status of provenance studies for obsidian artefacts from Palaeolithic sites in Korea has recently been summarised (Kim *et al.* 2007). At six major obsidian-containing Palaeolithic sites, 75 artefacts were analysed. Forty-seven of them were identified as PNK1 type; 17 as PNK2 type; four of Japanese origin; and the remaining seven were from unknown sources.

Since that time, one more major Palaeolithic site called Kigok, was studied. It is located in the coastal area of the East [Japan] Sea (coordinates 37°35′41″ N; 129°04′53″ E; see Figure 11.1). This site was excavated in 2001–3 by the Gangwon Research Institute of Cultural Properties (2005), and produced 7890 artefacts; the ^{14}C dates range from ca. 10,200 BP to older than 48,000 BP. Recently, Cho and Choi (2010) published a paper on the geochemical analysis of the Kigok obsidians. Ten trace elements (Sm, La, Ce, Sc, Cs, Hf, Lu, Rb, Tb, and Sb) were analysed by the ICP–MS method. A Principal Component Analysis (hereafter – PCA), including data from the Sangmuyongri, Suyanggae, and Janghungri Palaeolithic sites (in addition to the Kigok site), shows that there are five to six major groups, and among them only one group is assigned to the Paektusan source (Cho and Choi 2010, Figure 5). However, our re-analysis of the same data using the *StatistiXL* software (available commercially from Microsoft Co.) shows that nearly all Palaeolithic obsidian artefacts fall within the three groups of Paektusan origin as shown on Figure 11.4.

One of the reasons for the difference in the interpretation of the same data is the significantly high concentration of Rb in Kigok obsidians, while other element concentrations are identical to those of the PNK1 and PNK2 groups. This could be caused by several factors: instrumental error; error caused by the standard sample; and/or weathering of the sample's surface. In fact, it has been observed that perlitisation by weathering can significantly increase the concentration of Rb (Takahashi *et al.* 2003). These factors should be investigated first, before assigning an obsidian sample to a new source group. The result of cluster analysis (dendrogram) using StatistiXL is shown in Figure 11.5. It confirms the conclusion that Kigok obsidian also belongs to the PNK groups.

In the application of StatistiXL for the cluster analysis, we have chosen Euclidean Distance as a Similarity/Distance measure, and Nearest Neighbour as the Cluster Method. In the cluster analysis, some obsidian samples from Kyushu Island (Japan) are also included, to see how well two major obsidian sources from the Korean Peninsula — the Paektusan and Japanese localities — differ from each other. The Japanese samples used are 'geological' obsidian materials from six sources on Kyushu Island. They were measured using the Seoul National University ICP–MS Facility, and the results are presented in Table 11.3.

The Neolithic Sites

Previously, only a small number of obsidian artefacts excavated from the Neolithic sites in Korea were analysed for provenance purposes. In 1982, trace elements of obsidian excavated at the Osanri [Osanni] site, which is located on the east coast of Korea, were analysed using the XRF, and assigned to the Paektusan source (Higashimura 1984). Later, obsidian artefacts from the Dongsamdong [Tongsamdong] and Beombang [Bonbang] sites in southern Korea were analysed by a Japanese–Korean team (Takahashi *et al.* 2003), and the following conclusion was drawn. Of 22 samples from the Beombang site, 14 were assigned to the Koshidake source, six to the Himeshima [Himejima]/Ushinodake source, and one to the Hariojima source (all located on Kyushu Island); and one of unknown origin. Of 12

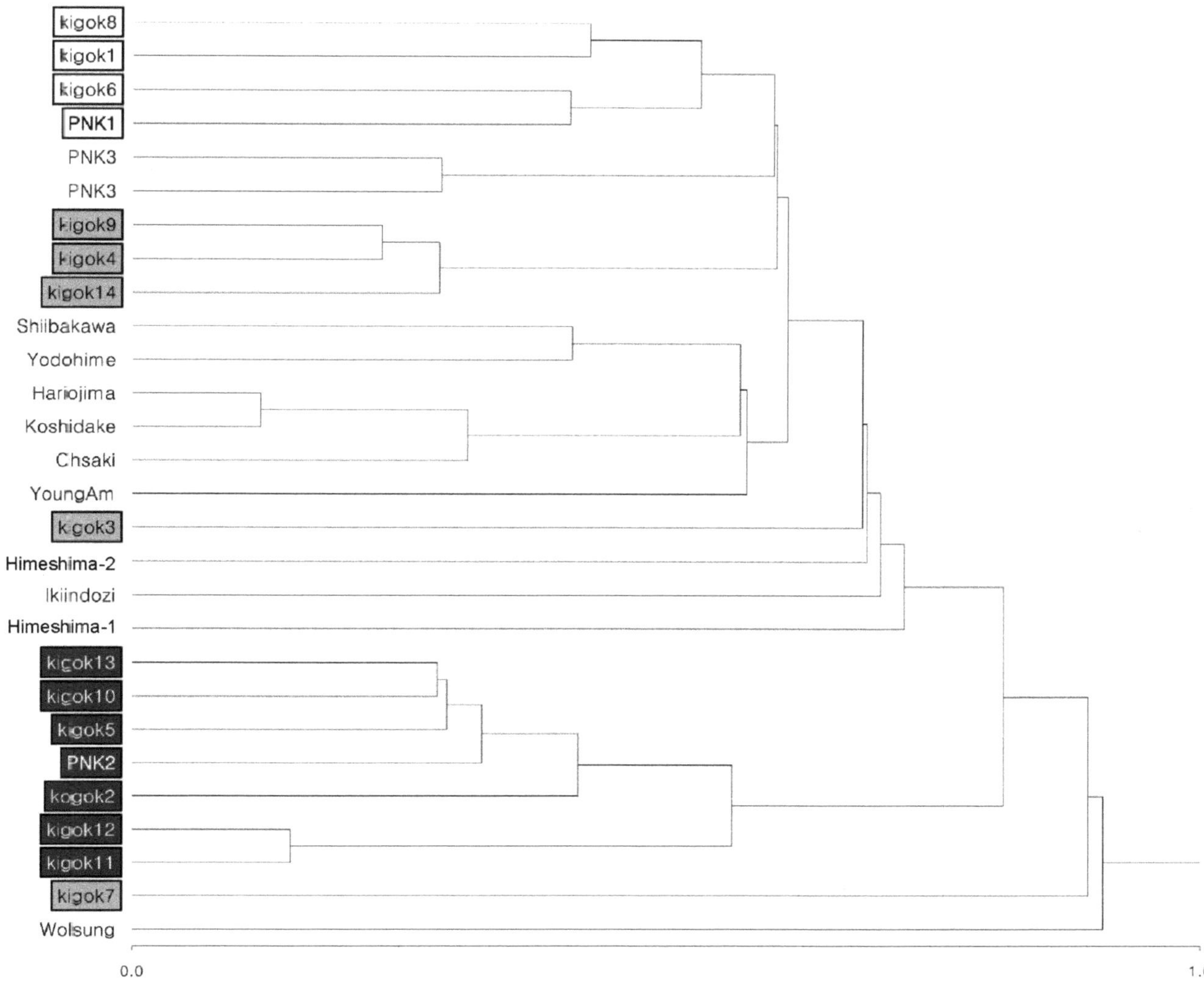

Figure 11.5. Dendrogram of obsidians from the Kigok site obtained by cluster analysis using StatisXL. Rectangles: Kigok8, Kigok1, and Kigok6 (PNK1 group); dark shaded rectangles: Kigok13, Kigok10, Kigok5, Kigok2, Kigok12, and Kigok11 (PNK2 group); and light shaded rectangles: Kigok 9, Kigok4, Kigok14, Kigok3, and Kigok7 (outliers)

obsidian artefacts from the Donsamdong site, nine were assigned to the Koshidake source, one to the Hariojima source, and two to an unknown source (see Takahashi *et al.* 2003).

More recently, Cho *et al.* (2006) reported NAA results on the content of ten trace elements (Sm, La, Ce, Sc, Cs, Hf, Lu, Rb, Tb, and Sb) for 28 obsidian samples from five Neolithic sites in the southern coastal region of Korea, together with some Palaeolithic obsidian artefacts. A cluster analysis of this dataset was conducted, taking into account that only part of the results belong to the Neolithic sites (see Figure 11.1). For this examination, data for the Kyushu sources was added from Kim *et al.* (2007) – the Hariojima, Koshidake, Yodohime, Ohsaki, Shibakawa, Ikiindozi, Himeshima-1, and Himeshima-2 sources (see Table 11.3); the PNK data were also included. The dendrogram resulting from the cluster analysis is presented in Figure 11.6. It shows the current status of provenance research for Neolithic obsidian in Korea quite well, despite the fact that two different analyses — ICP–MS and NAA — were employed.

Figure 11.6 shows that the Paektusan obsidian is clustered in the top and bottom portions of the dendrogram, indicating that it is distinctly different from obsidian geochemical groups of Japanese origin identified at the Korean Neolithic sites. In the central portion of the dendrogram, the Korean Neolithic specimens are intermingled with obsidians from sources in the Kyushu region of Japan. This shows the complexity of provenancing, indicating that more accurate and intensive geochemical analyses are required. Very limited use of Paektusan obsidian in the Neolithic is evident from the PNK1 group sample at the Yokjido site (Figure 11.6).

Cho *et al.* (2006) concluded that there are two obsidian sources identified for the Neolithic sites. One group is composed of part of the Dongsamdong site and of the Yeosu site which is of Koshidake origin; the other group is from the Yondaedo, Yokjido, and Sangnodaedo sites (see Figure 11.1), and part of the Dongsamdong site which is less related to the Koshidake source. This conclusion is well represented in the dendrogram (Figure 11.6).

Table 11.3. The results of ICP–MS analysis (in ppm) of the obsidian source samples from Kyushu Island, Japan

Groups (Sources)	Dy	Yb	Lu	Zr	Hf	Ta	Rb	Sr	Cs	Ba
Koshidake	3.70	2.09	0.30	74.88	3.92	14.96	184.3	42.94	12.45	229.94
Hariojima	3.64	2.03	0.29	53.83	3.27	13.79	167.7	37.92	12.79	218.61
Ohsaki	3.84	2.4	0.36	56.45	4.49	8.82	155.66	27.01	9.22	388.14
Yodohime	2.47	1.5	0.23	155.5	4.59	11.46	136.77	103.37	6.59	580.93
Himeshima-1	1.61	0.59	0.08	27.6	1.70	9.35	85.69	60.04	2.77	867.35
Himeshima-2	2.35	1.08	0.16	72.77	3.44	6.4	55.98	594.79	1.85	759.15
Shibakawa	1.91	1.00	0.15	56.68	3.26	12.14	122.82	179.28	10.77	638.53
Ikiindozi	9.94	5.74	0.87	230.76	14.84	27.85	196.20	4.13	7.05	14.95
PNK3	8.01	3.42	0.52	600.21	13.8	9.46	132.57	16.9	1.42	65.59
PNK3	6.77	2.79	0.42	506.48	11.12	6.91	111.39	29.83	1.17	117.87
PNK2	22.43	9.67	1.36	1961.2	47.69	25.19	351.97	3.82	5.34	9.16
PNK1	10.38	4.01	0.7	260	9.98	6.75	226	28	3.71	143

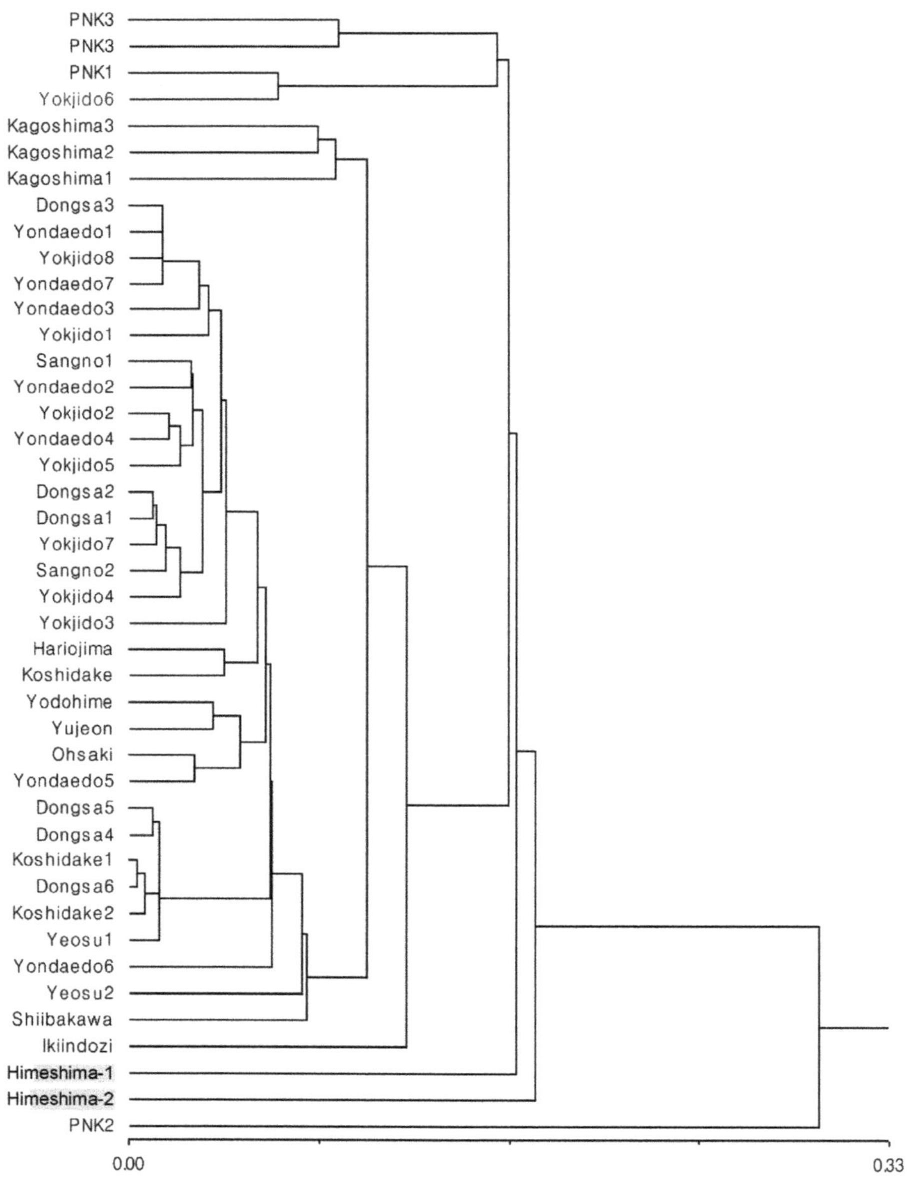

Figure 11.6. Dendrogram of obsidians from Neolithic sites in the southern part of Korea based on data in Cho et al. *(2006) obtained by cluster analysis*

Table 11.4. Results of the present LA–ICP–MS analysis of obsidian from the Little Glass Buttes and Sierra de Pachuca by the Korean Basic Science (KBS) Facility (in ppm, unless otherwise indicated), compared with the previous ICP–MS results at the Orleans Laboratory (Glascock 1999; Gratuze 1999)

Element	Little Glass Buttes		Sierra de Pachuca	
	KBS (n = 1)	*Orleans (n = 3)*	*KBS (n = 1)*	*Orleans (n = 3)*
Li	49	33	73	63
Ti	471	595	1025	1190
V	21	1.4	4.2	4.4
Mn	411	269	1139	1048
Fe (%)	21.78	0.68	32.69	1.66
Co	0.6	–	–	–
Zn	–	26	–	219
Ga	112	–	29.2	–
Rb	129	97	204	203
Sr	59	52	2.00	1.86
Y	22	18	102	111
Zr	94	83	892	1058
Nb	12	9.00	89.9	116
Sb	0.2	0.2	0.3	0.23
Cs	4.7	3.2	3.9	3.88
Ba	1215	1080	9.7	9.3
La	25	21	38	38
Ce	55	43	96	91
Pr	4.7	15.4	10.7	10.5
Nd	18.5	15.4	35	38
Sm	3.4	2.87	10.1	10.1
Eu	2.7	0.56	1.6	1.6
Gd	4.3	3.51	11.3	11.6
Tb	0.87	0.42	2.2	2.3
Dy	3.14	3.00	15.8	16.4
Ho	0.74	0.56	3.5	3.7
Er	2.57	2.00	11.4	12
Tm	0.36	0.28	1.8	1.8
Yb	2.6	2.4	13.3	13.2
Lu	0.42	0.38	2.00	2.00
Hf	3.94	3.00	25.8	29
Ta	1.14	1.09	5.7	6.5
Pb	–	–	31.6	35
Th	9.2	8.1	19.6	21
U	5.00	3.9	7.00	7.5

THE LA–ICP–MS TEST MEASUREMENT OF TWO REFERENCE SAMPLES

Archaeological sites in Korea have produced relatively small amounts of obsidian artefacts, and good preservation is generally required after geochemical analysis. Therefore, non-destructive analyses are preferred to destructive ones. One of the non-destructive techniques is PIXE, but it has a limitation: only a small number of elements can be measured with the necessary accuracy. A new LA–ICP–MS Facility at the Korea Basic Science Institute (Daejeon [Taejon], Republic of Korea), is a very promising laboratory for non-destructive obsidian studies of Korean prehistoric artefacts. The reproducibility of this LA–ICP–MS system for the trace element analyses of the USGS GSE standard is claimed to be better than 10% at the 2-sigma level, and the repeatability is better than 3% of the relative standard deviation at the 2-sigma level (Kil *et al.* 2011). As for its non-destructive feature, it is claimed that the depth of penetration for laser shots is less than ca. 400 μm, and cross-sections of the laser holes are less than 50 μm in diameter.

To examine the performance of this facility, obsidian samples from two sources, Little Glass Buttes and Sierra de Pachuca, were measured. They became available from the Archaeometry Group, Missouri University Research Reactor Center (see Glascock 1999), with a data sheet showing the previous LA–ICP–MS measurements in the Orleans Laboratory (Glascock 1999; Gratuze 1999). The results, except for Fe, are in good agreement with the previous data as shown in Table 11.4. No visible damage to the obsidian sample was found after the measurement.

DISCUSSION

According to Smith (1979), even a single volcanic eruption larger than 1 km^3 in volume is almost always compositionally zoned. A recent study of the composition of pyroclastic deposits resulting from the tenth century AD explosion of the Paektusan Volcano (the so-called "Millennium Eruption"; see the latest age determinations: Yatsuzuka *et al.* 2010; Yin *et al.* 2012), which ejected more than 78 km^3 of tephra (Machida *et al.* 1990; Horn and Schmincke 2000), showed that there were several lava flows, such as comendite, trachyte, latite, and trachybasalt (Dunlap 1996). More than 80% by volume was composed of comendite (a peralkaline rhyolite); the next rock in terms of volume was trachyte. The trace element composition of these two kinds of lava is nearly identical to the PNK2 and PNK3 groups. In particular, REE are in perfect agreement with the PNK obsidians as shown in Figure 11.3. The Janghungri Palaeolithic site (Choi *et al.* 2000), where PNK2 obsidian artefacts were excavated (see Popov *et al.* 2005), is ^{14}C-dated to ca. 24,000 BP. Therefore, we can infer that the source of the Paektusan magma from Millennium Eruption was the same as for the explosion which took place at least 24,000 years ago. The trace element composition of the PNK1 obsidian (if the PNK1 is also associated with the Paektusan magma) is a bit elusive. When the magma evolves to the higher concentration of SiO_2 (e.g., from trachytic to comenditic magma), the concentration of trace elements such as REE and incompatible elements like Zr increases (Horn and Schmincke 2000). However, in case of the PNK1 group the increase in SiO_2 was met with much lower concentrations of these trace elements. Therefore, it is certain that the magma source of the PNK1 obsidian developed quite separately from the magma responsible for the PNK2 and PNK3 obsidians. There is an opinion that the Chongjin obsidian may indicate the existence of a yet unknown source of high quality volcanic glass in North Korea other than Paektusan. The ca. 100 km distance between Chongjin and Paektusan seems too big for the former to be a secondary obsidian quarry of the Paektusan Volcano. However, the Mule Creek obsidian quarry mentioned by Glascock *et al.* (1998) was about 80 km from the source volcano, which is a similar distance.

So far, among the obsidian artefacts studied at the Palaeolithic sites in Korea there are numerous pieces for which sources are unknown. Sometimes it is hard to distinguish obsidian from unusual rocks at archaeological sites, as shown by Cho and Choi (2010) where seven out of 25 submitted "obsidians" turned out to be non-obsidian rocks. Therefore, it would be worthwhile to perform a petrological investigation to check whether the outliers are obsidian or not, before proceeding to provenance studies. On the other hand, in the region where there are not so many obsidian sources, like the Korean Peninsula, the chances are higher for finding 'exotic' obsidian than on a terrain with multiple sources. Exotic obsidian will provide a very valuable clue for the study of ancient exchange networks.

CONCLUSIONS

As a result of the 2007 Paektusan fieldtrip the source problem of the PNK2 obsidian, which is one of the most frequently excavated varieties at Palaeolithic sites in Korea, has successfully been settled. Although the primary source of the PNK1 obsidian has not yet been located, in view of the accidental finding of Chongjin obsidian boulders with the PNK1 geochemical signature it would be worthwhile to conduct fieldwork in the future at distant places from the Paektusan Volcano, and to go further downstream on the Tumen or Yalu [Yalujiang] rivers (Figure 11.1). The conclusion derived from the present multivariate statistical analysis based on the ICP–MS data is in agreement with that of the previous bivariate statistical study which was based on a small number of elements from PIXE data, in which it was stated that about 85% of Palaeolithic obsidians in Korea originated from the Paektusan Volcano.

The test LA–ICP–MS measurements of two reference obsidian samples in the new Korean laboratory has shown satisfactory results, so this method can now be safely applied to non-destructive trace element analysis of obsidian artefacts.

Multi-element and non-destructive analysis would be very valuable for provenancing the Neolithic obsidians from the Korean Peninsula, because it could deal with the multiple volcanic sources in the Japanese Archipelago. Also, obsidian artefacts from the Palaeolithic site of Shinbuk [Sinbuk], which according to a previous study originated from the Kyushu region of Japan separated from the utilisation site by the wide Korea [Tsushima] Strait (Figure 11.1; see Kim *et al.* 2007), raise the challenging problem of palaeo-seafaring.

Acknowledgements

I am grateful to Dr M. D. Glascock for providing obsidian samples from Little Glass Buttes and Sierra de Pachuca, and to Dr M.-Y. Hong (Seoul, Korea) for obsidian from the Chongjin locality. Thanks to Dr V. K. Popov for the description of the 2007 fieldtrip samples, and to Dr Y. V. Kuzmin for help in organising this trip. My sincere thanks go to Prof. A. Ono for inviting me to participate in the 2011 Nagano Workshop "Methodological Issues of Obsidian Provenance Studies and the Standardisation of Geologic Obsidian", and for financial support. Finally, I am grateful to Drs Y. V. Kuzmin, M. D. Glascock, and S. G. Keates, and to Ms C. C. Lindsey for help with correcting the grammar of the original manuscript.

References

CHO, N.-C. 2005. Classification of Obsidian Artifacts Found in Korean Peninsula Based on the Chemical Composition, Texture and Magnetic Property. Unpublished PhD Dissertation. Kangwon National University, Chuncheon, Korea (in Korean with English Summary).

CHO, N.-C., and S. Y. CHOI. 2010. Miryang-Seongbun ul Riyonghan TongHae Kigok Guseokgi Yujeok Chulto Hukyoseok SanJi Yeonku [Provenance Study of Obsidian from Kigok Palaeolithic site in Donghae-si Using Trace Elements]. *Hanguk Sanggosa Hakbo* 70, 5–20.

CHO, N.-C., H. T. KANG, and K. Y. CHUNG. 2006. Miryang-Seongbun mit Strontium Tongwi wonso-bi rul Riyonghan Han-Bando Hukyoseok-je SeogKi ui SanJi Choojeong [Provenance Study of Obsidian Artefacts Found in the Korean Peninsula Based on Trace Elements and Strontium Isotope Ratios]. *Hanguk Sanggosa Hakbo* 53, 5–21.

CHO, N.-C., J.-C. KIM, and H.-T. KANG. 2010. Provenance Study of Obsidian Artefacts Excavated from Palaeolithic Sites on the Korean Peninsula. In *Crossing the Straits: Prehistoric Obsidian Source Exploitation in the North Pacific Rim* (B.A.R. International Series 2152), edited by Y. V. Kuzmin and M. D. Glascock, 73–87. Oxford, BAR Publishing.

CHOI, B.-K., S. Y. CHOI, and C. Y. CHOI. 2000. Chulwon [A Study of Jangheungri Upper Palaeolithic Site at Chulwon]. *Hanguk Guseokgi Hakbo* 3, 1–23.

DAVIS, M. K., T. L. JACKSON, M. S. SHACKLEY, T. TEAGUE, and J. H. HAMPEL. 1998. Factors Affecting the Energy Dispersive X-ray Fluorescence (EDXRF) Analysis of Archaeological Obsidian. In *Archaeological Obsidian Studies: Method and Theory*, edited by M. S. Shackley, 159–179. New York & London, Plenum Press.

DOELMAN, T., N. KONONENKO, V. POPOV, G. SUMMERHAYES, R. TORRENCE, A. BONETTI, A. GUGLIELMETTI, A. MANZONI, and M. ODDONE. 2004. Acquisition and Movement of Volcanic Glass in the Primorye Region of Far Eastern Russia. *Rossiya i ATR* 4(46), 112–125.

DUNLAP, C. E. 1996. Physical, Chemical, and Temporal Relations among Products of the 11th Century Eruption of Baitoushan, China/North Korea. Unpublished PhD Dissertation. University of California–Santa Cruz, Santa Cruz, CA, USA.

Gangwon Research Institute of Cultural Properties and Korea Highway Corporation. 2005. *Donghae Kigok Yujeok* [The Donghae Kigok Palaeolithic Site] (Research Report of Gangwon Research Institute of Cultural Properties Vol. 30). Chuncheon, Gangwon Research Institute of Cultural Properties.

Geologiya Korei [The Geology of Korea]. 1993. Pyongyang, Foreign Languages Books Publishing House.

GLASCOCK, M. D. 1999. An Inter-Laboratory Comparison of Element Compositions of Two Obsidian Sources. *International Association for Obsidian Studies Bulletin* 23, 13–15.

GLASCOCK, M. D., G. E. BRASWELL, and R. H. COBEAN. 1998. A Systematic Approach to Obsidian Source Characterization. In *Archaeological Obsidian Studies: Method and Theory*, edited by M. S. Shackley, 15–65. New York & London, Plenum Press.

GRATUZE, B. 1999. Description of the Procedures for LA–ICP–MS Analysis at Orleans, France. *International Association for Obsidian Studies Bulletin* 23, 18.

HIGASHIMURA, T. 1984. Osanri Yuzeok Chulto Hukyoseok ui Hyong-gwang Bunseok [X-ray Fluorescence Analysis of Obsidians Excavated at Osanni Site]. In *Osanri Yuzeok*, edited by H.-J. Im and H.-S. Kwon, 69–73. Seoul, Seoul National University Museum Press.

HORN, S., and H.-U. SCHMINCKE. 2000. Volatile Emission during the Eruption of Baitoushan Volcano (China/North Korea) ca. 969 AD. *Bulletin of Volcanology* 61, 537–555.

IZUHO, M., and W. HIROSE. 2010. A Review of Archaeological Obsidian Studies on Hokkaido Island (Japan). In *Crossing the Straits: Prehistoric Obsidian Source Exploitation in the North Pacific Rim* (B.A.R. International Series 2152), edited by Y. V. Kuzmin and M. D. Glascock, 9–25. Oxford, BAR Publishing.

JIA, P. W., T. DOELMAN, C. CHEN, H. ZHAO, S. LIN, R. TORRENCE, and M. D. GLASCOCK. 2010. Moving Sources: A Preliminary Study of Volcanic Glass Artifact Distributions in Northeast China Using pXRF. *Journal of Archaeological Science* 37, 1670–1677.

JIA, P. W., T. DOELMAN, R. TORRENCE, and M. D. GLASCOCK. 2013. New Pieces: The Acquisition and Distribution of Volcanic Glass Sources in Northeast China during the Holocene. *Journal of Archaeological Science* 40, 971–982.

KIL, Y., H.-S. SHIN, H.-Y. OH, J.-S. KIM, M.-S. CHOI, H.-J. SHIN, and S.-C. PARK. 2011. In-Situ Trace Element Analysis of Clinopyroxene on Thin Section by Using LA–ICP–MS. *Geosciences Journal* 15, 177–183.

KIM, J.-C., D. K. KIM, M. YOON, C. C. YUN, G. PARK, H. J. WOO, M.-Y. HONG, and G. K. LEE. 2007. PIXE Provenancing of Obsidian Artefacts from Paleolithic Sites in Korea. *Bulletin of the Indo-Pacific Prehistory Association* 27, 122–128.

KUZMIN, Y. V., V. K. POPOV, M. D. GLASCOCK, and M. S. SHACKLEY. 2002. Sources of Archaeological Volcanic Glass in the Primorye (Maritime) Province, Russian Far East. *Archaeometry* 44, 505–515.

LEE, D. Y. 1998. JiJil Hak Jeok Jaryo ui Bunseok [Analysis of Geological Samples]. In *Kogohak Yeonku Bangbeopron*, edited by M. R. Choi, S. R. Choi and S. J. Shin, 257–299. Seoul, Seoul National University Press.

MACHIDA, H., H. MORIWAKI, and D.-C. ZHAO. 1990. The Recent Major Eruption of Changbai Volcano and Its Environmental Effects. *Geographical Reports of Tokyo Metropolitan University* 25, 1–20.

NISHITANI, M. 1982. Chosen Hanto no Kokuyoseki Nitsuite [On Obsidians in Korean Peninsula]. In *Kagawa Mitsuo Sensei Kanreki Kinen Ronbun Shu*, edited by the Committee for the Proceedings of the 60th Anniversary of Prof. Mitsuo Kagawa, 129–138. Beppu, Japan, Committee for the Proceedings of the 60th Anniversary of Prof. Mitsuo Kagawa.

PARK, Y.-H., K.-S. LEE, H.-Y. LEE, I. SON, and J.-R. LEE (eds). 2000. *Atlas of Korea*. Seoul, Sungji Munhwa Publishers.

POPOV, V. K., V. G. SAKHNO, Y. V. KUZMIN, M. D. GLASCOCK, and B.-K. CHOI. 2005. Geochemistry of Volcanic Glasses of the Paektusan Volcano. *Doklady Earth Sciences* 403, 254–259.

RI, D. 1996. Paektu Volcano. In *Geology of Korea*, edited by R. J. Paek, K. H. Gap and G. P. Jon, 325–344. Pyongyang, Foreign Book Publishing House.

SAKHNO, V. G. 2007. Chronology of Eruptions, Composition, and Magmatic Evolution of the Paektusan Volcano: Evidence from K-Ar, $^{87}Sr/^{86}Sr$, and $\delta^{18}O$ Isotope Data. *Doklady Earth Sciences* 412, 22–28.

SAKHNO, V. G. 2008. *Noveishyi i Sovremenny Vulkanizm Yuga Dalnego Vostoka (Pozdnepleistotsen – Golotsenovy Etap)* [The Latest and Recent Volcanism in the Southern Far East (Late Pleistocene – Holocene Stage)]. Vladivostok, Dalnauka Press.

SMITH, R. L. 1979. Ash-Flow Magmatism. In *Ash-Flow Tuffs* (Geological Society of America Special Paper 180), edited by C. E. Chapin and W. E. Elston, 5–27. Boulder, CO, Geological Society of America.

SOHN, P. 1993. *Sokchang-ni Prehistoric Site*. Seoul, Yonsei University Press (in Korean with English Abstract).

SUMMERHAYES, G. R., J. R. BIRD, R. FULLAGAR, C. GOSDEN, J. SPECHT, and R. TORRENCE. 1998. Application of PIXE–PIGME to Archaeological Analysis of Changing Patterns of Obsidian Use in West New Britain, Papua New Guinea. In *Archaeological Obsidian Studies: Method and Theory*, edited by M. S. Shackley, 129–158. New York & London, Plenum Press.

TAKAHASHI, Y., I. HA, and H. OBATA. 2003. Hyong-goang X-son Bunsoku e Uihan Tongsam-dong/Bonbang Yujok Churto Hugyosok Sanji Chujong [Sources Identification of Obsidian Tools from Tongsamdong and Bonbang Sites by X-ray Fluorescence Analysis]. *Hanguk Shinsokki Yongu* 6, 83–99.

THOMPSON, R. N., M. A. MORRISON, A. P. DICKIN, and G. L. HENDRY. 1983. Continental Flood Basalts ... Arachnids Rule OR? In *Continental Basalts and Mantle Xenoliths*, edited by C. J. Hawkesworth and M. J. Norry, 158–185. Cambridge, MA, Shiva Publishing.

WEI, H., Y. WANG, J. JIN, L. GAO, S.-H. YUN, and B. JIN. 2007. Timescale and Evolution of the Intracontinental Tianchi Volcanic Shield and Ignimbrite–Forming Eruption, Changbaishan, Northeast China. *Lithos* 96, 315–324.

YATSUZUKA, S., M. OKUNO, T. NAKAMURA, K. KIMURA, Y. SETOMA, T. MIYAMOTO, K. H. KIM, H. MORIWAKI, T. NAGASE, X. JIN, B. L. JIN, T. TAKAHASHI, and H. TANIGUCHI. 2010. ^{14}C Wiggle-Matching of the B–TM Tephra, Baitoushan Volcano, China/North Korea. *Radiocarbon* 52, 933–940.

YIN, J., A. J. T. JULL, G. S. BURR, and Y. ZHENG. 2012. A Wiggle-Match Age for the Millennium Eruption of Tianchi Volcano at Changbaishan, Northeastern China. *Quaternary Science Reviews* 47, 150–159.

Appendix
DESCRIPTION OF SAMPLES COLLECTED DURING THE 2007 JOINT KOREAN–RUSSIAN FIELDTRIP TO THE PAEKTUSAN VOLCANO (COMPILED BY DR V. K. POPOV)

The general geological situation with sources of volcanic glass on the Paektusan Volcano can be summarised as follows:

1. The volcanic glasses of trachytes, trachydacites, pantellerites, and rhyolithes are related to the caldera and post-caldera explosive eruptions of the Paektusan Volcano, which occurred from 200,000 years ago until before AD 964. In some sources (see Ri 1996), the older ages are presented; for example, the age of the Puksollen Suite is 2.2 Ma (K–Ar method).
2. The rock names for samples studied and their suggested geological age are shown in Table 1. Most of volcanic glass is in the form of small fragments (glassy debris) in the explosion deposits. The pumices which created these deposits are 'juvenile' (from the magma reservoir), and they reflect the composition of the magma. These are pumices of the Float Stone Forest and Gorge (Canyon) of the Mount Paektu. Glass in these pumices could be either juvenile or consists of xenoliths (taken during the eruption process). Thus, the composition of the B14 pumice sample from the Float Stone Forest is identical to the glass of B12-1 sample (they both belong to the PNK1 group); but the glass sample B12-2 belongs to the PNK3 group. In general, they are all trachytic rocks.

Short description of samples taken from different parts of the Paektusan Volcano follows here: 1) the caldera rim on the northern and western sides of the volcano; 2) stream deposits near the Erdaobaihe River headwater and the Changbaek Waterfall; and 3) the canyon of the Float Stone Forest (see Figure 11.2).

Northern Part of the Caldera's Rim

Pumice of the PNK2 group (samples B1 and B2) belongs to the pantellerite. It possibly occurs in the pumice deposits created during the Millennium Eruption in AD 969 (Horn and Schmincke 2000) or AD 1024 (Wei *et al.* 2007). The geological age of the glass in this group is most probably ca. 0.065–0.095 Ma (Sakhno 2007). It is important that artefacts belonging to the PNK2 group are found at Palaeolithic sites in South Korea (Kim *et al.* 2007), and the geological age of the PNK2 group is therefore much older than tenth century AD.

Glassy debris and bomb fragment of the PNK3 group (samples B5-1, B5-2, B7, and B8) are glasses (obsidians) of trachytic composition, and they constitute debris and volcanic bombs in the eruption deposits of the Paektusan Volcano dated to ca. 40,000 years ago (Sakhno 2007).

Tephra and glassy melt (?) of the PNK1 group (samples B4 and B10) are also of trachytic composition, and are possibly related to the AD 1688 eruption (Wei *et al.* 2007) if the tephra sample B4 was taken from the layer near the Black Wind Mouth [Heifengkou] outcrop.

Glassy debris of the PNK4 group (sample B11) belongs to trachyte (trachydacite), and its age is 0.04–0.2 Ma (Wei *et al.* 2007).

Changbaek Waterfall and the Erdaobaihe River Headwaters

Glassy rock (intrusive?) of the PNK2 group (sample B26-1), lava (B26-2), and ignimbrite (B20 and B21). Their age is perhaps 0.065–0.095 Ma (Sakhno 2007).

Float Stone Forest

Glassy debris and pumice of the PNK1 group (samples B12-1 and B-13) are close enough to the trachytes of the Black Wind Mouth locale, but most probably belong to the older generation (dated to 0.04–0.2 Ma ago). Glassy debris of the PNK3 group (sample B12-2) is also trachyte, and its geological age is probably 0.04–0.2 Ma (Wei *et al.* 2007).

Western Part of the Caldera Rim

Glassy debris, bomb fragment, tephra and rock (intrusive) of the PNK4 group (samples B15, B17, B24, and B31): these rocks could have been formed during the trachytic eruption in AD 1668. Glassy debris and pumice (samples B16 and B23): according to Wei *et al.* (2007), there are deposits of the Millennium Eruption on the western side of the volcano rim. The light comendite pumices with debris of black glassy trachytes are typical for these deposits.

General Comments

The pumices of the Float Stone Forest and Gorge of Paektu Mount were formed at ca. 45,000–65,000 years ago according to Sakhno (2008). In other papers (Ri 1996; Wei *et al.* 2007), the younger age of these deposits is indicated. The glassy debris or at least part of it are xenoliths, and they are older (Table 1). Thus, trachytic obsidians and obsidian rhyoliths ('obsidianites'), and pumice-like perlites of the Ch'onji Suite are distributed on the southeastern slope of the Paektusan Volcano, east of the Jianggun Peak (peaks of Sanmunchjige, Hebal, and Tangel or Tanger) [these terms are very approximate transliterations of the Russian text of *Geologiya Korei* (1993); there is nothing similar in the *Atlas of Korea* (Park *et al.* 2000, 103)]. The absolute age of these rocks is from 0.1 to 0.0087 Ma (K–Ar dating) (Ri 1996).

The ages of the volcanic glass as presented in Table 1 are preliminary. In order to make a final conclusion, isotope dating of representative samples from each group should be undertaken. I tried to correlate the absolute age of rocks and the Rb/Zr ratio in the acidic rocks of the Paektusan Volcano, using geochronological and geochemical studies by Sakhno (2007, 2008). The data obtained testify in favour of old ages of the trachytic glass (PNK1, PNK3, and PNK4 groups): 0.065–1 Ma. The age of the PNK2 obsidians is unclear. Rocks of such compositions (pantellerites and comendites) were dated to 0.140–0.545 Ma and AD 969–1903.

Table 1. Pyroclastic debris from the Paektusan Volcano collected during the 2007 fieldtrip (see Figure 11.2 for sampling locations)

Location*	Field No.	Facies type	Geochemical group by PIXE**	Chemical composition	Age
Northern part of the caldera rim	B1 B2	pumice pumice	PNK2	pantellerite	Probably AD 969 (Horn and Schmincke 2000) or AD 1024 (Wei *et al.* 2007), or fragments of an eruption dated to 0.065–0.095 Ma ago.
	B5-1 B5-2 B7 B8	glassy debris glassy debris bomb fragment glassy debris	PNK3	trachyte	Fragments are products of an eruption dated to 0.04 Ma ago.
	B4 B10	tephra glassy melt	PNK1	trachyte	Fragments are products of an eruption dated probably to AD 1688
	B11	glassy debris	PNK4	trachydacite	AD 1668? 0.04–0.095 Ma ago?
Changbaek Waterfall and the Erdaobaihe River headwater	B26-1 B26-2 B20 B21	rock (intrusive?) lava glassy rock (ignimbrite?) glassy rock (ignimbrite?)	PNK2	pantellerite	Probably fragments of an eruption dated to AD 969 (Horn and Schmincke 2000), or AD 1024 (Wei *et al.* 2007), or to 0.065–0.095 Ma ago.
Float Stone Forest	B12-1 B13	glassy debris pumice	PNK1	trachyte	AD 1668? Fragments of an eruption dated to 0.04–0.2 Ma ago?
	B12-2	glassy debris	PNK3	trachyte	AD 1668? Fragments of an eruption dated to 0.04–0.2 Ma ago?
Western part of the caldera rim	B15 B17 B24 B31	glassy debris bomb fragment tephra rock (intrusive?)	PNK4	trachydacite	Probably fragments of an eruption dated to 0.04–0.095 Ma ago or AD 1668
	B16 B23	glassy debris pumice	PNK2	pantellerite	Probably fragments of an eruption dated to either AD 969 (Horn and Schmincke 2000) or AD 1024 (Wei *et al.* 2007).

* Only general sampling locations are described here and shown in Figure 11.2. However, the B4 sample was collected very near the Heifengkou Outcrop mentioned by Jia *et al.* (2013, 976, Figure 8).

** After analysis of these samples by ICP–MS, the PNK1 group turned out to be PNK3, and the PNK4 group to be PNK2 (see main text, Table 11.2).

INDEX

www.ingramcontent.com/pod-product-compliance
Lightning Source LLC
LaVergne TN
LVHW071208210926
848482LV00015B/755